National Center for Construction Education and Research

Pipeline Corrosion Control Level Two

Annotated Instructor's Guide

Prentice Hall
Upper Saddle River, NJ
Columbus, OH

contren™ Learning Series

American Petroleum Institute
Pipeliner Training and Assessment Program (PTAP)

This information is general in nature and intended for training purposes only. Actual performance of activities described in this manual requires compliance with all applicable operating, service, maintenance, and safety procedures under the direction of qualified personnel. References in this manual to patented or proprietary devices do not constitute a recommendation of their use.

---

10 9 8
ISBN 0-13-046687-5

Preface

This volume was developed by the National Center for Construction Education and Research (NCCER) in response to the training needs of the construction, maintenance, and pipeline industries. It is one of many in NCCER's *Contren™ Learning Series*. The program, covering training for close to 40 construction and maintenance areas, and including skills assessments, safety training, and management education, was developed over a period of years by industry and education specialists.

NCCER also maintains a National Registry that provides transcripts, certificates, and wallet cards to individuals who have successfully completed modules of NCCER's *Contren™ Learning Series*, when the training program is delivered by an NCCER Accredited Training Sponsor.

The NCCER is a not-for-profit 501(c)(3) education foundation established in 1995 by the world's largest and most progressive construction companies and national construction associations. It was founded to address the severe workforce shortage facing the industry and to develop a standardized training process and curricula. Today, NCCER is supported by hundreds of leading construction and maintenance companies, manufacturers, and national associations, including the following partnering organizations:

PARTNERING ASSOCIATIONS

- American Fire Sprinkler Association
- American Petroleum Institute
- American Society for Training & Development
- American Welding Society
- Associated Builders & Contractors, Inc.
- Association for Career and Technical Education
- Associated General Contractors of America
- Carolinas AGC, Inc.
- Carolinas Electrical Contractors Association
- Citizens Democracy Corps
- Construction Industry Institute
- Construction Users Roundtable
- Design-Build Institute of America
- Merit Contractors Association of Canada
- Metal Building Manufacturers Association
- National Association of Minority Contractors
- National Association of State Supervisors for Trade and Industrial Education
- National Association of Women in Construction
- National Insulation Association
- National Ready Mixed Concrete Association
- National Utility Contractors Association
- National Vocational Technical Honor Society
- North American Crane Bureau
- Painting & Decorating Contractors of America
- Portland Cement Association
- SkillsUSA-VICA
- Steel Erectors Association of America
- Texas Gulf Coast Chapter ABC
- U.S. Army Corps of Engineers
- University of Florida
- Women Construction Owners & Executives, USA

Some features of NCCER's *Contren™ Learning Series* are:

- An industry-proven record of success
- Curricula developed by the industry for the industry
- National standardization providing portability of learned job skills and educational credits
- Credentials for individuals through NCCER's National Registry
- Compliance with Apprenticeship, Training, Employer, and Labor Services (ATELS) requirements for related classroom training (CFR 29:29)
- Well-illustrated, up-to-date, and practical information

Acknowledgments

This curriculum is a product of the farsightedness and leadership of the following sponsors:

Bass Engineering Company
Enhance Co., Inc.
Green Equipment Company
Marathon Ashland Pipeline
TEPPCO
Enbridge Energy Partners
Equistar Pipeline
Koch Pipeline Company
Phillips Pipeline Company
Valero Logistics Operations

This curriculum would not exist were it not for the dedication and unselfish energy of those volunteers who served on the Authoring Team. A sincere thanks is extended to:

Clay Brelsford
Roger Floyd
Wayne Guild
Kimberly Harris
David Selvage
Keith Wooten
Allen Carlile
Dan Franzese
Greg Guy
Tom Pickthall
Stephen Waguespack
Will Wright

A final note: This book is the result of a collaborative effort involving the production, editorial, and development staff at Prentice-Hall, Inc., and the National Center for Construction Education and Research. Thanks to all of the dedicated people involved in the many stages of this project.

Contents

Module 61201-02

Install Cathodic Protection Systems

(CT 9.2, 9.3, and 9.4)

Install Cathodic Protection Systems
Annotated Instructor's Guide

Module 61201-02

MODULE OVERVIEW

This module introduces trainees to corrosion prevention through the use of cathodic protection (CP). Trainees will learn how to select system components, describe installation techniques, and perform rectifier installation.

PREREQUISITES

Please refer to the Course Map in the Trainee Module. Prior to training with this module, it is recommended that the trainee shall have successfully completed the following:

Core Curriculum; Pipeline Core, Modules 66101-02 and 66102-02; Pipeline Corrosion Control Level One

OBJECTIVES

Upon completion of this module, the trainee will be able to:

1. Describe the information that must be gathered when planning a cathodic protection system.
2. Explain how to select the system components.
3. Describe installation techniques for galvanic and impressed current systems (CT 9.2 and 9.4).
4. Perform rectifier installation (CT 9.3).

PERFORMANCE TASKS

Under the supervision of the instructor, the trainee should be able to:

1. Correctly install galvanic anodes (CT 9.2 and 9.4).
   - Install anode.
   - Attach leads to structure and test station.
2. Install impressed current ground beds (CT 9.4).
3. Properly install rectifiers (CT 9.3).
   - Connect positive cable and negative header cables to proper output terminals.

NCCER STANDARDIZED CRAFT TRAINING PROGRAM

The National Center for Construction Education and Research (NCCER) provides a standardized national program of accredited craft training. Key features of the program include instructor certification, competency-based training, and performance testing. The program provides trainees, instructors, and companies with a standard form of recognition through a National Craft Training Registry. The program is described in full in the *Guidelines for Accreditation*, published by the NCCER. For more information on standardized craft training, contact the NCCER by writing us at P.O. Box 141104, Gainesville, FL 32614-1104; calling 352-334-0911; or e-mailing info@nccer.org. More information may be found at our Web site, www.nccer.org.

HOW TO USE THIS ANNOTATED INSTRUCTOR'S GUIDE

Each page presents two sections of information. The larger section displays each page exactly as it appears in the Trainee Module. The narrow column ties suggested trainee and instructor actions to each page and provides icons to call your attention to material, safety, audiovisual, or testing requirements. The bottom of each page includes space for your notes.

If you see the Teaching Tip icon, that means there is a teaching tip associated with this section. Also refer to any suggested teaching tips at the end of the module.

SAFETY CONSIDERATIONS

Ensure that the trainees are equipped with appropriate personal protective equipment. Emphasize the importance of following all safety precautions and procedures when working with power tools.

PREPARATION

Before teaching this module, you should review the Module Outline, Objectives, Performance Tasks, and the Materials and Equipment List. Be sure to allow ample time to prepare your own training or lesson plan and gather all required equipment and materials.

MATERIALS AND EQUIPMENT LIST

Overhead projector and screen
Transparencies
Felt-tip markers/chalk
Blank acetate sheets
Transparency pens
Pencils and scratch paper
Module Examinations*
Performance Profile Sheets*
Copies of Performance Verifications for Covered Task 9**
Copies of Quick Quiz†
Whiteboard/chalkboard
Appropriate personal protective equipment
Copies of *49 CFR Part 192* (Gas) and/or *49 CFR Part 195* (Liquid)
Copies of your company policy and procedures manual
Magnesium anodes
Knives suitable for cutting magnesium anodes
Ribbon (strip) anodes
Vertical anodes
Horizontal anodes
Header cables
Practice anodes
Lead wires
Junction boxes
Tools to connect anodes
Tools to splice header cables to the anode lead wires
Rectifiers
AC power feed wires
Conduit
Tools to connect the specified conduits to the rectifier
Header cables
Single thermite weld
Tools to separate the cable strands and connect them to a junction box
Manufacturer's instructions and wiring diagrams

* Located in the Test Booklet.
** Available for download from www.nccer.org.
† Located at the end of this module.

TEACHING TIME FOR THIS MODULE

An outline for use in developing your lesson plan is presented below. Note that each Roman numeral in the outline equates to one session of instruction. Each session has a suggested time period of 3 hours. This includes 10 minutes at the beginning of each session for administrative tasks and one 10-minute break during the session. Approximately 18 hours are suggested to cover *Install Cathodic Protection Systems (CT 9.2, 9.3, and 9.4)*. You will need to adjust the time required for hands-on activity and testing based on your class size and resources.

Topic	Planned Time
Session I. Introduction and Gathering Information	
A. NACE Standard	___________
B. Gathering Information	___________
Session II. Installation Techniques, Part One	
A. Anode Inspection	___________
B. Installing Galvanic Anode Ground Beds	___________
1. Laboratory	___________
C. Performance Testing	___________
Session III. Installation Techniques, Part Two	
A. Installing Impressed Current Surface Ground Beds	___________
1. Laboratory	___________
B. Performance Testing	___________
Session IV. Installation Techniques, Part Three	
A. Installing Header Cable	___________
B. Installing Rectifiers	___________
1. Laboratory	___________
Session V. Installation Techniques, Part Four	
A. Input and Output Voltages	___________
B. Negative Circuits	___________
Session VI. Review and Testing	
A. Cathodic Protection Review	___________
B. Summary	
1. Summarize module	___________
2. Answer questions	___________
C. Module Examination	___________
1. Trainees must score 70% or higher to receive recognition from the NCCER.	
2. Record the testing results on Craft Training Report Form 200 and submit the results to the Training Program Sponsor.	
D. Performance Testing	___________
1. Trainees must perform each task to the satisfaction of the instructor to receive recognition from the NCCER.	
2. Record the testing results on Craft Training Report Form 200 and submit the results to the Training Program Sponsor.	

Module 61201-02

Install Cathodic Protection Systems

(CT 9.2, 9.3, and 9.4)

Instructor's Notes:

Instructor's Notes:

Introduction to the Pipeline Operator Qualification Rule, API, and NCCER

It could be said that petroleum products keep America running. Petroleum products like oil and natural gas are the principal sources of energy in the United States and are used in many manufacturing and chemical processes. Petroleum products are primarily transported by pipeline. There are more than a million miles of liquid and natural gas pipeline in this country, ranging in diameter from two inches to four feet. These pipelines bring crude oil and natural gas from production fields to processing plants and refineries—and ultimately to the consumer.

Pipeline Operator Qualification

Petroleum products are considered hazardous, so pipeline personnel must take extra precautions and follow specific procedures when working with them. To help ensure the safe handling and transportation of petroleum products, the United States Department of Transportation (DOT) mandates safe practices as federal regulations.

One such regulation, the Operator Qualification (OQ) rule, requires pipeline operators to develop a written qualification plan to evaluate their personnel and their contract personnel (see *49 CFR Part 192 Subpart N—Gas,* and *49 CFR Part 195 Subpart G—Liquid*). The evaluation hinges on the operator's ability to perform certain covered tasks and to be able to recognize and respond to abnormal operating conditions that may be encountered while performing these covered tasks.

Covered Tasks

A pipeline performance task is deemed *covered* if the activity passes the following four-part test:

1. Is it performed on a pipeline facility?
2. Is it an operations or maintenance task?
3. Is it performed as a requirement of *49 CFR Part 195 (Liquid)* or *49 CFR Part 192 (Gas)*?
4. Does it affect the operation or integrity of the pipeline facility?

If the answer to all four questions is *yes*, then the activity is considered *covered* and must figure into the evaluation process of pipeline personnel.

American Petroleum Institute

The OQ rule and its related covered tasks provided the catalyst to develop a unified pipeline training and assessment program—so that pipeline operators and owners have ready access to the assessment and training tools they need. This effort was the direct result of the vision and leadership provided by The American Petroleum Institute (API) and API members.

Efforts to unify the industry are nothing new to API. Indeed, the impetus for forming the Institute in 1919 was the need to standardize engineering specifications for drilling and production equipment. Today, the most pressing issues revolve around public perceptions and government policies toward the industry. Speaking with one voice on these issues has become as essential as having interchangeable parts in the field. API's members determine the petroleum industry's positions on public policy as well as the standards that govern its day-to-day operations worldwide.

Consortium on Operator Qualification

API assembled a focus group whose primary purpose was to identify covered tasks and provide compliance recommendations to pipeline operators.

The Consortium on Operator Qualification (COOQ) became the guiding force for the identification and development of covered tasks for the liquid pipeline industry. Their work product, API publication 1161, *Guidance Document for the Qualification of Liquid Pipeline Personnel*, formed the basis for the resulting Pipeliner Training and Assessment Program.

The covered task list created in API publication 1161 has now evolved to include gas covered tasks and is maintained by NCCER. The most current covered task list can be downloaded at www.nccer.org.

National Center for Construction Education and Research

With the guidance and support of API, its members, and the Consortium on Operator Qualification, the National Center for Construction Education and Research (NCCER) was able to design and implement the Pipeliner Training and Assessment Program.

NCCER is a not-for-profit 501(c)(3) education foundation established in 1995 by the world's largest and most progressive construction and maintenance contractors and national contractor associations. NCCER was founded to address the

severe workforce shortage facing the construction and maintenance industries and to develop a standardized training process and curricula. Today, NCCER is supported by hundreds of leading construction and maintenance companies, manufacturers, and national associations and developed more than 35 construction and maintenance titles.

NCCER also maintains a National Registry, which contains the records of trainees and instructors who have received NCCER training. All completion information for every trainee is sent to the NCCER and filed within the Registry. The Registry can then confirm training and skills for workers as they move from company to company, state to state, or even within their own company.

Standardized Curriculum (NCCER's Contren™ Learning Series)

NCCER curricula are developed using the expertise of industry subject matter experts (SMEs). The use of SMEs as an integral part of every new development or revision ensures that the curriculum is current, industry-driven, and indicative of industry practices. The development and/or revision process is facilitated from start to finish by NCCER project managers, who work closely with the SMEs and a select team of technical writer(s) on the development and presentation of the curriculum.

Pipeline: Five Training Paths

The pipeline curriculum focuses on essential skills and competencies within five skill areas:

- Corrosion Control
- Electrical and Instrumentation
- Maintenance
- Mechanical
- Operations (Field and Control Center)

SMEs from the pipeline industry have determined the skills and competencies needed based on API's covered tasks and the consensus of the group/industry. Specific company information is generally not incorporated into the curriculum unless the SMEs determine that the information is the industry standard.

Training Modules and Covered Tasks

The smallest unit of instruction in the curriculum is a *module*. Each curriculum is composed of modules (the equivalent of stand-alone chapters) that are logically sequenced into complete *levels* of instruction. Every module is task-oriented and requires the successful completion of a written (knowledge-based) examination and a performance (hands-on) test. Modules may be individually selected and/or combined for efficient task training. Modules in the pipeline curricula, to the extent possible, will align with API-developed covered tasks. While not all modules will relate to covered tasks, all covered tasks will relate to at least one module.

Benefits of Training to the Pipeline Marketplace

Applying NCCER's standardized training process allows pipeline owners and contractors in the United States to establish the benchmark for acceptable skill and competency levels within the pipeline industry. This process will assist training organizations in developing and/or improving standardized pipeline training programs that effectively match their resources and operating conditions while responding to regulatory demands and to the needs of the industry. The process will also recognize training organizations that meet the minimum standards of competence, warranting regulatory, public, and professional confidence. This in turn will foster national unity within the pipeline industry for the mutual benefit of individual technicians, owners, contractors, and ultimately pipeline-users and society.

Additional Resources

To review the Pipeliner Training and Assessment Program Implementation and Resource Guide and the latest version of pipeline covered tasks, visit NCCER's Web site at www.nccer.org, or contact:

National Center for Construction Education and Research
PO Box 141104
Gainesville, FL 32614-1104
(888) 622-3720
Web site: www.nccer.org

For more information on API, please contact:

American Petroleum Institute
1220 L Street, Northwest
Washington, D.C. 20005
(202) 682-8000
Web site: www.api.org

To order *API Pub 1161*, contact Global Engineering Documents by calling 1-800-854-7149 or visiting their Web site at www.global.his.com.

Additional sources of information on the pipeline operator qualification regulation include:

DOT – Office of Pipeline Safety
www.ops.dot.gov

Transportation Safety Institute
www.tsi.dot.gov

Instructor's Notes:

COURSE MAP

This course map shows all of the modules in the second level of the Pipeline Corrosion Control curriculum. The suggested training order begins at the bottom and proceeds up. Skill levels increase as you advance on the course map. The local Training Program Sponsor may adjust the training order.

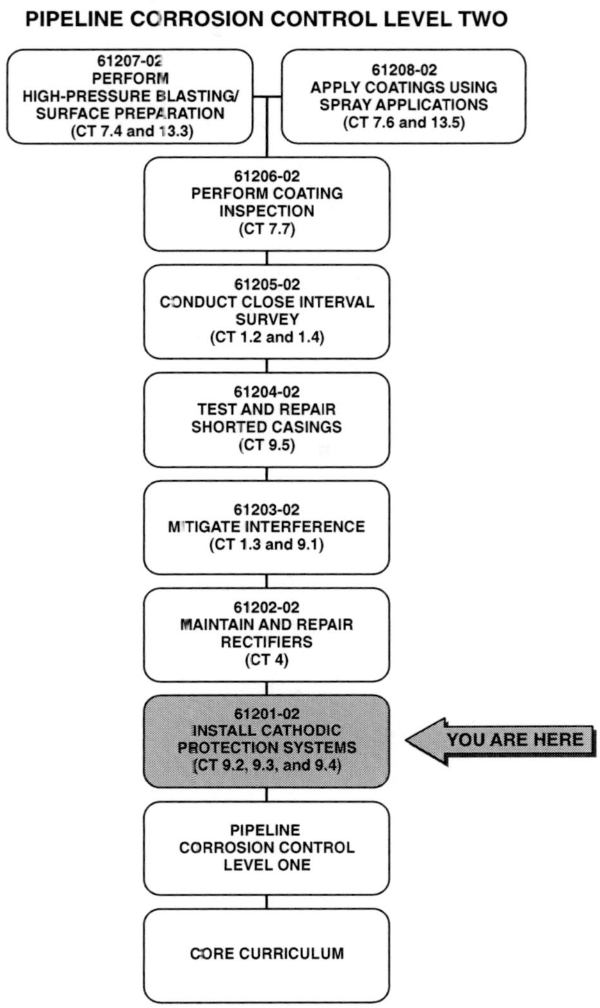

201CMAP.EPS

Homework

Assign reading of Module 61201-02.

Instructor's Notes:

MODULE 61201-02 CONTENTS

Figures

Instructor's Notes:

MODULE 61201-02

Install Cathodic Protection Systems (CT 9.2, 9.3, and 9.4)

Objectives

When you have completed this module, you will be able to do the following:

1. Describe the information that must be gathered when planning a cathodic protection system.
2. Explain how to select the system components.
3. Describe installation techniques for galvanic and impressed current systems (CT 9.2 and 9.4).
4. Perform rectifier installation (CT 9.3).

Prerequisites

Before you begin this module, it is recommended that you successfully complete the following: Core Curriculum; Pipeline Core, Modules 66101-02 and 66102-02; Pipeline Corrosion Control, Level One.

Required Trainee Materials

1. Pencil and paper
2. Appropriate personal protective equipment

1.0.0 ◆ INTRODUCTION

The *Code of Federal Regulations* mandates cathodic protection (CP) for underground transmission and distribution lines as well as some gathering pipelines. Corrosion prevention is essential to the safety and service of a pipeline.

Corrosion occurs when certain elements are brought together to form an electrochemical cell consisting of four components. There must be a positive area, called the **anode**, and a negative area, called the **cathode**. There must also be a path for the ionic current flow, called the **electrolyte**, and a path for the electronic current flow, called the metallic path. The four conditions can involve two separate structures or can be present on one structure. If any one of the four conditions is eliminated, the corrosion can be prevented. CP is the application of a specially designed electrochemical cell in such a way that the structure is the cathode and a separate expendable (sacrificial) material is chosen for the anode.

Current always flows away from the anodic (positive) metal, through the electrolyte, to the cathodic (negative) metal; this causes the anodic metal to corrode, but the cathodic metal does not corrode. The potential that causes the current to flow is always due to a difference between the anode and the cathode, such as between two different metals. The positive area, the anode, is the one that corrodes and loses metal. The metal that is lost is directly proportional to the amount of current flow.

The corrosion individual must have a thorough understanding of corrosion and the two types of CP to install and maintain either galvanic or impressed current systems. Both galvanic and impressed current CP systems send the current from the anode through the electrolyte onto the structure.

The rate of corrosion can be reduced if all of the exposed metal on the surface of a pipeline can be made to collect current instead of lose it. To make the surface of the structure negative, or cathodic, so that it will collect current, direct current is forced onto all surfaces of the pipeline. This shifts the potential of the pipeline in the negative direction, resulting in a reduction in the corrosion rate of the metal. When the amount of applied direct current is adjusted properly, it overpowers the

Materials

Ensure that you have everything required to teach the course. Check the Materials and Equipment List at the front of this module.

Classroom

Explain that terms shown in bold are defined in the Glossary at the back of this module.

Audiovisual

Show Transparency 1, Course Objectives.

Show Transparency 2, Performance Tasks.

Teaching Tip

See the general Teaching Tip at the end of this module.

Classroom

Discuss the four elements that form an electrochemical cell. Explain how cathodic protection (CP) is applied to prevent corrosion on a pipeline.

Explain how galvanic and impressed current CP systems reduce the rate of corrosion. Answer any questions trainees may have.

Teaching Tip

See the Teaching Tip for Section 1.0.0 at the end of this module.

Classroom

Review the National Association of Corrosion Engineers' *Standard RP0169*. Stress the importance of following the latest version of this standard.

Review the considerations for installing a CP system. Stress that environmental improvements can reduce the amount of protective current that is needed over time.

Homework

Have trainees review Sections 3.0.0–3.2.3 for the next session.

Materials

Ensure that you have everything required for the laboratory and testing during this session.

corrosion current that is discharging from the anodic areas on the pipeline. The entire surface then becomes a cathode, and the rate of corrosion is reduced.

1.1.0 NACE Standard

National Association of Corrosion Engineers (NACE) *Standard RP0169, Section 6*, states that a negative (cathodic) potential of at least 850 millivolts and/or a minimum of 100 millivolts polarization shift, as measured between the structure surface and a saturated copper-copper sulfate reference electrode contacting the electrolyte, will provide the corrosion control criteria used by most judicious pipeline companies. This voltage is determined with the protective current applied.

> **NOTE**
> Always verify that you are using the latest version of the standard.

2.0.0 ◆ GATHERING INFORMATION

Before any CP system can be installed, a number of factors must be determined about the structure to be protected and its environment, such as the following:

- *Physical size of the structure* – length, width, depth, height, diameter
- *Condition of the existing structure* – coated or uncoated, corroded or in good condition
- *Condition of any new structures* – type of coating
- *Past corrosion of existing structures* – abnormal conditions, environment
- *Foreign structures in the area* – stray current interference
- *Surveys*
  - Electrolyte (soil/water) resistivity
  - Electrolyte (soil/water) pH
  - Structure-to-electrolyte potential
- *Current requirements* – amount of current per square foot
- *Electrical continuity/isolation*

When testing the environment, be sure to consider any other possible improvements, such as increasing the drainage or mixing in another type of soil. For example, sand could be added to heavy clay to improve the drainage quality. In some cases, altering the soil type could help to reduce the corrosion rate. However, before doing this, local, state, and federal restrictions must be considered; some areas may be protected, allowing only minimal disturbances of the environment. When feasible, environmental improvements could be a one-time effort that would reduce the amount of protective current needed for many years.

3.0.0 ◆ INSTALLATION TECHNIQUES

Efficient operation of any cathodic protection system is dependent on proper installation. The system elements must be selected on the basis of careful calculations. The installation plan must include all the details, and the proper procedures must be used to install the system. Attention to these details helps ensure that the system will provide the protection for which it was designed for a long time.

Compared to an impressed current system, galvanic anode systems are relatively easy to install.

3.1.0 Anode Inspection

All anodes should be inspected before they are installed. Be sure that the packaging has not been torn or punctured, and that the **backfill** material is intact. Store the anodes in a dry environment, and keep the plastic or waterproof shipping materials on the anodes until they can be installed. Remove any waterproof material just before installation.

Inspect the lead wire carefully for any damage to the insulation or for any defects. Take care to avoid damaging the wire insulation when handling the anode, such as when you are storing it or placing it in the hole. Repair or replace any anodes with damage or defects in the lead wire or the insulation.

3.2.0 Installing Galvanic Anode Ground Beds

Galvanic anode systems are relatively simple once all the calculations are done. If the calculations were done correctly and the elements are properly installed, galvanic anode systems are self-regulating and require little maintenance other than monitoring. The elements of a galvanic anode system are the anodes, the lead wires and connections, and the test stations. *Figure 1* shows a typical single packaged anode installation without a test station.

Galvanic anodes are installed with the top of the anode below the structure. The anodes can be installed at about 3' out, in vertical or horizontal positions, from the sides of the structure.

Classroom

Explain how to properly inspect anodes before installation.

Audiovisual

Show Transparency 3 (Figure 1). Review the elements of a galvanic anode system.

Instructor's Notes:

The negative cable can be connected to the structure by either thermite welding or a mechanical connection, whichever is specified in the installation plan. Be careful to avoid damaging the insulation on the cables.

To reduce the resistance to earth, multiple anodes in a single bed should be placed in a straight line, either perpendicular or parallel to the pipeline. If the right-of-way boundaries are a consideration, the parallel line may be the most feasible. Both zinc and magnesium anodes can be placed as close to the structure as 5'; a minimum of 10' is preferred. Always consult company specifications for the proper distance.

Before installing a **ground bed** for a coated pipeline, inspect the **header cable** for any damage to the insulation. If the cable is in good condition, connect it securely to the header cable. Then, before the ground bed installation takes place, connect the header cable to the multiple anodes; the anodes may be either zinc or magnesium. Connect the header cable to a test point; the test point connection allows the CP system to be monitored periodically. Current output measurements can also be taken through the test point connections. After completing the installation, cover the entire connection area with waterproof insulation material. *Figure 2* shows a detail of an anode lead wire to header cable connection.

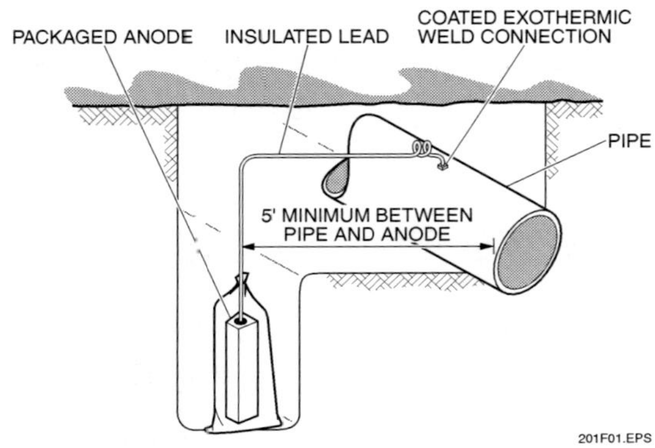

Figure 1 ◆ Typical single packaged anode installation without the test station.

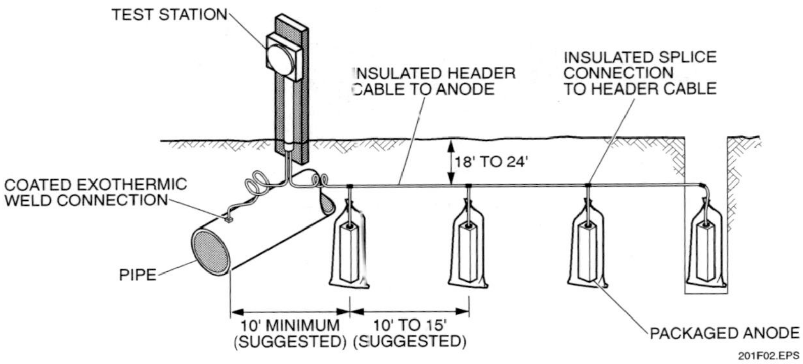

Figure 2 ◆ Detail of an anode lead wire to header cable connection.

Classroom

Explain how to install galvanic anodes. Go over any questions trainees may have.

Explain how installing anodes in a straight line reduces the resistance to earth.

Audiovisual

Show Transparency 4 (Figure 2). Explain how to connect the header cable to the multiple anodes. Have a trainee explain the importance of covering the connection with waterproof insulation.

Classroom

Review the advantages of using long ribbon anodes where sections of pipeline are bare or poorly coated.

Laboratory

Set up stations with magnesium anodes and knives suitable for cutting the anodes. Ask trainees to practice exposing the steel wire core on magnesium anodes. Ask a trainee to explain how to expose a zinc anode.

3.2.1 Installing Ribbon or Strip Anodes

Where sections of pipeline are bare or poorly coated, long ribbon (also called strip) anodes are sometimes plowed or trenched into the ground to lie parallel with the pipeline. Ribbon anodes, which have a steel wire core, provide continuous protection; they can be made of either zinc or magnesium. *Figure 3*, an image of a continuous galvanic anode installation, also shows a typical ribbon anode installation.

The steel core may remain active after the anode material is consumed, which usually occurs in areas of lower soil resistivity. To complete the CP circuit, the anode core should be connected to the pipeline at relatively short distances, not more than 1,000', preferably at test points so that readings can be taken. If the distances between connections are too far apart and as the anode material is consumed, the voltage drop will increase proportionately.

Strip anodes are furnished bare, and although backfill material is not required, its absence increases the risk of insufficient current and anode passivation. It is safer to use backfill in the recommended amount of 70 pounds per 100' of strip anode. With or without backfill, the anode should be buried deeply enough that it is always buried at least 2' in moist soil; in dry climates anodes must be buried deeper to reach needed moisture.

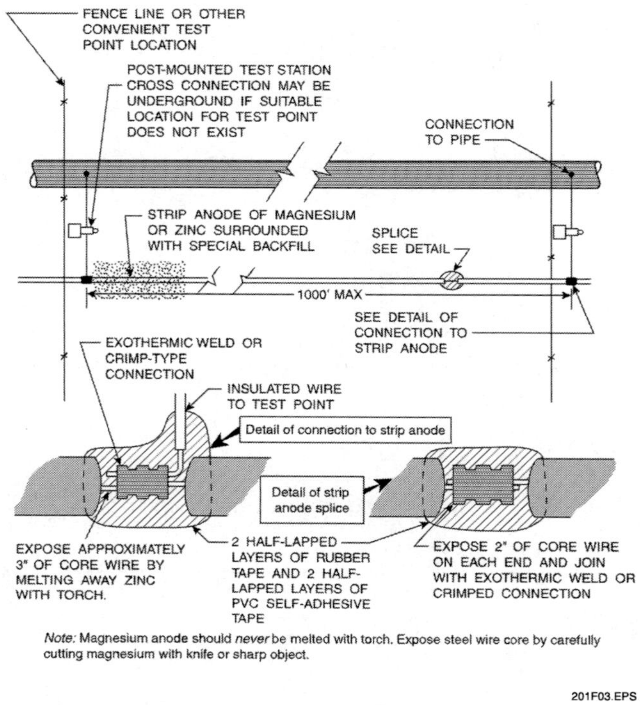

Figure 3 ◆ Continuous galvanic anode installation.

Instructor's Notes:

3.2.2 Installing Distributed Anodes

Distributed anodes, which are used with both galvanic and impressed current systems, are often installed to minimize the effects of stray currents in congested areas with a number of nearby structures. Distributed anodes also provide good current distribution. *Figure 4* shows a typical distributed **anode ground bed** installation. *Figure 5* shows a typical distributed anode grounded for a wharf.

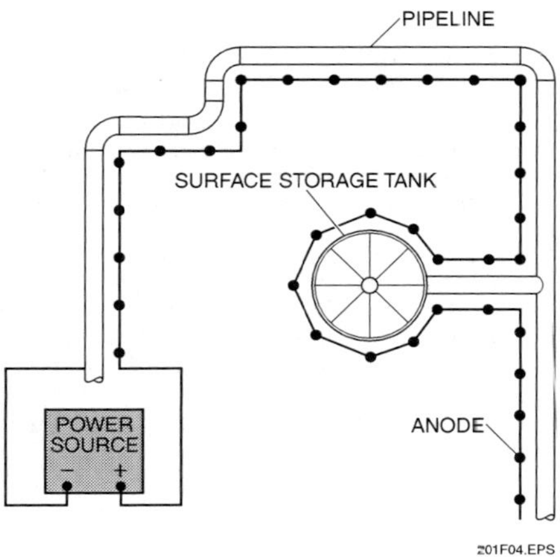

Figure 4 ◆ Distributed anode groundbed installation.

3.2.3 Test Stations

Test stations can be used to monitor the potential of the structure as well as the current out of the anode or anodes when used in connection with a shunt.

The structural cable can be connected to the structure by either thermite welding or a mechanical connection, whichever is specified in the installation plan. Be careful to avoid damaging the insulation on the structural cables.

3.3.0 Installing Impressed Current Surface Ground Beds

All connections must be in good condition, and both the connections and the wiring must be sealed against moisture. This includes the connection between anode and the anode lead, the anode lead wire insulation, the header cable insulation, and the insulation around the connections for the anode lead and header cable.

If the insulation is not maintained, current will be discharged through the **holidays**, and the wires will corrode and break apart, disabling the cathodic protection ground bed. When the connections between the header cable and the lead wires are broken, the system will cease to function.

3.3.1 Following Specifications

The ground bed and the **rectifier** must be installed exactly as specified in the installation plans; otherwise, the system may fail in short order.

All the connections must be secure mechanically. The connections to the rectifier must be electrically conductive. The negative lead wire must

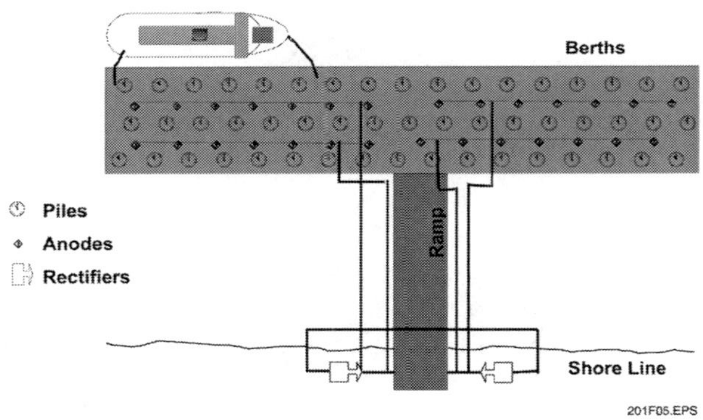

Figure 5 ◆ Distributed anode groundbed protecting wharf piling.

Classroom

Ask a trainee to explain when ribbon anodes are preferable to ground beds. Ask a trainee to explain at what depth ribbon anodes should be buried.

Audiovisual

Show Transparency 5 (Figure 4). Explain how distributed anodes are able to provide good current distribution. Ask whether trainees have any questions.

Performance Testing

Have each trainee perform Performance Task 1 to your satisfaction. Fill out a Performance Profile Sheet for each trainee.

Homework

Assign the reading of Sections 3.3.0–3.3.5 for the next session.

Materials

Ensure that you have everything required for the laboratory and testing during this session.

Classroom

Discuss problems that result if the insulation is not maintained when installing impressed current surface ground beds.

Classroom

Stress the importance of following the installation plans when installing the ground bed and the rectifier.

Ensure that trainees understand how to verify proper polarity when connecting the anodes.

Audiovisual

Show Transparency 6 (Figure 6). Explain how the anodes are grouped in a remote surface ground bed.

be properly connected to the structure. Verify that the connections between the header cable and the anode leads are secure and conductive. Always be sure the negative conductor is connected to the structure to be protected and the positive conductor is connected to the anodes before energizing power. Verify proper polarity after energizing the power. Always refer to company specifications to check polarity.

If underground **splices** on the header cable cannot be avoided, use the fewest number possible. Buried or submerged connections must be completely sealed against moisture to isolate them electrically from the surrounding environment.

Leave all wiring with some slack to prevent straining them. Be careful to avoid damaging buried cable header; the insulation must remain intact or the system will fail from electrolytic corrosion within a short time. Before backfilling, sift any reused backfill or native soil to remove any rocks or debris that could cause damage to the wiring or insulation. Follow all guidelines and use extra care when plowing in a cable.

When anodes are installed from about 5' to 15' below ground level, they are referred to as surface ground beds. The anodes may be installed either vertically or horizontally. When there are multiple anodes in the bed, the lead wires are usually connected to a header cable of sufficient size. The header cable is connected to the positive terminal of the rectifier.

Some of the more specific installation procedures are explained in the following sections.

3.3.2 Installing Remote Surface Ground Beds

On transmission pipelines and places where few structures exist to be affected by stray current, remote surface ground beds are often used. In these ground beds, the anodes are grouped together in a single location, and the current flows from the ground bed directly to the structure. *Figure 6* shows a typical remote surface ground bed installation.

3.3.3 Installing Vertical Anodes in a Surface Ground Bed

Drill holes according to the specifications on the plan. If bare anodes are being installed, put some **carbonaceous** backfill in the bottom of the hole and tamp it down. Carefully lower the anode into the hole.

Place the anode in the center of the hole and fill around it with backfill. Tamp down the backfill even with the top of the anode. Then fill in above the anode to the specified height with more backfill, being careful to avoid damaging the lead wire or the insulation. Fill the remaining space to the bottom of the header cable with soil and tamp it down.

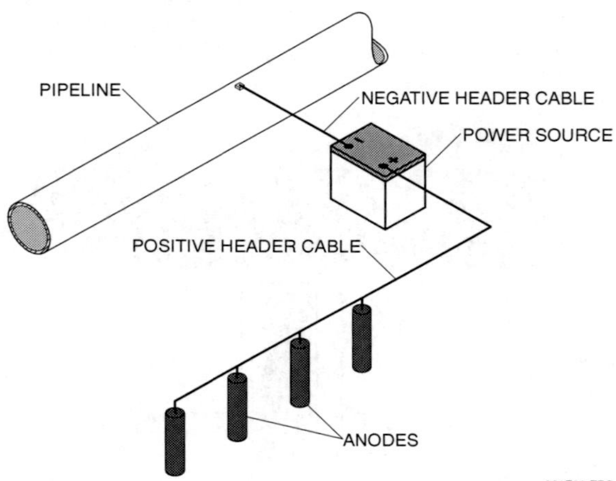

Figure 6 ◆ Remote surface ground bed installation.

Instructor's Notes:

CAUTION
Do not hold the anode by its lead wire. Either the connection or the wire could break or be damaged.

Use the same method for installing packaged anodes. Plans usually call for additional carbonaceous backfill to be placed around the package as well as above and below it. Be sure the installation meets the plan specifications. *Figure 7* shows a typical vertical impressed current anode installation.

The dimensions shown in *Figure 7* are typical dimensions of a vertical surface ground bed. Always use the dimensions on your installation plan.

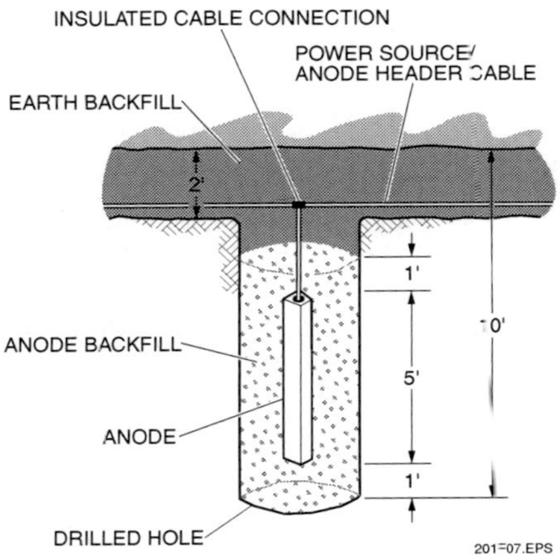

Figure 7 ◆ Vertical impressed current anode installation.

3.3.4 Installation of Horizontal Anodes

Installing horizontal anodes is similar to installing vertical anodes. Horizontal installations can use either bare or packaged anodes and carbonaceous backfill.

Check the spacing of the anodes in the installation plan. If the spacing of the anodes is less than 10', digging a single, long trench that can accommodate both the anodes and the header cable is usually recommended. However, when the spacing exceeds 10', a separate hole is prepared for each anode. Installation plans should have all the specifications. *Figure 8* shows a typical horizontal anode installation.

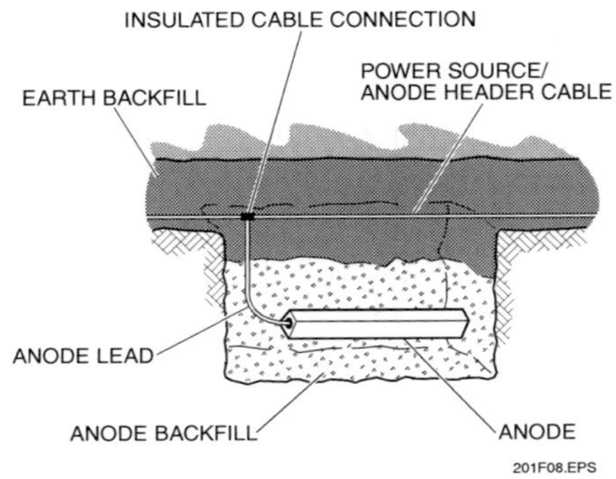

Figure 8 ◆ Horizontal anode installation.

3.3.5 Installing Deep Anode Ground Beds

Anode depth is another important consideration. Some anode beds may be hundreds of feet in the ground. In these deep anode installations, a machine auger, usually truck-mounted, drills the hole; the anode is placed at the bottom. Holes for deep anodes are usually drilled to a specified depth and diameter. The rotary drills can use water-based drilling mud to preserve the integrity of the hole; they also remove the contents from the hole. Water-based drilling muds (gels) are used to preserve the integrity of the hole when the soil type would allow the walls to collapse. When the anodes are deeper (as opposed to surface ground beds) in the earth, the potential gradient effects and seasonal output variations from moisture are reduced. There is also less likelihood of damage to the connecting wires from other excavations in the area. *Figure 9* shows a typical deep anode installation.

Conduct an electrical log of the hole to locate the area of lowest resistivity, and measure the circuit resistance. To find the most advantageous depth for the anode, a log is kept of these measurements.

The equipment used in deep anode installations may include a compressed air circulating system, portable rotary drill rig, mud circulation system, water truck, mud pit to contain the drill mud and cuttings, and various support systems and equipment. *Figure 10* shows a typical portable rotary mud drill rig.

Anodes may be lowered by their leads, but this must done very carefully. A centering device keeps the anodes centered in the hole.

Safety

Discuss the consequences of holding an anode by its lead wire.

Classroom

Refer to Figure 7. Review the procedures for installing bare and packaged vertical anodes. Answer any questions trainees may have.

Audiovisual

Show Transparency 7 (Figure 8). Compare the procedures for installing horizontal anodes with the procedures for installing vertical anodes.

Laboratory

Using sample anodes and header cables, ask a trainee to demonstrate the procedure for installing vertical and horizontal anodes. Have other trainees determine which procedure the trainee is demonstrating.

Classroom

Discuss considerations for determining the appropriate anode depth.

Audiovisual

Show Transparency 8 (Figure 9). Explain how the depth of an anode affects the likelihood of damage to the connecting wires.

Classroom

Refer to Figure 10. Review the types of equipment that may be used for deep anode installation. Go over any questions trainees may have.

Teaching Tip

See the Teaching Tip for Section 3.3.0 at the end of this module.

Performance Testing

Have each trainee perform Performance Task 2 to your satisfaction. Fill out a Performance Profile Sheet for each trainee.

Homework

Assign the reading of Sections 3.4.0–3.5.1 for the next session.

Materials

Ensure that you have everything required for the demonstration and laboratory during this session.

Classroom

Emphasize the importance of reviewing the specifications before installing header cable.

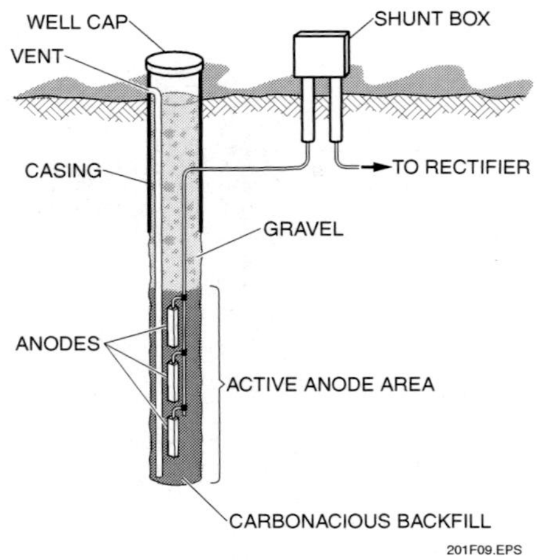

Figure 9 ◆ Deep anode installation.

Compressed air circulation systems may be advantageous in limited situations in which downhole formations allow their use. Installing a vent tube in the hole allows any gases to escape to the atmosphere. The tube is usually a length of slotted or perforated PVC or polyethelyne with a screened top at ground level. If gases are trapped in the hole, they can increase the resistance of the anode and interfere with its performance.

The integrity of the upper portion of the hole is sometimes maintained by a length of surface casing. After the final backfill has been added, the surface casing is pulled or capped.

When all the components are in place, carbonaceous backfill is added to the level specified and dropped on the top of the lead wire or the cable header. Above the backfill, the hole may be filled with soil, gravel, or bentonite, or may be cased.

Each of the anode lead wires is connected to a separate terminal in a junction box. Each lead should be marked for identification in the junction box. The terminals are sometimes connected through a shunt to a positive bus bar to allow the monitoring of the current output to each of the individual anodes. The bus bar is then connected by cable to the positive terminal in the rectifier.

State and/or local codes govern deep anode installations. Be sure to obtain the proper permits before beginning the work and then adhere to all applicable regulations.

3.4.0 Installing Header Cable

Always begin by checking the specifications in the installation plan. If a trench is required for a header cable or for anodes, backhoes or trenchers usually dig them. After the trench is prepared and the anodes (if applicable) are placed, the header cable should be carefully laid on about 3" of sand

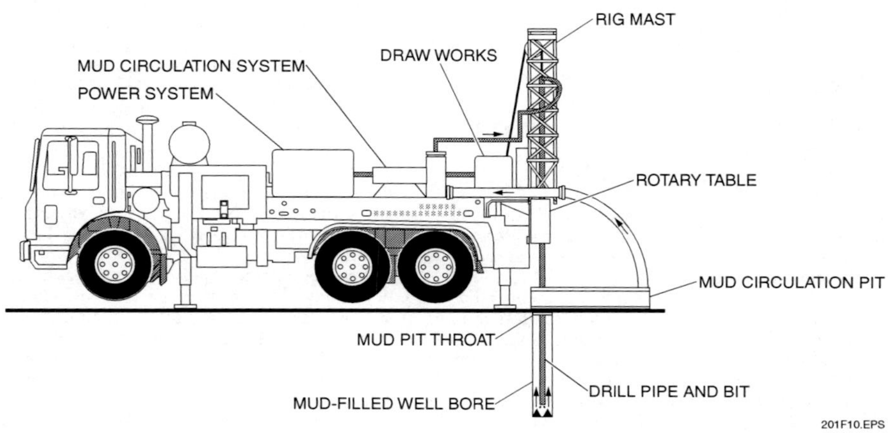

Figure 10 ◆ Portable rotary mud drill rig.

Instructor's Notes:

or other approved backfill material. Be careful to avoid damaging the cable insulation. Occasionally the cable may be plowed in, but this requires great care to avoid damaging the insulation.

After the cable is laid, the anode lead wires are spliced in. The splice is the most critical part of the process. To prevent corrosion, each splice must be insulated according to the installation plan, using the specified insulation material or method. All wiring should have plenty of slack.

When the splices are complete and the insulation has been properly applied, the trench is backfilled, again using the material specified in the installation plan. Cover the cable with approximately the same amount of sand, soil, or backfill material as was used below the cable. Be careful to remove any rocks or debris from the backfill material. Finish filling in the trench with soil.

If more digging or excavating is expected in the area, it may be wise to include a plastic warning tape about a foot below the surface, to warn of the presence of the cable and the anodes.

When splicing the header cable to the anode lead wires, a heavy-duty compression splice or similar splice is recommended. The splice must be completely enclosed and waterproofed by the epoxy coating and must be suitable for direct burial. It would be better to avoid underground splices altogether, but if they must be used, be sure to follow the manufacturer's directions carefully.

3.5.0 Installing Rectifiers

The rectifier or other power source should be installed so that the possibility of damage or vandalism is minimized. All wiring should meet the applicable codes and requirements and any other applicable standards. Before installing the rectifier, a fused switch box with an external disconnect should be installed so that the power can be turned off before work begins on the rectifier.

The rectifier case must be properly grounded before being put into operation. *Figure 11* shows a typical air-cooled rectifier.

Figure 12 shows a wooden utility pole-mounted rectifier.

Figure 13 shows a pad-mounted rectifier for mounting air-cooled (pedestal type), oil-filled, and explosionproof rectifiers.

Some units, referred to as constant current, will automatically adjust the voltage to meet the specified output by means of an internal sensor. A constant potential unit works in a similar manner, but they use a reference cell to obtain the desired voltage. Manually controlled rectifiers with built-in air-cooling systems are often used with smaller impressed current systems.

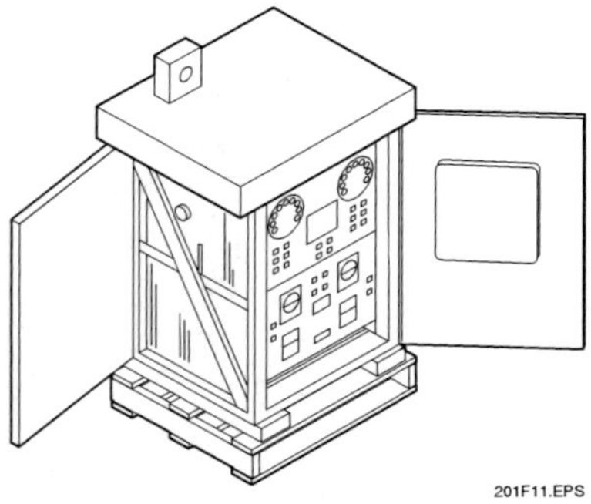

201F11.EPS

Figure 11 ◆ Air-cooled rectifier.

Rectifiers should be installed in safe locations, usually well away from traffic. They can be located inside a protective structure or outside, as long as they are rated for the environment. Good air circulation is essential to prevent overheating.

3.5.1 Rectifier Installation

Sometimes rectifiers are not repaired through component replacement and are replaced. To replace a rectifier it is important to use proper installation procedures and to follow all safety rules. Follow these steps to install rectifiers:

Step 1 Mount the rectifier securely at the appropriate location.

> **NOTE**
> Rectifiers may be mounted on poles, posts, walls, panels, and concrete pads, and must be mounted securely using appropriately sized fasteners.

Step 2 Connect AC power feed wires through appropriately sized conduit in accordance with the appropriate sections of the *National Electric Code®*, the National Electric Safety Code, and local electric and building codes.

Step 3 Terminate AC feed wires at terminal board, circuit breaker, or AC input connection wires.

Step 4 Connect specified conduits to the rectifier.

Step 5 Install DC cables from the anode ground bed and the structure in their respective conduits and terminate on their respective terminals.

Demonstration

Demonstrate how to splice header cables to the anode lead wires. Explain problems that could result from underground splices.

Classroom

Review the different types of rectifiers. Explain how to properly ground the rectifier case.

Review the procedure for installing a rectifier. Emphasize that trainees must comply with the hazardous classifications and installation specifications when working with rectifiers.

Laboratory

Ask trainees to practice connecting AC power feed wires to the appropriate conduit and the specified conduits to the rectifier. Go over any questions trainees may have.

Safety

Explain problems that can result if wires are crossed.

Teaching Tip

See the Teaching Tip for Section 3.5.1 at the end of this module.

Homework

Have trainees review Sections 3.5.2–3.5.3 for the next session.

Materials

Ensure that you have everything required for the demonstration during this session.

Safety

Emphasize the importance of preventing electrical sparks and other sources of ignition.

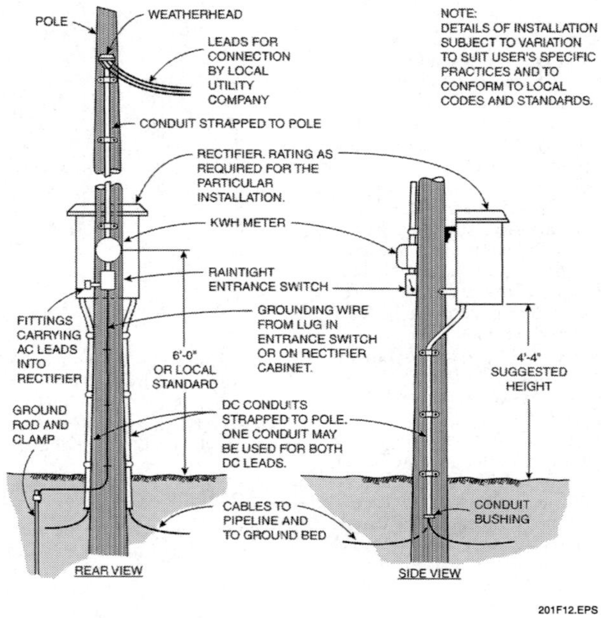

Figure 12 ◆ Pole mounted rectifier.

> **WARNING!**
> Termination at the proper terminal is essential. Crossing the wires (connecting the anode ground bed to the negative and the structure to the positive) can have disastrous consequences.

Step 6 Document installation and make company notifications as applicable.

3.5.2 Input and Output Voltages

The input voltage for the rectifier may range from 115 volts to as high as 480 volts. The rectifier may require single- or three-phase power or both. The wiring connections should comply with the manufacturer's installation specifications.

Conduit should be used for any aboveground wiring. Ensure that all wiring, seals, enclosures, and other electrical equipment meet any hazardous classification requirements. The presence of gas or vapors in an atmosphere in sufficient quantities to ignite or explode makes the location a Class I hazard, according to the types defined by the NEC.

> **WARNING!**
> Any time flammable gases or vapors are present in the atmosphere, the potential for explosion exists; an electrical spark or other source of ignition could ignite the gases or vapors.

Instructor's Notes:

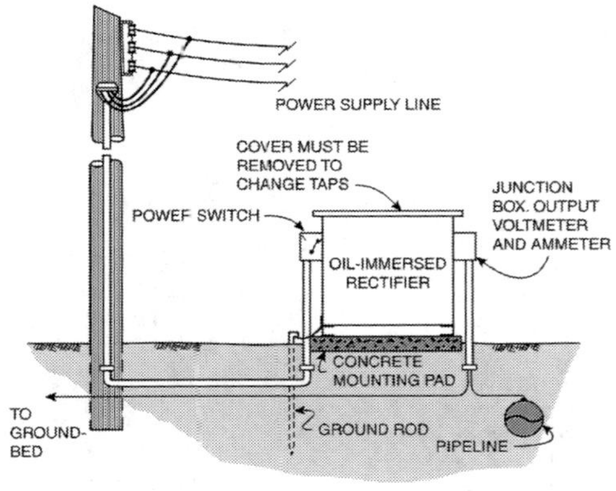

Figure 13 ◆ Pad mounted rectifier.

All Class I locations require Class I hazardous location equipment. Many locations within the petroleum industry fall under Class I.

Pipeline operators are subject to various standards for rectifier installations, depending on location, conditions, and type of pipeline. Always check which regulations apply and ensure that all wiring meets the applicable standards. *Figures 12* and *13* do not represent an attempt to standardize rectifier installation practices. They are shown only to serve as a guide for installing different types of rectifiers. In the installation process, always follow your company plans and policies.

3.5.3 Negative Circuits

With a rectifier, the positive lead is connected to an anode bed and the negative lead is connected to the structure to be protected. If the positive terminal were connected to the pipeline, it would increase or accelerate the corrosion to the pipeline instead of reducing it. The terminals are sometimes labeled to help prevent such errors. Always follow the manufacturer's installation instructions for connecting the rectifier.

If two or more structures are being protected by a single rectifier, it may be necessary to have more than one negative circuit. The negative cables can be connected to separate terminals in a junction box with added resistors. Consult the wiring diagram to identify correct terminals for the negative circuits.

The header cables range from No. 8 AWG to No. 2/0 AWG or larger and may be too large to connect with a single thermite weld. The largest charge that can be used on a steel pressure pipeline is 15 grams. In this case, use one of the following methods:

- Separate the cable strands into smaller groups and then use the proper sized charge for each group.
- Weld a lug onto the structure and connect the lug using a mechanical connector or by **exothermic welding**.

After the connection is made, coat the negative connection according to the installation instructions.

Classroom

Ensure that trainees understand how to select the appropriate equipment for the rectifier's voltage and for the location's class designation.

Emphasize the importance of following appropriate regulations and company policies.

Explain when it may be necessary to have more than one negative circuit. Review the methods for connecting the header cables.

Demonstration

Bring in header cables and a single thermite weld. Demonstrate how to separate the cable strands into smaller groups. Ask whether trainees have any questions.

Classroom

Distribute sample manufacturers' instructions and wiring diagrams. Ask trainees to review the diagrams and identify the correct terminals for negative circuits.

Teaching Tip

See the Teaching Tip for Section 3.5.3 at the end of this module.

Homework

Ask trainees to complete the Review Questions and prepare for the Module Examination and testing during the next session.

Materials

Ensure that you have everything required for testing during this session.

Classroom

Go over the Review Questions for this module, and answer any questions trainees may have.

Summary

Before a CP system can be installed, certain information must be gathered about the area and the structure to be protected. The size and condition of the structure and the environmental conditions of the surrounding area must be considered when planning and laying out the system.

NACE standards for the amount of protection and the type of test measurements must be factored into the plan.

Various installation techniques can be used to help ensure the efficiency of the CP system. The components, including the anodes, wiring, and insulation, must be inspected before installation. The location and depth of the ground bed must be determined; the location will depend on whether a galvanic anode or an impressed current system is used. The type of anode and the type and location of the test stations are also factors to be considered.

If impressed current ground beds are being installed, you must determine the power source, whether the anodes should be vertical or horizontal, and whether the ground bed should be surface level or deep. Each of these factors requires special installation techniques. The header cable must be installed according to plan and with care to avoid damage to the insulation. The rectifier must be installed in such a way as to minimize the possibility of vandalism or damage. Wiring must be in compliance with the NEC and must be done according to the manufacturer's specifications. Any conditions imposed by hazards in the area must be considered, and all equipment and wiring must meet NEC Class requirements.

Review Questions

1. Corrosion occurs when certain elements are brought together to form an electrochemical cell consisting of _____ components.
   a. three
   b. four
   c. five
   d. six

2. Current always flows away from the _____.
   a. cathodic metal
   b. metallic path
   c. electrolyte
   d. anodic metal

3. When the amount of applied direct current is adjusted properly, it overpowers the _____.
   a. pipeline potential
   b. galvanic system
   c. corrosion current
   d. CP system

4. NACE states that a negative potential of at least _____ mV and/or a minimum of _____ mV polarization shift will provide corrosion control criteria.
   a. 100; 850
   b. 200; 750
   c. 750; 200
   d. 850; 100

5. In some cases, altering the _____ could help to reduce the corrosion rate.
   a. current requirements
   b. soil type
   c. structure size
   d. ground bed

6. Efficient operation of any cathodic protection system is dependent on _____.
   a. structure size
   b. current requirements
   c. proper installation
   d. regular surveys

7. Galvanic anodes are installed with the top of the anode below the structure, usually about _____ out from the sides of the structure.
   a. 1'
   b. 2'
   c. 3'
   d. 4'

Instructor's Notes:

8. To reduce the resistance to earth, multiple anodes in a single bed should be placed in a _____.
   a. circle
   b. rectangular pattern
   c. cluster
   d. straight line

9. Where sections of pipeline are bare or poorly coated, _____ anodes are sometimes plowed into the ground.
   a. distributed
   b. bare
   c. long ribbon
   d. short flat

10. _____ anodes are often installed to minimize the effects of stray currents in congested areas.
   a. Distributed
   b. Bare
   c. Long ribbon
   d. Short flat

11. Test stations can be used to monitor the _____ of the structure.
   a. currents
   b. corrosion
   c. integrity
   d. potential

12. If the insulation is not maintained, current will be discharged through the _____.
   a. anodes
   b. holidays
   c. header cable
   d. ground bed

13. Always be sure the _____ conductor is connected to the _____.
   a. negative, structure
   b. positive, structure
   c. negative, anodes
   d. positive, rectifier

14. When anodes are installed from about 5' to 15' below ground level, they are referred to as _____.
   a. shallow ground beds
   b. buried ground beds
   c. surface anodes
   d. surface ground beds

15. If bare anodes are being installed, put some _____ backfill in the bottom of the holes.
   a. sand
   b. dirt
   c. carbonaceous
   d. gravel

Examination

Administer the Module Examination. Be sure to record the results of the Exam on Craft Training Report Form 200 and submit the results to the Training Program Sponsor.

Performance Testing

Have each trainee perform Performance Task 3 to your satisfaction. Fill out a Performance Profile Sheet for each trainee.

Ensure that all Performance Tests have been completed and Performance Profile Sheets for each trainee are filled out. Be sure to record the results of the testing on Craft Training Report Form 200 and submit the results to the Training Program Sponsor.

GLOSSARY

Glossary

This glossary will help you learn some terms specific to installing cathodic protection systems. Trade Terms introduced in this module appear in italic text.

**Anode:* The electrode of an electrochemical cell at which oxidation occurs. Electrons flow away from the anode in the external circuit. Corrosion usually occurs and metal ions enter the solution at the anode.

Anode ground bed: Two or more anodes installed in a specific configuration where the anodes are connected together by a header cable/wire.

**Backfill:* Material placed in a hole to fill the space around the anodes, vent pipe, and buried components of a cathodic protection system.

Carbonaceous: Relating to or composed of carbon.

**Cathode:* The electrode of an electrochemical cell at which reduction is the principal reaction. Electrons flow toward the cathode in the external circuit.

**Electrolyte:* A chemical substance containing ions that migrate in an electric field.

Exothermic weld: Connection made to structure using exothermic welding equipment and techniques, usually through the semi-controlled firing of a powered material.

**Groundbed:* One or more anodes installed below the earth's surface for the purpose of supplying cathodic protection.

Header cable/wire: A cable or wire to which the anode lead wire is connected.

**Holiday:* A discontinuity in a protective coating that exposes unprotected surface to the environment.

Rectifier: A device used to convert AC current to DC current. In an impressed cathodic protection system, it supplies current to the structure in the soil.

Splice connection: The electrical connection between the anode lead wire and the header cable/wire.

Ultrasonic tester: An instrument that determines the thickness of a structure.

* Term published in NACE *Glossary of Corrosion-Related Terms.* Reproduced with permission from NACE International, Houston, TX. For a copy of the complete *Glossary of Corrosion-Related Terms*, contact NACE International, 1440 South Creek Dr., Houston, TX 77084. The Glossary is revised periodically, so information in an outdated version may not be accurate.

Instructor's Notes:

REFERENCES & ACKNOWLEDGMENTS

Additional Resources

This module is intended to present thorough resources for task training. The following reference works are suggested for further study. These are optional materials for continued education rather than for task training.

Peabody's Control of Pipeline Corrosion, 2nd Edition, 2001. A. W. Peabody. Houston, TX: NACE International.

Cathodic Protection Level 1 Training Manual, September 2001. NACE International.

Handbook of Cathodic Corrosion Protection, 3rd Edition, June 1995. W. von Baeckmann, W. Schwenk, and W. Prinz. Houston, TX: Gulf Publishing Company.

Pipeline Corrosion and Cathodic Protection, 3rd Edition, June 1995. Marshall E. Parker, Edward G. Peattie. Houston, TX: Gulf Professional Publishing.

Figure Credits

Figure 3 courtesy of NACE International

(Source: *Peabody's Control of Pipeline Corrosion, Second Edition*, Pg 197, Fig 9.8, Continuous galvanic anode installation) ©NACE International – The Corrosion Society, 1440 South Creek Drive, Houston, Texas 77084. Additional publications on corrosion control technology are available from www.nace.org or phone (281) 228-6200.

Figure 5 courtesy of NACE International

(Source: *NACE Cathodic Protection Level 1 Training Manual*, Pg 8:26, Fig 8.23 Distributed anode groundbed protecting wharf piling) ©NACE International – The Corrosion Society, 1440 South Creek Drive, Houston, Texas 77084. Additional publications on corrosion control technology are available from www.nace.org or phone (281) 228-6200.

Figure 12 courtesy of NACE International

(Source: *Peabody's Control of Pipeline Corrosion, Second Edition*, Pg 165, Fig 8.5, Pole mounted rectifier) ©NACE International – The Corrosion Society, 1440 South Creek Drive, Houston, Texas 77084. Additional publications on corrosion control technology are available from www.nace.org or phone (281) 228-6200.

Figure 13 courtesy of NACE International

(Source: *Peabody's Control of Pipeline Corrosion, Second Edition*, Pg 166, Fig 8.6, Pad mounted rectifier) ©NACE International – The Corrosion Society, 1440 South Creek Drive, Houston, Texas 77084. Additional publications on corrosion control technology are available from www.nace.org or phone (281) 228-6200.

MODULE 61201-02 — TEACHING TIPS

The following are suggested activities or instructional methods to help you teach the material in this AIG.

General

When you call on someone to answer a question, the rest of the class relaxes or even tunes out because they expect that the question and answer will take place only between you and the trainee you called on. Instead, use this technique to involve more trainees in answering questions and to keep them on their toes.

1. Ask trainees to define a term or explain a concept.
2. After one trainee has answered, ask a trainee seated nearby if the answer is right. Then ask whether a trainee in the back of the room agrees.
3. Ask trainees to explain why they think an answer is right or wrong.
4. Use the session to clear up incorrect ideas and encourage trainees to learn from their mistakes.

Section 1.0.0 *Introduction*

This exercise will reinforce trainees' understanding of corrosion and the four components of an electrochemical cell. Trainees will need their Trainee Guides and pencils and paper to take notes for discussion. Allow 20 minutes for this exercise.

1. Divide the class into small groups. Ask each member of a group to select one component of an electrochemical cell: anode, cathode, electrolyte, or metallic path.
2. Have trainees explain their chosen component's function. Ask trainees to illustrate the flow of current and explain the flow.
3. Ask trainees to answer the following questions:
   - Explain how to reduce the rate of corrosion (*eliminate one of the conditions*).
   - Identify which component loses metal and corrodes (*anode*).
   - Identify the objective of CP (*making the entire surface a cathode*).
4. Repeat the questions to each group, as time permits. Answer any questions the trainees may have about the four components of an electrochemical cell.

Section 3.3.0 *Installing Impressed Current Surface Ground Beds*

This exercise will give trainees practice connecting anodes. For this exercise, you will need sample anodes, lead wires, junction boxes, and the tools to connect anodes. Trainees will need their Trainee Guides and pencils and paper to take notes for discussion. Allow between 20 and 30 minutes for this exercise.

1. Set up several stations with anodes, lead wires, junction boxes, and the tools to connect anodes.
2. Divide the class into small groups according to the number of stations.
3. Ask trainees to practice connecting the anode lead wires to separate terminals in the junction box. Ensure that trainees identify each lead.
4. Review and assess trainees' work. Answer any questions trainees may have about installing impressed current surface ground beds.

Section 3.5.1 *Rectifier Installation*

This exercise will help trainees memorize the procedure for installing rectifiers. The exercise will also reinforce safety warnings. Trainees will need their Trainee Guides and pencils and paper to take notes for discussion. Allow 15 minutes for this exercise.

1. Ask trainees to refer to *Figure 12* and to review the procedures for installing rectifiers.
2. For each step, have trainees identify the location on *Figure 12* where this action would likely take place. Ask trainees to label the appropriate step at the selected location.
3. Have trainees note the relevant safety considerations for each step.
4. Review and assess trainees' work. Answer any questions they may have about rectifier installation.

Section 3.5.3 *Negative Circuits*

This Quick Quiz will familiarize trainees with the details of negative circuits. You will need photocopies of the quiz provided on the following page. Trainees will need pencils. If you allow trainees to use the Trainee Guide, decrease the amount of time you give them to complete the quiz.

1. Make as many photocopies as you will need of the quiz.
2. Give trainees between 5 and 10 minutes to complete the quiz.
3. Go over the answers to the quiz.
4. Ask trainees whether they have questions about negative circuits.

Answers to Quick Quiz

1. c. accelerate the corrosion
2. d. separate the cable strands into smaller groups
3. a. an anode bed
4. c. the structure to be protected
5. d. more than one negative circuit

Quick Quiz *Negative Circuits*

Circle the correct answer.

1. If you connect the positive terminal to the pipeline, it will _____.

a. make it necessary to add another positive circuit

b. make it necessary to add another negative circuit

c. accelerate the corrosion

d. reduce the corrosion

2. If header cables are too large to connect with a single thermite weld, you can _____.

a. use more than one positive circuit

b. weld a lug onto the anode bed

c. use more than one negative circuit

d. separate the cable strands into smaller groups

3. With a rectifier, the positive lead is connected to _____.

a. an anode bed

b. the pipeline

c. a negative circuit

d. a negative cable

4. With a rectifier, the negative lead is connected to _____.

a. the pipeline

b. an anode bed

c. the structure to be protected

d. a junction box

5. If two or more structures are being protected by a single rectifier, it may be necessary to have _____.

a. more than one anode bed

b. larger cables

c. more than one positive circuit

d. more than one negative circuit

MODULE 61201-02 — ANSWERS TO REVIEW QUESTIONS

Answer	Section
1. b	1.0.0
2. d	1.0.0
3. c	1.0.0
4. d	1.1.0
5. b	2.0.0
6. c	3.0.0
7. c	3.2.0
8. d	3.2.0
9. c	3.2.1
10. a	3.2.2
11. d	3.2.3
12. b	3.3.0
13. a	3.3.1
14. d	3.3.1
15. c	3.3.3

CONTREN™ LEARNING SERIES — USER UPDATES

The NCCER makes every effort to keep these textbooks up-to-date and free of technical errors. We appreciate your help in this process. If you have an idea for improving this textbook, or if you find an error, a typographical mistake, or an inaccuracy in NCCER's Contren™ textbooks, please write us, using this form or a photocopy. Be sure to include the exact module number, page number, a detailed description, and the correction, if applicable. Your input will be brought to the attention of the Technical Review Committee. Thank you for your assistance.

Instructors – If you found that additional materials were necessary in order to teach this module effectively, please let us know so that we may include them in the Equipment/Materials list in the Instructor's Guide.

Write: Curriculum Revision and Development Department
National Center for Construction Education and Research
P.O. Box 141104, Gainesville, FL 32614-1104

Fax: 352-334-0932

E-mail: curriculum@nccer.org

Craft ______________________ Module Name ______________________

Copyright Date __________ Module Number __________ Page Number(s) __________

Description

____

____

____

____

(Optional) Correction

____

____

____

(Optional) Your Name and Address

____

____

____

Module 61202-02

Maintain and Repair Rectifiers (CT 4)

MODULE OVERVIEW

This module discusses the regular maintenance program for cathodic protection rectifier systems, which supply the DC power to the anodes to protect the pipeline. Trainees will learn how to troubleshoot and repair a rectifier.

PREREQUISITES

Please refer to the Course Map in the Trainee Module. Prior to training with this module, it is recommended that the trainee shall have successfully completed the following:

Core Curriculum; Pipeline Core, Modules 66101-02 and 66102-02; Pipeline Corrosion Control Level One; Pipeline Corrosion Control Level Two, Module 61201-02

OBJECTIVES

Upon completion of this module, the trainee will be able to:

1. Describe the characteristics of cathodic protection rectifiers.
2. Describe the functions of typical rectifier components.
3. Describe troubleshooting techniques for rectifier and bond connections (CT 4.1).
4. Describe the common repair process for a rectifier (CT 4.2).
5. Describe the adjustment procedures for a rectifier (CT 4.3).

PERFORMANCE TASKS

Under the supervision of the instructor, the trainee should be able to:

1. Troubleshoot rectifier bond connections (CT 4.1).
2. Properly repair and replace defective rectifier components (CT 4.2).
3. Correctly adjust rectifier components for proper output (CT 4.3).

NCCER STANDARDIZED CRAFT TRAINING PROGRAM

The National Center for Construction Education and Research (NCCER) provides a standardized national program of accredited craft training. Key features of the program include instructor certification, competency-based training, and performance testing. The program provides trainees, instructors, and companies with a standard form of recognition through a National Craft Training Registry. The program is described in full in the *Guidelines for Accreditation*, published by the NCCER. For more information on standardized craft training, contact the NCCER by writing us at P.O. Box 141104, Gainesville, FL 32614-1104; calling 352-334-0911; or e-mailing info@nccer.org. More information may be found at our Web site, www.nccer.org.

HOW TO USE THIS ANNOTATED INSTRUCTOR'S GUIDE

Each page presents two sections of information. The larger section displays each page exactly as it appears in the Trainee Module. The narrow column ties suggested trainee and instructor actions to each page and provides icons to call your attention to material, safety, audiovisual, or testing requirements. The bottom of each page includes space for your notes.

Teaching Tip

If you see the Teaching Tip icon, that means there is a teaching tip associated with this section. Also refer to any suggested teaching tips at the end of the module.

SAFETY CONSIDERATIONS

Ensure that the trainees are equipped with appropriate personal protective equipment. Emphasize the importance of following all safety precautions and procedures when working with power tools.

PREPARATION

Before teaching this module, you should review the Module Outline, Objectives, Performance Tasks, and the Materials and Equipment List. Be sure to allow ample time to prepare your own training or lesson plan and gather all required equipment and materials.

MATERIALS AND EQUIPMENT LIST

Overhead projector and screen

Transparencies

Felt-tip markers/chalk

Blank acetate sheets

Transparency pens

Pencils and scratch paper

Module Examinations*

Performance Profile Sheets*

Copies of Performance Verifications for Covered Task 4**

Whiteboard/chalkboard

Appropriate personal protective equipment

Copies of *49 CFR Part 192* (Gas) and/or *49 CFR Part 195* (Liquid)

Copies of your company policy and procedures manual

Rectifying cells

Bridge circuits

Rectifiers

A variety of meters, including the following
- d'Arsonval meters
- Voltmeters
- Ammeters
- Volt-ammeters
- Digital meters

Tools to connect meters to rectifiers

Copies of *Figure 1* with the callouts covered

Several ohmmeters

Test leads

Tools to connect leads

Fuses

Rectifier components, including
- Meters
- AC breakers
- AC fuses
- Transformers
- Stacks

Three-phase circuits

Tools to test three-phase circuits

Tools to adjust the fine tap settings

Electronic control cards

* Located in the Test Booklet.
** Available for download from www.nccer.org.

TEACHING TIME FOR THIS MODULE

An outline for use in developing your lesson plan is presented below. Note that each Roman numeral in the outline equates to one session of instruction. Each session has a suggested time period of 3 hours. This includes 10 minutes at the beginning of each session for administrative tasks and one 10-minute break during the session. Approximately 18 hours are suggested to cover *Maintain and Repair Rectifiers (CT 4).* You will need to adjust the time required for hands-on activity and testing based on your class size and resources.

Topic	Planned Time
Session I. Rectifier Systems, Part One	
A. Transformer	______
B. Lightning Arresters	______
Session II. Rectifier Systems, Part Two	
A. Circuit Breakers	______
B. Rectifying Cells and Bridge Circuits	______
C. Meters	______
1. Laboratory	______
Session III. Troubleshooting, Part One	
A. Test Equipment	______
B. Safety	______
1. Laboratory	______
C. Testing Single-Phase Circuits	______
1. Laboratory	______
Session IV. Troubleshooting, Part Two	
A. Testing Three-Phase Circuits	______
1. Laboratory	______
B. Troubleshooting Rectifier Problems	______
C. Rectifier Troubleshooting Procedure	______
D. Troubleshooting Rectifier Bond Connections	______
1. Performance Testing	______
Session V. Repair and Replace Defective Rectifier Components	
A. Replace Rectifier Meter	______
B. Replace AC Breakers	______
C. Replace Rectifier AC Fuse(s)	______
D. Replace the Transformer	______
E. Repair or Replace the Rectifier Stack	______
1. Laboratory	______
2. Performance Testing	______
Session VI. Adjust the Rectifier	
A. Adjustment of the Rectifier	______
B. Setting Up Electronic Control Cards	______
C. Summary	
1. Summarize module	______
2. Answer questions	______
D. Module Examination	______
1. Trainees must score 70% or higher to receive recognition from the NCCER.	
2. Record the testing results on Craft Training Report Form 200 and submit the results to the Training Program Sponsor.	

E. Performance Testing

1. Trainees must perform each task to the satisfaction of the instructor to receive recognition from the NCCER.
2. Record the testing results on Craft Training Report Form 200 and submit the results to the Training Program Sponsor.

Module 61202-02

Maintain and Repair Rectifiers

(CT 4)

Instructor's Notes:

Instructor's Notes:

Introduction to the Pipeline Operator Qualification Rule, API, and NCCER

It could be said that petroleum products keep America running. Petroleum products like oil and natural gas are the principal sources of energy in the United States and are used in many manufacturing and chemical processes. Petroleum products are primarily transported by pipeline. There are more than a million miles of liquid and natural gas pipeline in this country, ranging in diameter from two inches to four feet. These pipelines bring crude oil and natural gas from production fields to processing plants and refineries—and ultimately to the consumer.

Pipeline Operator Qualification

Petroleum products are considered hazardous, so pipeline personnel must take extra precautions and follow specific procedures when working with them. To help ensure the safe handling and transportation of petroleum products, the United States Department of Transportation (DOT) mandates safe practices as federal regulations.

One such regulation, the Operator Qualification (OQ) rule, requires pipeline operators to develop a written qualification plan to evaluate their personnel and their contract personnel (see *49 CFR Part 192 Subpart N—Gas,* and *49 CFR Part 195 Subpart G—Liquid*). The evaluation hinges on the operator's ability to perform certain covered tasks and to be able to recognize and respond to abnormal operating conditions that may be encountered while performing these covered tasks.

Covered Tasks

A pipeline performance task is deemed *covered* if the activity passes the following four-part test:

1. Is it performed on a pipeline facility?
2. Is it an operations or maintenance task?
3. Is it performed as a requirement of *49 CFR Part 195 (Liquid)* or *49 CFR Part 192 (Gas)*?
4. Does it affect the operation or integrity of the pipeline facility?

If the answer to all four questions is *yes*, then the activity is considered *covered* and must figure into the evaluation process of pipeline personnel.

American Petroleum Institute

The OQ rule and its related covered tasks provided the catalyst to develop a unified pipeline training and assessment program—so that pipeline operators and owners have ready access to the assessment and training tools they need. This effort was the direct result of the vision and leadership provided by The American Petroleum Institute (API) and API members.

Efforts to unify the industry are nothing new to API. Indeed, the impetus for forming the Institute in 1919 was the need to standardize engineering specifications for drilling and production equipment. Today, the most pressing issues revolve around public perceptions and government policies toward the industry. Speaking with one voice on these issues has become as essential as having interchangeable parts in the field. API's members determine the petroleum industry's positions on public policy as well as the standards that govern its day-to-day operations worldwide.

Consortium on Operator Qualification

API assembled a focus group whose primary purpose was to identify covered tasks and provide compliance recommendations to pipeline operators.

The Consortium on Operator Qualification (COOQ) became the guiding force for the identification and development of covered tasks for the liquid pipeline industry. Their work product, API publication 1161, *Guidance Document for the Qualification of Liquid Pipeline Personnel*, formed the basis for the resulting Pipeliner Training and Assessment Program.

The covered task list created in API publication 1161 has now evolved to include gas covered tasks and is maintained by NCCER. The most current covered task list can be downloaded at www.nccer.org.

National Center for Construction Education and Research

With the guidance and support of API, its members, and the Consortium on Operator Qualification, the National Center for Construction Education and Research (NCCER) was able to design and implement the Pipeliner Training and Assessment Program.

NCCER is a not-for-profit 501(c)(3) education foundation established in 1995 by the world's largest and most progressive construction and maintenance contractors and national contractor associations. NCCER was founded to address the

severe workforce shortage facing the construction and maintenance industries and to develop a standardized training process and curricula. Today, NCCER is supported by hundreds of leading construction and maintenance companies, manufacturers, and national associations and developed more than 35 construction and maintenance titles.

NCCER also maintains a National Registry, which contains the records of trainees and instructors who have received NCCER training. All completion information for every trainee is sent to the NCCER and filed within the Registry. The Registry can then confirm training and skills for workers as they move from company to company, state to state, or even within their own company.

Standardized Curriculum (NCCER's Contren™ Learning Series)

NCCER curricula are developed using the expertise of industry subject matter experts (SMEs). The use of SMEs as an integral part of every new development or revision ensures that the curriculum is current, industry-driven, and indicative of industry practices. The development and/or revision process is facilitated from start to finish by NCCER project managers, who work closely with the SMEs and a select team of technical writer(s) on the development and presentation of the curriculum.

Pipeline: Five Training Paths

The pipeline curriculum focuses on essential skills and competencies within five skill areas:

- Corrosion Control
- Electrical and Instrumentation
- Maintenance
- Mechanical
- Operations (Field and Control Center)

SMEs from the pipeline industry have determined the skills and competencies needed based on API's covered tasks and the consensus of the group/industry. Specific company information is generally not incorporated into the curriculum unless the SMEs determine that the information is the industry standard.

Training Modules and Covered Tasks

The smallest unit of instruction in the curriculum is a *module*. Each curriculum is composed of modules (the equivalent of stand-alone chapters) that are logically sequenced into complete *levels* of instruction. Every module is task-oriented and requires the successful completion of a written (knowledge-based) examination and a performance (hands-on) test. Modules may be individually selected and/or combined for efficient task training. Modules in the pipeline curricula, to the extent possible, will align with API-developed covered tasks. While not all modules will relate to covered tasks, all covered tasks will relate to at least one module.

Benefits of Training to the Pipeline Marketplace

Applying NCCER's standardized training process allows pipeline owners and contractors in the United States to establish the benchmark for acceptable skill and competency levels within the pipeline industry. This process will assist training organizations in developing and/or improving standardized pipeline training programs that effectively match their resources and operating conditions while responding to regulatory demands and to the needs of the industry. The process will also recognize training organizations that meet the minimum standards of competence, warranting regulatory, public, and professional confidence. This in turn will foster national unity within the pipeline industry for the mutual benefit of individual technicians, owners, contractors, and ultimately pipeline-users and society.

Additional Resources

To review the Pipeliner Training and Assessment Program Implementation and Resource Guide and the latest version of pipeline covered tasks, visit NCCER's Web site at www.nccer.org, or contact:

National Center for Construction Education and Research
PO Box 141104
Gainesville, FL 32614-1104
(888) 622-3720
Web site: www.nccer.org

For more information on API, please contact:

American Petroleum Institute
1220 L Street, Northwest
Washington, D.C. 20005
(202) 682-8000
Web site: www.api.org

To order *API Pub 1161*, contact Global Engineering Documents by calling 1-800-854-7149 or visiting their Web site at www.global.his.com.

Additional sources of information on the pipeline operator qualification regulation include:

DOT – Office of Pipeline Safety
www.ops.dot.gov

Transportation Safety Institute
www.tsi.dot.gov

Instructor's Notes:

COURSE MAP

This course map shows all of the modules in the second level of the Pipeline Corrosion Control curriculum. The suggested training order begins at the bottom and proceeds up. Skill levels increase as you advance on the course map. The local Training Program Sponsor may adjust the training order.

PIPELINE CORROSION CONTROL LEVEL TWO

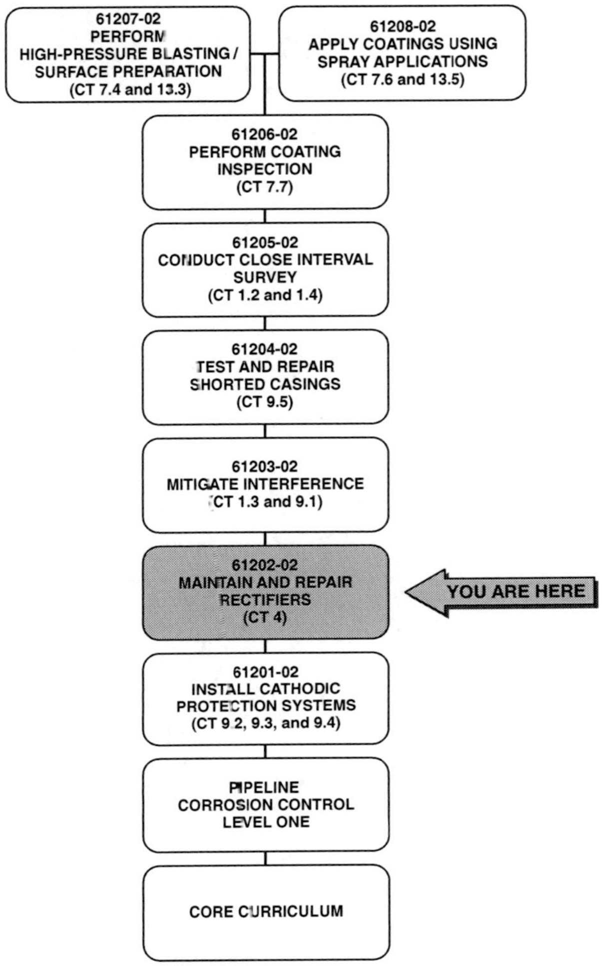

202CMAP.EPS

Homework

Assign reading of Module 61202-02.

Instructor's Notes:

MODULE 61202-02 CONTENTS

MODULE 61202-02 CONTENTS (Continued)

Instructor's Notes:

Figures

Tables

Instructor's Notes:

MODULE 61202-02

Maintain and Repair Rectifiers (CT 4)

Objectives

When you have completed this module, you will be able to do the following:

1. Describe the characteristics of cathodic protection rectifiers.
2. Describe the functions of typical rectifier components.
3. Describe troubleshooting techniques for rectifier and bond connections (CT 4.1).
4. Describe the common repair process for a rectifier (CT 4.2).
5. Describe the adjustment procedures for a rectifier (CT 4.3).

Prerequisites

Before you begin this module, it is recommended that you successfully complete the following: Core Curriculum; Pipeline Core, Modules 66101-02 and 66102-02; Pipeline Corrosion Control Level One; Pipeline Corrosion Control Level Two, Module 61201-02.

Required Trainee Materials

1. Pencil and paper
2. Appropriate personal protective equipment

1.0.0 ◆ INTRODUCTION

One way to protect a pipeline from corroding is to add electrical energy to the pipeline. The preferred way to provide this protection is with a constant supply of DC electrical **current**. This is because corrosion is a DC phenomenon. Unfortunately, only AC power is readily available. The solution is to convert or rectify the AC to DC, which is the purpose of the rectifier.

A rectifier is the device that supplies the DC power to the anodes to protect the pipeline. Supplying the ground bed or pipe with this converted **voltage** provides a constant source of energy.

Pipeline safety regulations relating to rectifiers are covered in Department of Transportation (DOT) *Office of Pipeline Safety Regulations, Part 192 and 195*. The cathodic protection rectifier is discussed in the following sections.

2.0.0 ◆ RECTIFIER SYSTEMS

Rectifiers used for cathodic protection are basically simple electrical circuits. The circuit contains a transformer, lightning arresters, **circuit breakers**, rectifying cells and bridge circuits, and meters. The system or circuit is housed in a rectifier cabinet with a control panel on the front and electrical connections to the structure to be protected. The following describes these components and their functions. *Figure 1* shows a simple rectifier circuit.

2.1.0 Transformer

The transformer's purpose in the system is to step down or step up voltage from the supply line to the voltage required in the application. Transformers are constructed with primary and secondary coils around a core. The supply current travels through the primary coil, creating a magnetic field around the coil. This field induces a current in the secondary coil that provides the power to the application.

Materials

Ensure that you have everything required to teach the course. Check the Materials and Equipment List at the front of this module.

Classroom

Explain that terms shown in bold are defined in the Glossary at the back of this module.

Audiovisual

Show Transparency 1, Course Objectives.

Show Transparency 2, Performance Tasks.

Teaching Tip

See the general Teaching Tip at the end of this module.

Classroom

Explain how providing a constant supply of DC electrical current can protect a pipeline from corroding.

Explain why it is necessary to use a rectifier to convert AC power to DC.

Audiovisual

Show Transparency 3 (Figure 1). Review the components in a rectifier circuit.

Classroom

Discuss the role of the transformer in a rectifier. Answer any questions trainees may have.

Audiovisual

Show Transparency 4 (Figure 2). Explain how a transformer operates.

Classroom

Review the various designs of lightning arresters and explain how they operate.

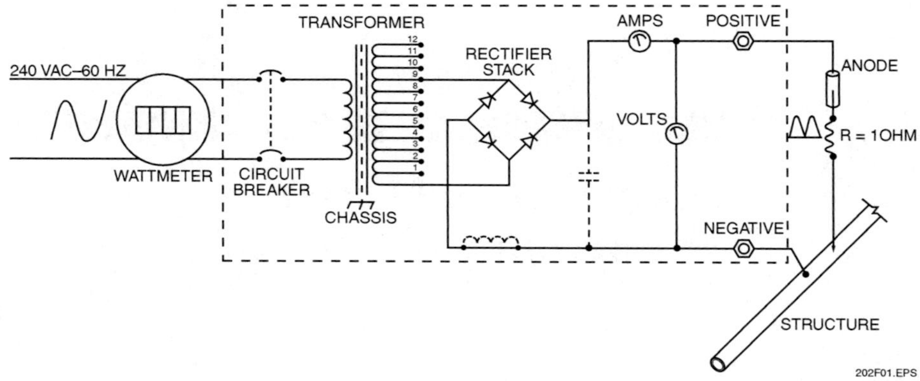

Figure 1 ◆ Simple rectifier circuit.

The amount of change to the supply power depends on the ratio of the number of windings in the primary coil to the number of windings in the secondary coil. Windings are the loops of the coils around the core. *Figure 2* shows the basics of a transformer.

If the primary coil has three loops or windings and the secondary coil has six, then the power will be stepped up or doubled in voltage when produced in the secondary coil. If the reverse is true, the power supplied to the transformer will be cut in half in the secondary coil, or stepped down. The ratio in the first case is 1:2; in the second case, the ratio is 2:1.

Transformers only work on AC power supplies. But since the coils are not connected electrically, the incoming AC power is isolated from the DC power in the application. *Figure 3* shows a typical rectifier unit transformer.

The voltage ratio produced in the transformer in *Figure 3* is under no-load conditions. When a load is applied, the voltage from the secondary to the load, or application, is reduced by the effects of wire resistance and lamination. This loss in voltage is commonly called transformer regulation and depends on the equipment and application.

The voltage produced in a transformer can be adjusted by the use of taps installed on the secondary windings. In effect, these taps change the number of windings by completing different circuits in the coil.

2.2.0 Lightning Arresters

To protect against voltage surges, lightning arresters are installed on both the supply and application sides of the rectifier. Lightning arresters come in a variety of designs. Some of the different types of lightning arresters used include the following.

2.2.1 Arc-Gap Arresters

An arc-gap arrester is simply designed, having two electrodes between which a high-voltage surge ionizes the air. This causes the current to arc across the gap between the electrodes as long as the higher voltage lasts. The arcing dissipates the incoming voltage before it can damage the circuit and equipment.

2.2.2 Expulsion Arresters

The expulsion arrester (see *Figure 4*) works much the same way as the arc-gap arrester. The expulsion arrester consists of two line electrodes and a ground electrode. The main difference is in the fiber discs, which act to separate the line electrodes from the ground. They also break up current arcs into smaller arcs, dissipating voltage faster. This helps to protect the arrester itself.

2.2.3 Valve Arresters

A resistive element in the valve arrester provides low impedance to the high voltage of lightning but a high impedance to normal line voltage. This acts to prevent normal line voltage from sustaining the arc caused by the lightning.

The valve has two metal electrodes, which are on either side of the element, and metal fingers, which are separated from the electrodes by an air gap that

Instructor's Notes:

becomes the spark gap. A surge of high voltage causes an arc to jump from the fingers to the nearest electrode. Since the valve element has low impedance to this surge voltage, it conducts current from one electrode to the other, dissipating the voltage. *Figure 5* shows the valve arrester construction.

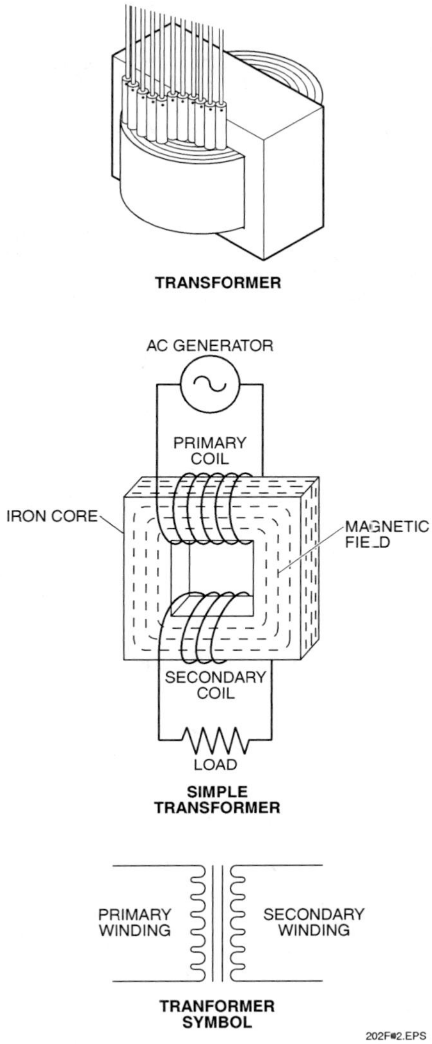

Figure 2 ◆ Basics of a transformer.

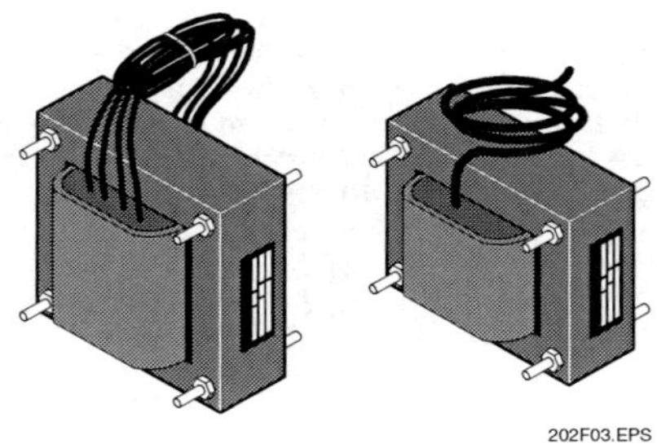

Figure 3 ◆ Typical rectifier unit transformer.

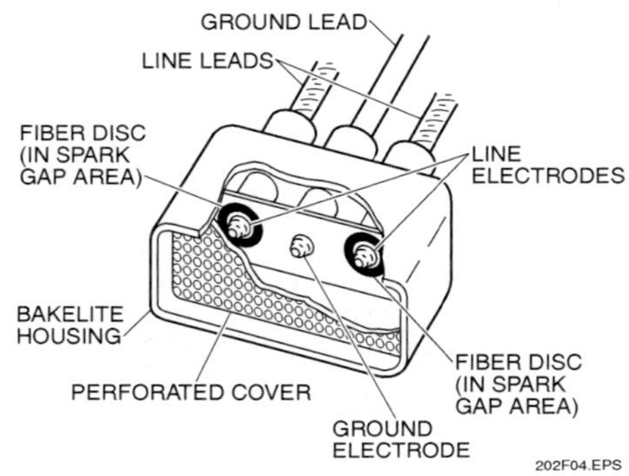

Figure 4 ◆ Expulsion arrestor.

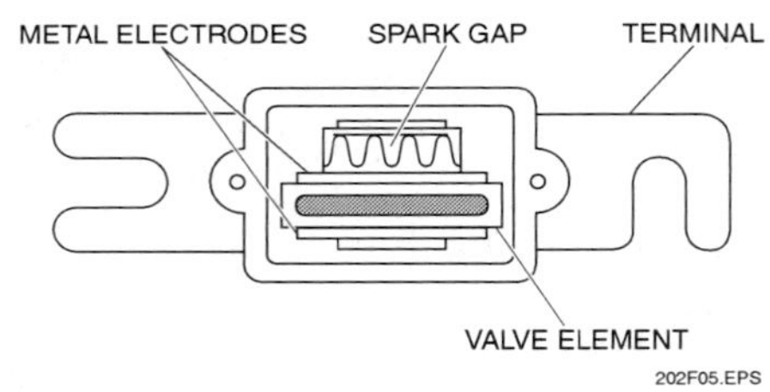

Figure 5 ◆ Valve arrester construction.

Classroom

Refer to Figure 5. Discuss the role of the valve element in a valve arrester.

Discuss the differences among arc-gap, expulsion, valve, gas, and silicon arresters.

Classroom

Have a trainee explain how a metal oxide varistor (MOV) works. Ask another trainee to identify the components that an MOV protects.

Ask trainees to identify each of the lightning arresters discussed in this section. Have trainees explain a suitable application for each of the lightning arresters. Review the advantages and/or disadvantages of each.

Homework

Have trainees review Sections 2.3.0–2.5.4 for the next session.

Materials

Ensure that you have everything required for the demonstration and laboratory during this session.

Classroom

Ask a trainee to explain why the handles on circuit breakers must be ganged.

Audiovisual

Show Transparency 5 (Figure 7). Discuss the role of the bimetallic element. Ask a trainee to explain what happens when an excessive current passes through the elements.

2.2.4 Gas Arresters

Inert gases are used in the gas arrester to conduct current from one electrode to the other. Argon and neon, normally insulators, are the gases most often used in gas arrestors. When a surge of voltage acts on the gases, they are ionized and become conductive to current. After the surge of voltage has passed, they return to their insulator state. Rare gas tubes can be connected in parallel with an arc gap to handle instantaneous surges. *Figure 6* shows a typical gas arrester.

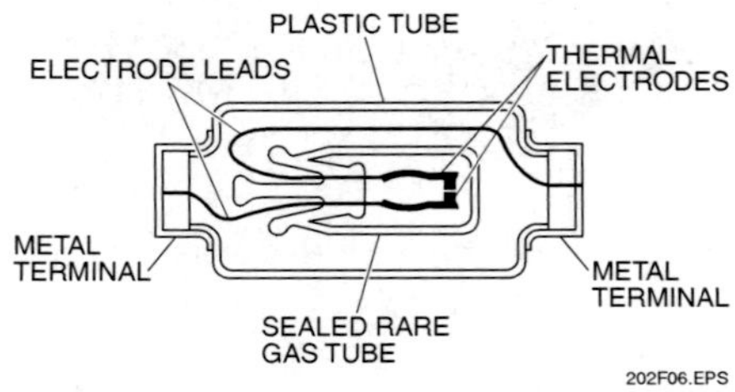

Figure 6 ◆ Typical gas arrester.

2.2.5 Silicon Arresters

The silicon arrester uses silicon oxide to surround two or more electrodes. The silicon oxide ionizes during a voltage surge, changing from a compound to silicon and allowing large amounts of current to pass through. After the surge, the silicon returns to its original state, which is no longer conductive to current. This type of arrester has no spark gap, no series resister for limiting current, and is not temperature sensitive.

2.2.6 Metal Oxide Varistors

Metal oxide **varistors** (MOVs) may be mounted across the AC power input and across the DC output to protect against voltage surges. The MOV is a voltage-sensitive **resistor** that reacts quickly to voltage surges, conducting large amounts of current for short periods of time. MOVs are used to protect **silicon diodes**, silicon controlled rectifiers (SCRs), and other sensitive electronic components from surges.

2.3.0 Circuit Breakers

Circuit breakers are used on the AC power input to turn the rectifier on and off. They also provide for overload protection. Breakers are placed in each hot AC supply line; with single-phase 120 volt AC, one breaker is placed in the hot line, and the ground or neutral line does not need one. Higher single-phase input requires breakers on both lines. Three-phase equipment must have breakers on each supply line. The handles on all circuit breakers must be **ganged** so that all input lines will open at the same time, cutting off all power.

Three basic types of circuit breakers are the fully thermal, the thermal-magnetic, and the fully magnetic. The differences between these circuit breakers are in their structure and their response to excessive current.

2.3.1 Fully Thermal Circuit Breakers

The fully thermal circuit breaker (*Figure 7*) operates by heat. It has bimetallic elements that carry current when the breaker is on. When an excessive current passes through these elements, the

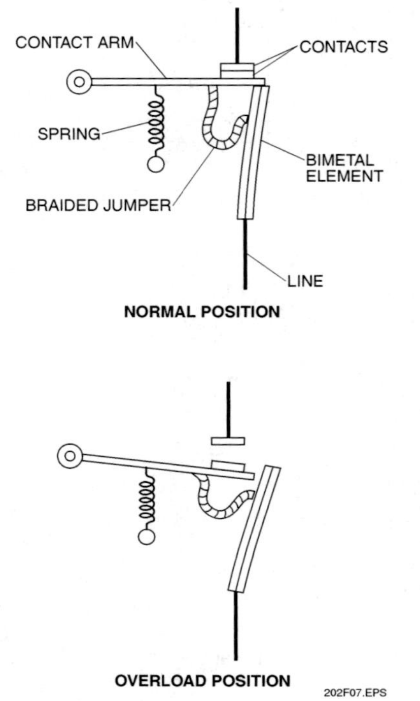

Figure 7 ◆ Thermal breaker.

Instructor's Notes:

resulting heat causes the two metals to expand according to their different heat coefficients. The difference in their expansion rates causes the elements to bend, releasing a moveable contact and opening the circuit. The nature of this breaker makes it sensitive to ambient temperature. It tends to open more quickly on warm days than cold and often must cool before resetting.

2.3.2 Thermal-Magnetic Circuit Breakers

The thermal-magnetic circuit breaker (*Figure 8*) is similar to the thermal breaker except for a magnetic plate attached to the bimetallic element. During heavy overload conditions, the plate will speed up the opening of the circuit contacts.

A magnetic field is set up around the magnetic plates during a heavy surge. This causes the plate to be attracted to an adjacent plate, opening the contacts. When no heavy overload exists, the magnetic plate has no function. These breakers may be used on rectifiers equipped with interrupters. The thermal-magnetic breaker has a slower reaction time than a fully magnetic breaker. This is an important quality for circuit interruption.

2.3.3 Fully Magnetic Circuit Breakers

The fully magnetic breaker responds only to overloads in current. This breaker is constructed with an iron core immersed in silicon fluid and retained by a compression spring. The core is encased in a sealed tube around which is wound a magnetic coil. It is the turns in this coil and the size of its wire that determines the current rating of the breaker. The iron core remains stationary as long as the current is within the rating limits. The fully magnetic breaker (*Figure 9*) operates independently of temperature.

During an overload, the magnetic action of the coil causes the core to be drawn toward the pole face on the end of the tube. As the core approaches the pole face, it increases a magnetic pull on the arm holding the contacts closed. At some point, the lever trips and opens the contacts. In a short-circuit condition, the magnetic action of the core itself trips the arm right away.

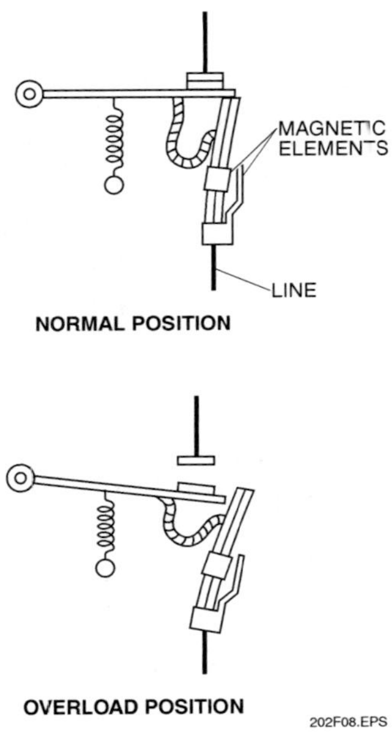

Figure 8 ◆ Thermal-magnetic breaker.

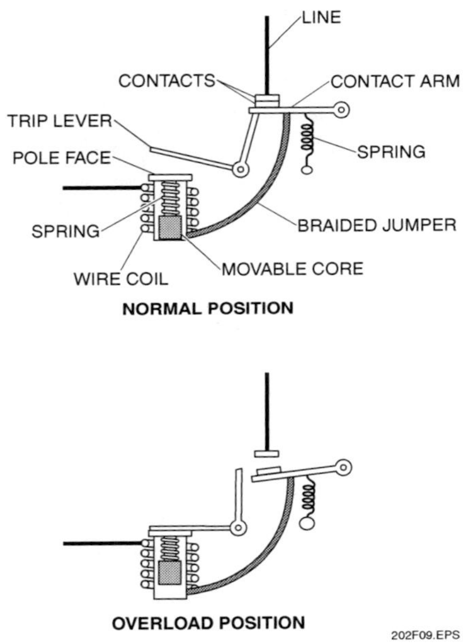

Figure 9 ◆ Fully magnetic breaker.

Classroom

Refer to Figure 8. Explain how a magnetic field causes the contacts to open.

Audiovisual

Show Transparency 6 (Figure 9). Explain what happens during an overload in the current.

Classroom

Discuss the differences among fully thermal, thermal-magnetic, and fully magnetic circuit breakers. Go over any questions trainees may have.

Audiovisual

Show Transparency 7 (Figure 10). Explain how a rectifying cell controls current flow.

Safety

Discuss the potential consequences of handling selenium.

Demonstration

Display several examples of rectifying cells and bridge circuits. Demonstrate how to properly handle highly toxic selenium cells.

2.4.0 Rectifying Cells and Bridge Circuits

A rectifying cell or element (*Figure 10*) within a rectifier will pass current in one direction but block current flow in the other direction. Rectifying cells are commonly made of selenium or silicon.

Within a rectifier, the current passes in one direction, with the amount of current dependent on the amount of voltage. When the **polarity** of the voltage changes, the amount of current flow decreases unless the applied voltage exceeds the reverse rating of the rectifier. This will cause the rectifier to fail, allowing current to flow in either direction, or to open, allowing no current to flow.

The schematic symbol for a rectifying cell includes an arrow to indicate the direction in which current will flow when the proper voltage is applied. Current will be blocked in the opposite direction. *Figure 11* shows a rectifying symbol.

2.4.1 Selenium Cell

A selenium cell (*Figure 12*) consists of an aluminum plate covered in selenium crystals covered by a metal film. The metal film is the contact area. A barrier layer on the selenium side of the plate prevents current from passing from the selenium to the aluminum.

CAUTION
Selenium is highly toxic. Always wash your hands carefully after handling a selenium cell.

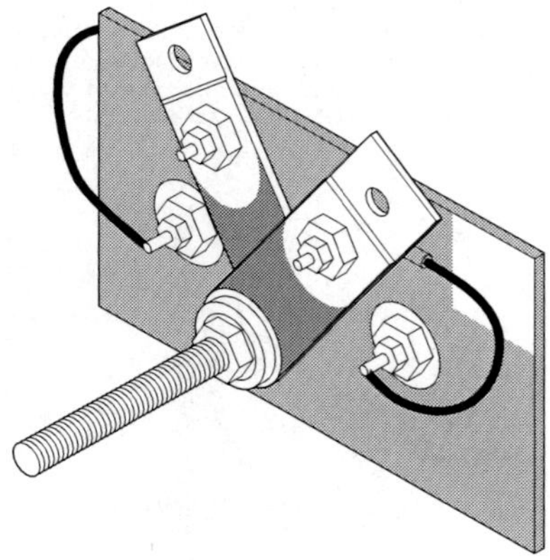

Figure 10 ◆ Typical rectifying cell or element.

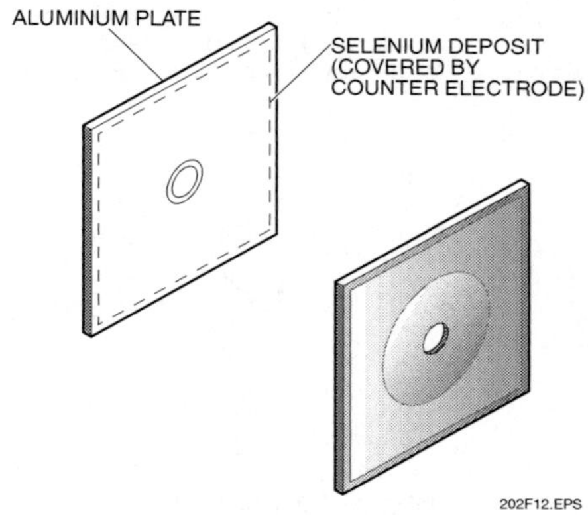

Figure 12 ◆ Selenium cell.

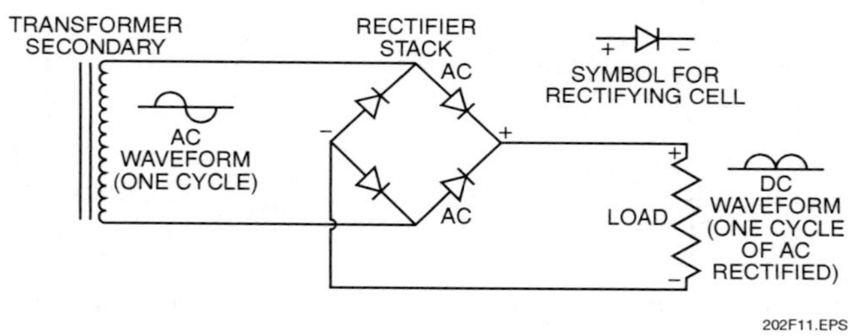

Figure 11 ◆ Rectifying symbol.

Instructor's Notes:

2.4.2 Silicon Diode

A silicon diode (*Figure 13*) is a single crystalline rectifying junction. One side of it is the anode, and the other side is the cathode. Because of its size and smaller rectifying area, the **diode** is mounted on a metal plate or heat sink that draws heat away from the junction area.

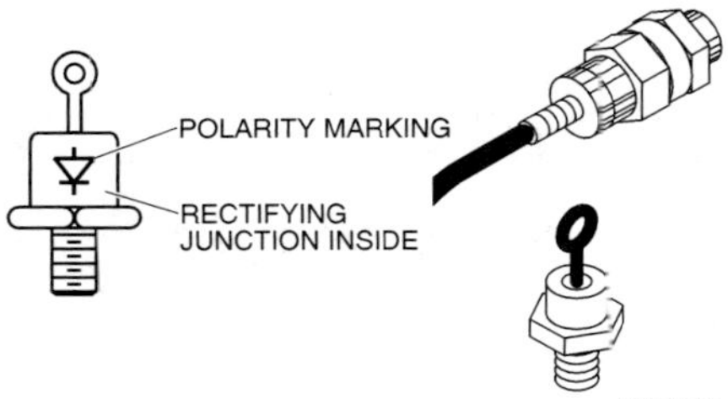

Figure 13 ◆ Silicon diode.

2.4.3 Silicon Controlled Rectifier

A silicon controlled rectifier (SCR), or thyristor, is an electronic switch, passing current in the forward direction when on and blocking current in the opposite direction, even with hundreds of applied volts. The SCR is similar to the diode electrically, but it has another lead that is used to turn it on and off.

2.4.4 Stack

A **stack** is the common name for an assembly of selenium plates or silicon diodes arranged to rectify AC current and produce DC output. Each stack is composed of one or more legs. A leg is a unidirectional path for current and may consist of one or more cells or elements.

A power module stack contains all the electronic components of a **rectifier stack** in one encapsulated package. They have lower current capabilities than equally rated individual rectifying elements because their construction makes them difficult to cool. Their small size and ability to isolate internal electrical components from the case are their advantages. They also have good serviceability.

2.5.0 Meters

Cathodic protection systems often use d'Arsonval meters. This type of meter consists of a coil of fine wire suspended so that it rotates within a permanent magnet. The coil rotates under the influence of current passing through it, moving a pointer across the face of a calibrated dial. *Figure 14* illustrates d'Arsonval meter movement.

The dial indicates an amount of deflection that is proportional to the amount of current. A spring restraint is used to oppose any torque generated by the current in the coil. In many cathodic protection rectifiers, the restraint is a helical spring, with the coil supported by a pivot and a jewel. Another type is the taut band suspension, in which a flat band held under tension provides support for the coil, resistance to twisting, and restraint.

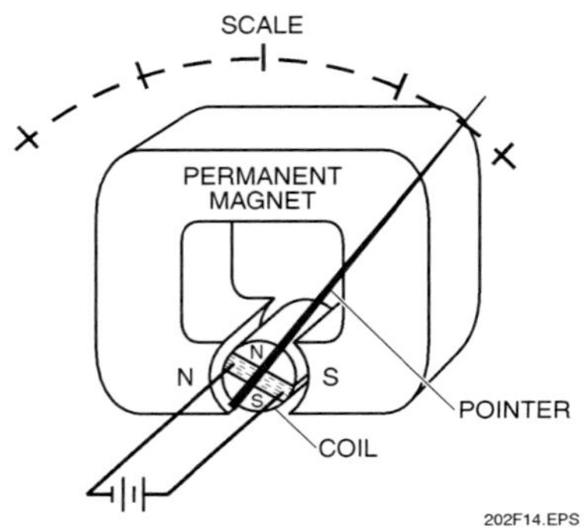

Figure 14 ◆ d'Arsonval meter movement.

2.5.1 Voltmeters

Voltmeters usually have 50-millivolt (mV) full-scale movements. They can be used with calibration resistors placed in series with the movement to prevent excess voltage and meter damage. A voltmeter must be connected across the rectifier output with correct polarity observed. *Figure 15* shows a voltmeter with a resistor.

2.5.2 Ammeters

A meter used as an ammeter is connected in series with either the positive or negative output of the rectifier. A **shunt** resistor is used to prevent full output current from passing through the meter coil. The shunt will drop the voltage when the current passes through it; the amount of the drop is proportional to the amount of current. Stamped on the shunt is a current rating that is also the rating of the ammeter.

Classroom

Compare a silicon diode with a silicon controlled rectifier.

Ask trainees to review the advantages and disadvantages of a rectifier stack.

Audiovisual

Show Transparency 8 (Figure 14). Review a d'Arsonval meter movement.

Classroom

Explain how voltmeters and ammeters are connected to rectifiers. Ask whether trainees have any questions.

Audiovisual

Show Transparency 9 (Figure 16). Ask a trainee to identify the distinct benefit of a volt-ammeter.

Laboratory

Set up stations with d'Arsonval meters, voltmeters, ammeters, volt-ammeters, and digital meters. Ask trainees to practice reading the meters. Have trainees identify the differences among the meters.

Teaching Tip

See the Teaching Tip for Sections 2.0.0–2.5.4 at the end of this module.

Homework

Assign the reading of Sections 3.0.0–3.3.7 for the next session.

Materials

Ensure that you have everything required for the laboratories and demonstration during this session.

Safety

Emphasize the importance of ensuring that a rectifier is not energized before touching it.

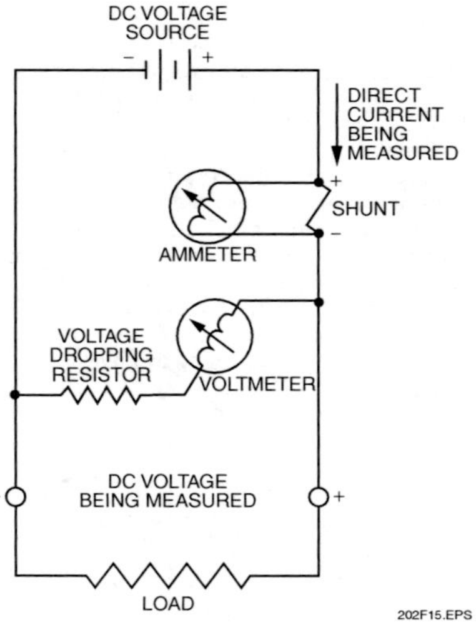

Figure 15 ◆ Voltmeter with a resistor.

2.5.3 Volt-Ammeter

Sometimes a meter is used as both ammeter and voltmeter. The meter used in this application can be larger and easier to read, since it takes up about the same space as two separate smaller meters. It also has a switch to remove it from the circuit until needed. This is a distinct benefit because a meter is one of the most sensitive components of a rectifier system and is therefore vulnerable to voltage surges. *Figure 16* shows a volt-ammeter in circuit.

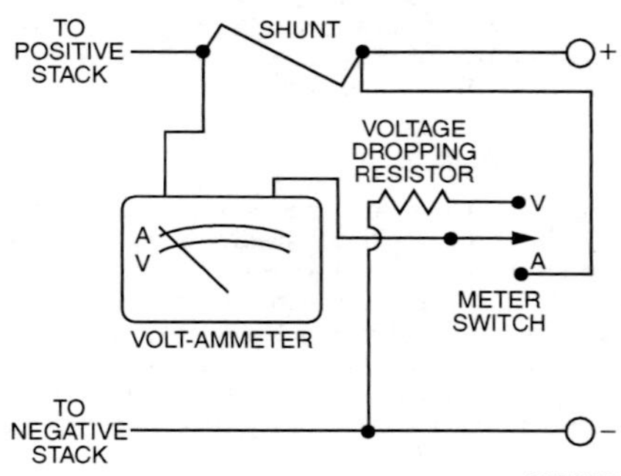

Figure 16 ◆ Volt-ammeter in circuit.

2.5.4 Digital Meters

Among the other meters that may be used are digital meters. They are used in the same way as d'Arsonval meters but have light-emitting diodes (LEDs) or liquid crystal displays (LCDs). In bright light such as sunlight, LEDs are difficult to read, whereas LCDs display well in bright light. However, LCDs cannot be read in the dark, and they require an external light source.

Digital meters can use a variety of input voltages and are available with a varying number of display digits. Multiplier ranges supplied with the meter make it possible to read volts and amperes.

3.0.0 ◆ TROUBLESHOOTING

Rectifier failures can often be prevented by routine maintenance and inspections. When failures do occur, the corrosion technician must be able to troubleshoot the problem. Safety procedures must always be followed when working with electrical equipment. This includes rectifiers.

> **WARNING!**
> Before touching a rectifier, always check that it is not energized.

Although the rectifier is grounded, there may be some malfunction causing it to still be energized. This could be a poor ground or other problem in the unit or the cabinet. A slight tap on the cabinet can de-energize the cabinet, lock it, and warn of the danger. The preferred method of checking is with a ground-fault detection device.

Once the unit is confirmed as safe, troubleshooting can begin. The techniques for troubleshooting a rectifier and **bond** connections are explained in the following sections.

3.1.0 Test Equipment

The equipment needed for troubleshooting a rectifier system is not elaborate. The list of equipment includes a digital or analog multimeter, test leads, and **fuses**.

A multimeter is used to read AC and DC voltages and DC current, to measure resistance, and to indicate short or **open circuits.** It can also check rectifier DC current by measuring the millivolt drop across the shunt.

Test leads for meters and meter fuses should be checked to ensure their serviceability before taking them to the field. However, it is still a good idea to always carry extra test leads and fuses for the meter being used.

Instructor's Notes:

3.2.0 Safety

A lot of pipeline rectifier troubleshooting may be performed in a remote area where assistance is not available. This makes safety consciousness even more critical. Always carry a means of communication when working in the field; even with contact, it could take some time before others can reach the site. In addition to the checks described earlier, always adhere to the following rules when working with rectifiers. This list only covers a few general safety items. Always consult your supervisor and/or company guidelines for specific information.

Step 1 Turn the rectifier off when handling components within the rectifier.

Step 2 Open the disconnect switch ahead of the rectifier as well as the internal circuit breakers.

Step 3 Follow company procedures when testing a rectifier in operation; this usually requires standing on a rubber mat and wearing rubber gloves.

Step 4 Consult the rectifier wiring diagram, if available, before starting work.

Step 5 Check that troubleshooting meters are properly connected:

- A voltmeter should be connected across or in parallel with the points to be measured.
- An ammeter should be connected in series with the circuit being tested.
- A multivoltmeter should be connected across the terminals of the shunt.

Step 6 Ensure that the correct polarity is observed when using DC instruments.

Step 7 Turn the rectifier off before using an ohmmeter.

3.3.0 Testing Single-Phase Circuits

The first step in testing a single-phase (or any other) rectifier circuit is to study the particular schematic for that circuit. The schematic is provided to display the internal workings of the rectifier and show how it is designed to work. *Figure 17* shows a typical single-phase rectifier circuit.

The circled letters on this schematic correspond to points on the control panel for this circuit. *Figure 18* shows a Good-All Electric, Inc. rectifier control panel. *Figure 19* shows another view of a Good-All single-phase control panel.

With an input of 240 VAC, a reading of 240 VAC should be obtained at the line side (A-A1) of the breaker and also at the load side (B-B1) with the circuit closed. Between both the high-and-low changeover and the B and low, a reading of 120 VAC should be measured. If the input voltage were decreased to 120 VAC, these readings would also be decreased by half. The DC output would be limited to one-half of the rated voltage.

3.3.1 Checking the Transformer

If the high-and-low changeover were connected to low, an input of 240 VAC would cause the voltage across the entire primary to be 480 VAC. This high voltage would cause heavy current through the primary winding, tripping the circuit breaker. It might, however, damage the transformer by saturating the lamination iron before the circuit break occurs.

To quickly check whether the transformer is working, take measurements at the tap change circle, points D and E, and at the coarse and fine taps. An equal AC voltage should exist between any two adjacent coarse taps and between any two adjacent fine taps, although the last voltage should be lower than the first.

3.3.2 No DC Output

When the circuit breaker is on, but there is no DC output from the rectifier, it may save time to first check the output from the transformer. This is done by measuring the voltage at the tap change circle (D and E or G and F).

The voltage measured at the center should be the voltage into the stack. This measurement is used to determine the approximate voltage out of the stack. The AC voltage measured into the stack should be greater than the DC voltage out of the stack because of the voltage drop associated with the rectifier elements. The difference will depend on the particular elements used. The DC voltage out of the stack is measured at points J-H or L-K.

If the measured DC voltage is only half of its expected value, then the stack is half-waving. This indicates that one leg of the stack is open.

3.3.3 Open Circuits

An ohmmeter can be used to check for open circuits. The disconnect switch must be open, and all power must be removed from the rectifier. With no power and the circuit breaker open, the resistance reading between the points A and B and between A1 and B1 should be infinity.

Classroom

Review the list of safety guidelines for working with rectifiers. Emphasize the importance of carrying an appropriate means of communication when working in the field.

Laboratory

Set up stations where trainees can practice the safety steps in turning off rectifiers, conducting testing, and connecting meters. Observe and assess the trainees' procedures. Answer any questions trainees may have.

Audiovisual

Show Transparency 10 (Figure 17). Identify the circuit breaker and explain how to assess various voltage measurements.

Classroom

Ensure that trainees understand how the circled letters on the schematic in Figure 17 correspond with the control panel in Figure 18.

Explain how to determine whether a transformer is working properly.

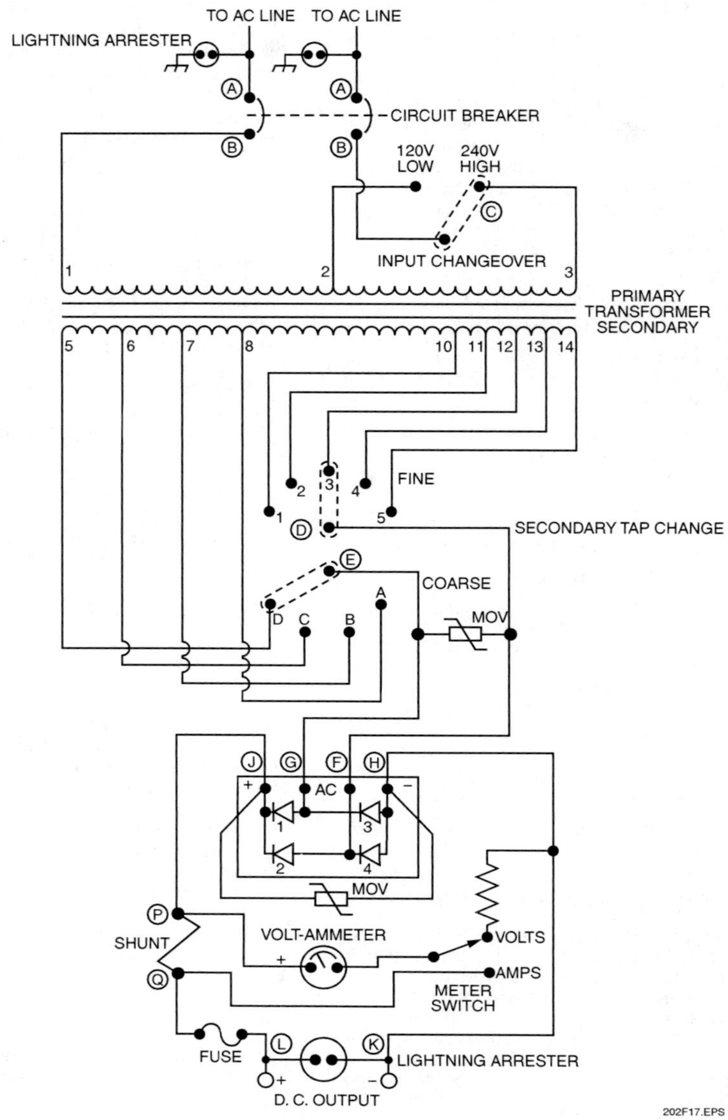

Figure 17 ◆ Typical single-phase rectifier circuit.

Instructor's Notes:

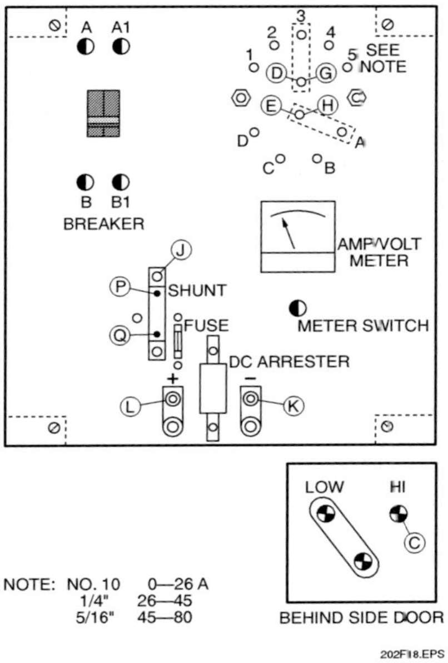

Figure 18 ◆ Good-All rectifier control panel.

With the breaker on, the measurements taken across each pole of breakers A-B and A1-B1 should yield a very low resistance. With the breaker still closed, the transformer primary can be checked by placing the ohmmeter probes on either side of the breaker. A low resistance should be measured since the primary wire is large with a resistance of 10 ohms or less per 1,000 feet.

NOTE

If the input changeover is on high, the entire transformer primary can be checked. If the input changeover is on low, only one half of the primary is measured.

3.3.4 *Transformer Resistance*

When checking transformer resistance, one link bar must be removed at the tap change circle (D or E). This will ensure that only transformer resistance is measured. Measuring between coarse taps D, C, B, and A, and fine taps 1, 2, 3, 4, and 5 should produce a low resistance reading. Taken in that order, the readings should increase progressively. With the ohmmeter probes connected between coarse D and fine 5, the entire resistance of the secondary can be read.

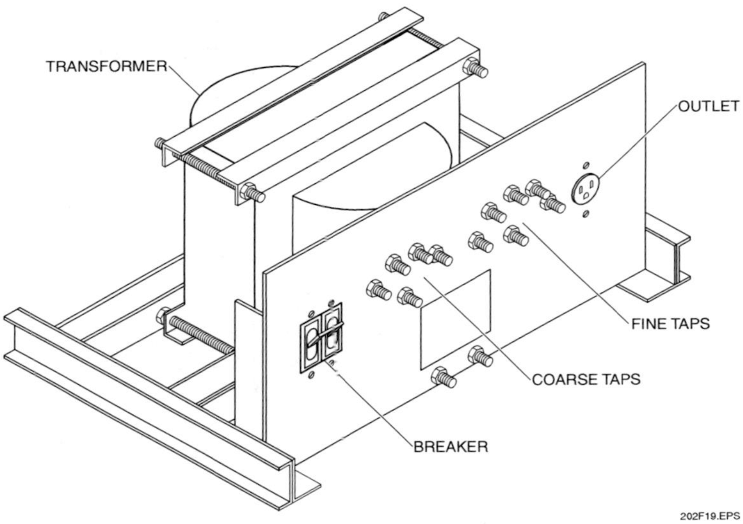

Figure 19 ◆ Good-All single-phase control panel.

Demonstration

Demonstrate how to use an ohmmeter. Explain how to measure resistance when the input changeover is on high versus when it is on low.

Audiovisual

Show Transparency 11 (Figure 20). Explain how to remove one link bar before checking the stack.

3.3.5 Checking the Stack

Before checking the stack, remove one link bar at the tap change circle and disconnect the load. *Figure 20* shows a typical rectifier stack. *Figure 21* shows a single-phase rectifier stack diagram.

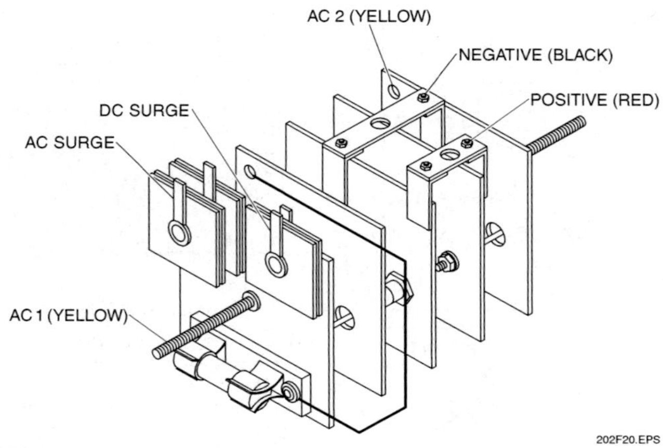

Figure 20 ◆ Typical rectifier stack.

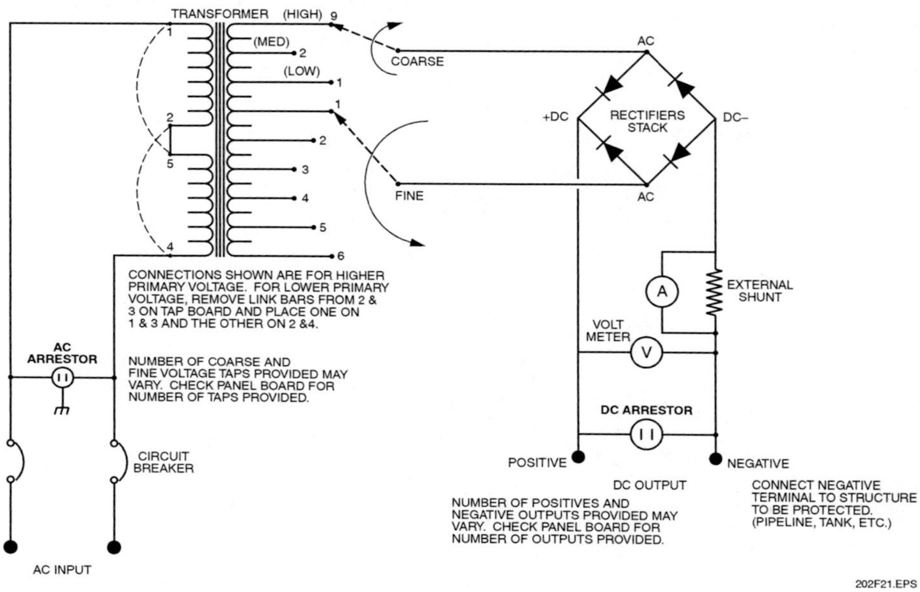

Figure 21 ◆ Single-phase rectifier stack diagram.

Instructor's Notes:

Follow these steps to check each diode:

Step 1 Set the ohmmeter on Rx1.

Step 2 Connect the ohmmeter positive lead (usually red) on the center of the tap change circle at E.

Step 3 Connect the ohmmeter common lead (usually black) on the positive output L.

Step 4 Check for a small resistance on diode 1 in the forward direction.

Step 5 Reverse the leads.

Step 6 Check for a large resistance on diode 1 in the reverse direction.

Step 7 Repeat the preceding steps to check each diode separately, using the connections shown in *Table 1*.

Step 8 If a diode has a low or high resistance in both directions, replace the diode and retest the stack.

3.3.6 Checking the Shunt

Determining the current rating of the rectifier shunt can usually be done quickly by one of two methods:

- Check the ammeter or volt-ammeter for the full-scale ampere rating (not the red-line rating).
- Check for a rating stamped on the shunt itself.

Most shunts are calibrated for a 50-millivolt drop when the rated current is flowing through, but some are rated differently.

The schematic should indicate which side of the shunt is positive and which is negative. When a shunt is in the negative side of the rectifier, the side connected to the stack negative is the negative of the shunt. When the shunt is in the positive side, the side connected to the stack positive is the positive of the shunt. Place the voltmeter probes on the appropriate shunt terminals to measure the millivolt drop across the shunt.

The resistance of a shunt is very low, usually 0.01 ohm or less. This is checked at the mounting studs with the load and meter disconnected. If the shunt has a high resistance, it is defective.

3.3.7 Other Checks

Fuses are easy to check, but they should always be removed from the circuit first. They should have a very low resistance if in good condition and an infinite resistance if bad.

Electrical connections to the meter and the switch are checked for open circuits with an ohmmeter. An ohmmeter should never be left connected directly across the rectifier terminals. This could cause the meter to exceed its rating and damage the meter.

The schematic should always be referred to before testing any part of the circuit, to avoid compounding a problem.

3.4.0 Testing Three-Phase Circuits

A three-phase rectifier circuit has three poles on the lightning arrester. The circuit breaker has three poles instead of two. A three-pole arrester and breaker are necessary since there are three hot lines to be protected. The transformer has three sections on the primary and the secondary, and the stack has six elements instead of four. *Figure 22* shows a typical three-phase rectifier circuit. *Figure 23* shows a three-phase control panel.

The transformer in the figure is connected delta-delta, although the three-phase transformer could also be connected in delta-wye or wye-wye configurations.

The same methods described for checking a single-phase circuit can be used on a three-phase circuit, except for the following:

- When measuring secondary voltages at the tap change circles, the AC voltage into the stack will be less than the DC voltage out of the stack.

Table 1 Diode Testing

Diode	Positive (Forward)	Common (Forward)	Forward Resistance	Reverse Resistance
1	E	L	Small	Large
2	D	L	Small	Large
3	E	K	Small	Large
4	D	K	Small	Large

202T01.EPS

Classroom

Refer to Table 1. Review the procedures for checking a diode. Ensure that trainees understand how to use the table for diode testing.

Laboratory

Bring in several ohmmeters. Ask trainees to practice setting the ohmmeters and connecting the leads. Ask a trainee to explain when a diode should be replaced.

Classroom

Explain how to determine the current rating of a rectifier shunt. Go over any questions trainees may have.

Emphasize the importance of referencing the schematic diagram before testing any part of a circuit.

Homework

Have trainees review Sections 3.4.0–3.7.0 for the next session.

Materials

Ensure that you have everything required for the laboratory and testing during this session.

Audiovisual

Show Transparency 12 (Figure 22). Explain why a three-pole arrester and breaker are necessary.

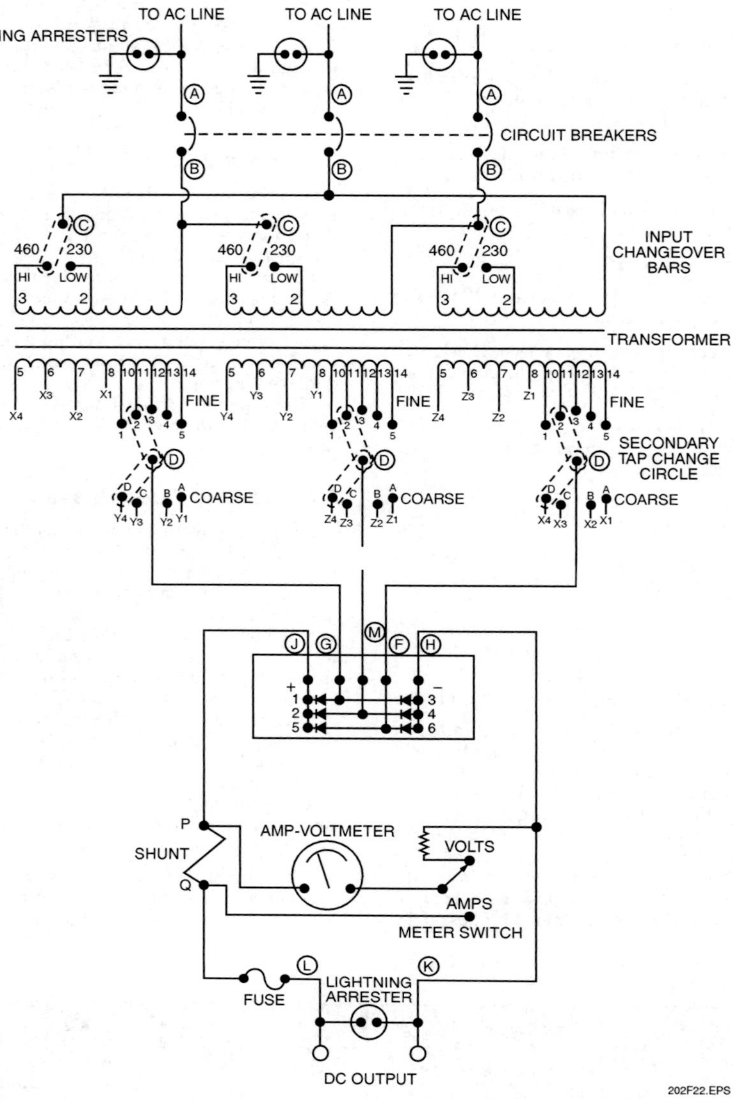

Figure 22 ◆ Typical three-phase rectifier circuit.

Instructor's Notes:

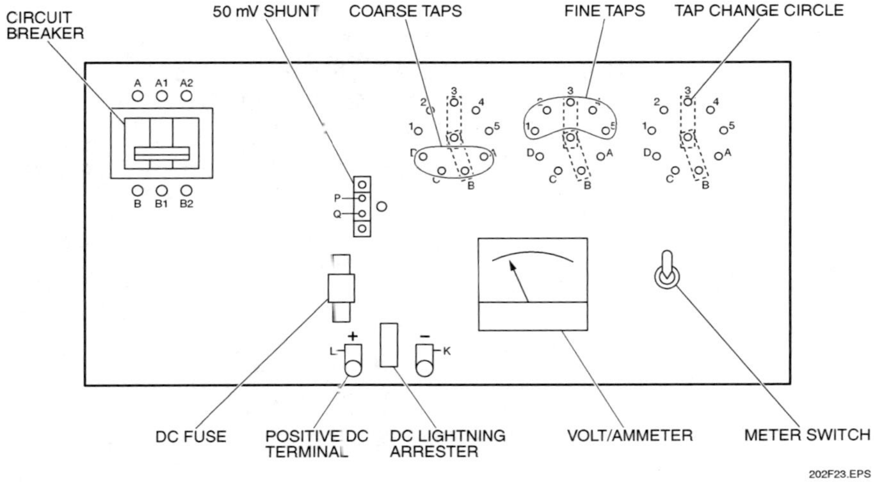

Figure 23 ◆ Typical three-phase control panel.

- When measuring the secondary AC voltage at the tap change circles, the value of the voltage between each of the studs should be approximately equal.
- There are two more diodes to be checked. The schematic should indicate test lead connections for the diodes in this circuit. *Figure 24* shows the method to test diodes.

> **CAUTION**
> The supply voltages on a three-phase circuit are usually higher than on a single-phase unit, and each line is hot.

3.5.0 Troubleshooting Rectifier Problems

Obvious signs of rectifier problems are loose connections, signs of arcing, and strange odors. These signs may indicate problems that do not require much testing. The most common problems with rectifiers are blown fuses, faulty meters, loose terminals, open ground bed leads, and lightning damage. The checks shown in *Table 2* may quickly eliminate many rectifier problems.

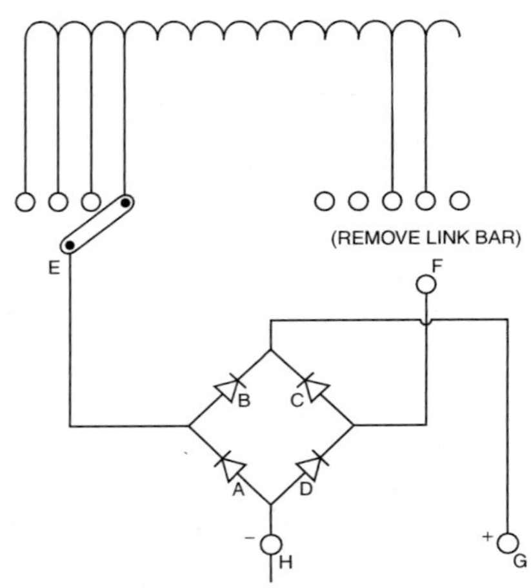

TO CHECK DIODE	PLACE LEADS BETWEEN
A	E AND H
B	E AND G
C	F AND G
D	F AND H

202F24.EPS

Figure 24 ◆ Method to test diodes.

Laboratory

Set up stations where trainees can practice the procedures for testing three-phase circuits. Have trainees review the differences between testing a single-phase circuit and a three-phase circuit.

Audiovisual

Show Transparency 13 (Figure 24.) Ask a trainee to explain how to correctly place the leads to check diodes A, B, C, and D.

Safety

Explain why the supply voltages on a three-phase circuit are usually higher than on a single-phase unit.

Classroom

Refer to Table 2. Review the possible causes of and solutions for common rectifier problems. Ask whether trainees have any questions.

Teaching Tip

See the Teaching Tip for Section 3.5.0 at the end of this module.

Table 2 Rectifier Problems

Condition	Possible Cause	Action
No output voltage or current	Blown fuse or tripped circuit breaker	Check fuses and circuit breakers.
	Open circuits	Check AC high-low selector, tap adjustments, and stack connections.
	Defective transformer	Check transformer; replace if defective.
	Voltage not applied	Check rectifier stacks with a voltmeter.
	Open stacks	Check with an ohmmeter.
Tripped circuit breaker or blown fuse	Steady circuit overload	Correct the cause of overload.
	Component short circuit	Determine which component is short-circuited.
	Temporary overload	Check to see if the cause is soil moisture, surges, or incorrect line voltage connections.
	Loose brackets or connections	Check and tighten.
	Effects of sun or temperature if a thermal breaker	Correct the situation.
Tripped circuit breaker	Defective transformer	Isolate the transformer from the DC circuit by removing the tap change link bars; if the circuit breaker continues to trip, the transformer is defective.
Tripped circuit breaker when a DC lead is connected to the stack, but not when DC lead is not connected	Short circuit in external ground bed or structure leads	Verify by removing one of the DC leads from the rectifier and turn the rectifier on again.
No AC line voltage	Service unavailable	Check for service.
AC line voltage, but no secondary current	Open secondary wiring	Check for audible hum; check secondary wiring.
	Faulty transformer	Isolate the transformer and check the winding resistance.
Maximum DC output voltage at rated DC current is only about one half of expected value	Incorrect rectifier connection	Check that the connection is for the actual voltage being used.
	Open plates or stacks	Turn off the unit and feel the plates in the stacks to determine if some are warm and others cold.
	Badly aged stacks	Check stacks; see note.
	Low line voltage	Check the line voltage.

NOTE: An ohmmeter check is not a valid test for determining whether selenium stacks are aged. If multiple stack arrangements are used, check the leads between the stacks for opens. On silicon stacks, disconnect each silicon diode and check individually with an ohmmeter for forward and reverse resistance. A bad silicon diode is either shorted or open.

202T02.EPS

Instructor's Notes:

3.6.0 Rectifier Troubleshooting Procedure

Whenever a rectifier problem occurs, the procedure is to systematically isolate and check the components until the defective part is found. Referring to the schematics shown in *Figures 17* and *22*, follow these steps to troubleshoot an inoperative rectifier:

Step 1 Check for voltage at the A points.

Step 2 Check for voltage at the B points.

Step 3 Check the taps for loose connections.

Step 4 Verify that the tap lead is set for the correct input voltage.

Step 5 Check for voltage in the secondary winding.

Step 6 Check the AC voltage supplied to the stack. If the circuit breaker trips, isolate the transformer from the DC circuit by removing the tap change link bars. If it continues to trip, the transformer is defective and should be replaced.

Step 7 Check the output voltage. If DC is present but much lower than expected, the stack should be tested. If the DC is about one half of what it should be, turn off the unit and feel the individual plates in the stack. If some parts are warm and others cold, the stack has an open circuit and is half-waving.

Step 8 Inspect the stack connection leads for opens by paralleling a suspect lead with a jumper cable. If the circuit breaker trips when a DC lead is connected to the stack but does not trip when it is removed, a short circuit may exist in the external ground bed or structure leads. To verify, remove one of the DC leads from the rectifier and turn the rectifier on again.

Step 9 Check the fuse, check for loose connections, and check for open leads between points J and L or H and K if DC voltage is present at the stack but not at the output terminals.

Step 10 Check for open external DC leads if there is DC voltage at the output terminals but no current is flowing.

Step 11 Ensure that meters are operating correctly and not giving false indications of rectifier problems.

Step 12 Ensure that filters (if present) are not faulty by placing a heavy jumper lead across the filter leads.

Step 13 Check the capacitor fuses in the interference filters. If the fuses are blown, replace them and check the capacitor.

Step 14 Check the capacitors in the interference filters by replacing the blown fuses and turning the unit on again. If the fuse blows again, the capacitor should be replaced.

NOTE

With its fuse open or removed, a capacitor is not in the circuit. The unit can be safely operated without it.

Step 15 Remove the faulty arresters to isolate them from the circuit. Replace them as soon as possible.

3.7.0 Troubleshooting Rectifier Bond Connections

Proper operation of bonds at rectifiers and crossings is essential to maintaining cathodic protection and mitigating stray current interference. Troubleshooting bond connections in a cathodic protection rectifier will include many of the checks and troubleshooting techniques already discussed. They will not be repeated here. Follow these steps to troubleshoot rectifier bond connections:

Step 1 Measure the pipe-to-soil potential at the major bond for the company pipeline structure.

Step 2 Measure and record the magnitude and direction of current flow through the bond.

Step 3 If no current flow exists, check all circuit connections, including, but not limited to, cable connections (including connections to the structure) and shunt connections.

Step 4 Examine the rectifier for any defects.

Step 5 Determine whether the rectifier needs adjusting. If not, continue to the next step. If it does, seek assistance from appropriate personnel for additional investigation and corrective actions before making any adjustments.

Step 6 Document the troubleshooting procedure and the results. Make company notifications as applicable.

Step 7 Identify potential abnormal operating conditions and determine the appropriate action to be taken.

Classroom

Explain that the procedure for troubleshooting a rectifier begins with systematically checking components until the defective part is found.

Refer to Figure 22. Ask trainees to review the steps for troubleshooting a rectifier. Ensure that trainees understand how to use the schematic to isolate the problem.

Classroom

Explain how to troubleshoot rectifier bond connections.

Performance Testing

Have each trainee perform Performance Task 1 to your satisfaction. Fill out a Performance Profile Sheet for each trainee.

Homework

Assign the reading of Sections 4.0.0–4.5.0 for the next session.

Materials

Ensure that you have everything required for the laboratory and testing during this session.

Classroom

Review the components of a rectifier that must be checked when determining the rectifier's condition.

Review the procedure for replacing a rectifier meter. Answer any questions trainees may have.

Explain how to replace AC breakers and AC fuses.

4.0.0 ◆ REPAIR AND REPLACE DEFECTIVE RECTIFIER COMPONENTS

The following sections cover the repair and replacement of defective rectifier components. Rectifier components include the rectifier meter, the AC fuse for the rectifier, AC breakers, the transformer, and the rectifier stack. Each component must be checked to determine its condition and then repaired or replaced as necessary.

4.1.0 Replace Rectifier Meter

Follow these steps to replace a rectifier meter:

Step 1 Switch the external disconnect to the OFF position.

Step 2 Place the internal power disconnects to the rectifier in the OFF position or remove the fuses, where applicable.

Step 3 Verify that there is no AC power to the rectifier, using a voltmeter.

Step 4 From the rectifier housing, remove the control panel cover holding the meter.

Step 5 Remove each wire from the meter and mark the wire to indicate the location of the wire on the meter.

Step 6 Remove the meter from the control panel cover.

Step 7 Inspect the new meter for damage and ensure that the rating of the meter matches the old meter.

Step 8 Install the new meter in the control panel cover.

Step 9 Install the wires as they were marked to the meter.

Step 10 Install the control panel cover to the rectifier housing.

> **NOTE**
> There may be a ground wire attached to one of the mounting screws. Ensure that the ground wire is installed to a clean surface.

Step 11 Place the internal power disconnects to the rectifier in the ON position or install the removed fuses, where applicable.

> **NOTE**
> Check the fuses for damage and corrosion before installing. Replace as required.

Step 12 Switch the external disconnect to the ON position.

Step 13 Check the rectifier for proper operation and correct voltage and amp readings.

Step 14 Complete the company's required documentation and notifications.

4.2.0 Replace AC Breakers

Proper operation of the rectifier's AC breaker is essential for protection of the rectifier components during power surges, electrical shorts, or component failure. Follow these steps to replace the AC breaker:

Step 1 Turn off the external AC supply to the rectifier.

Step 2 Disconnect the wires from the supply to the rectifier.

Step 3 Disconnect the wires from the breaker to the rectifier.

Step 4 Replace the defective breaker with a new breaker.

Step 5 Connect the wires from the breaker to the rectifier.

Step 6 Connect the wires from the AC supply to the breaker.

Step 7 Restore the power supply to the rectifier.

4.3.0 Replace Rectifier AC Fuse(s)

Proper operation of the rectifier's AC fuse(s) is essential for protecting rectifier components during power surges, electrical shorts, or component failures. Follow these steps to replace fuse(s) on a rectifier:

Step 1 Turn off the external AC supply to the rectifier.

Step 2 Check that there is no AC present at the fuse(s), with a voltmeter.

Step 3 Remove the fuse(s) from the rectifier.

Step 4 Install the new fuse(s) with the specified fuses.

Instructor's Notes:

NOTE
Inspect the new fuse for damage before installing it. If the new fuse is damaged or is not the correct size, replace it.

Step 5 Turn on the external AC power supply to the rectifier.

Step 6 Check the AC power across the fuse(s) to ensure that full AC power is supplied to the rectifier.

Step 7 Complete the company's required documentation and notifications.

4.4.0 Replace the Transformer

Transformers are required to reduce primary AC voltage to a lower adjustable AC voltage. Follow these steps to replace a transformer:

Step 1 Switch the external disconnect to the OFF position.

Step 2 Place the internal power disconnects to the rectifier in the OFF position or remove the fuses, where applicable.

Step 3 Verify that there is no AC power to the rectifier, using a voltmeter.

Step 4 Remove the control panel cover to gain access to the transformer.

Step 5 Disconnect the wires from the AC breaker to the transformer.

Step 6 Remove each wire from the transformer coarse and fine tap panels.

NOTE
Mark the positions of the wires as they are removed from the plate. Make sure the wire shows the locations on the coarse and fine tap locations, as these leads will need to be located in the same position on the new transformer.

Step 7 Remove the defective transformer from the transformer-mounting surface.

Step 8 Inspect the transformer-mounting surface for damage and corrosion. Repair as required.

Step 9 Inspect the new transformer for any signs of damage, such as loose insulation, loose or broken terminals, and any discoloration in the coil. Replace as required.

NOTE
The new transformer must have the same rating as the removed transformer.

Step 10 Connect the wires from the transformer to the coarse and fine tap panel as they were marked.

Step 11 Connect the wires from the transformer to the AC rectifier breaker as they were marked.

NOTE
Ensure that the wires are positioned correctly on the plate and that all connections are secured.

Step 12 Install the control panel cover.

Step 13 Place the internal power disconnect to the rectifier in the ON position or install the fuses.

Step 14 Switch the external disconnect to the ON position.

Step 15 Complete the company's required documentation and notifications.

4.5.0 Repair or Replace the Rectifier Stack

Rectifier stacks are required to change AC current to DC current. Follow these steps to repair or replace the rectifier stack:

Step 1 Switch the external disconnect to the OFF position.

Step 2 Place the internal power disconnects to the rectifier in the OFF position or remove the fuses, where applicable.

Step 3 Verify that there is no AC power to the rectifier, using a voltmeter.

Step 4 Remove the control panel cover to gain access to the rectifier.

Step 5 Disconnect the wires from the fine and coarse tap panels to the stack.

NOTE
Mark the wire locations on the fine and coarse tap panels.

Classroom

Emphasize the importance of inspecting new fuses for damage and corrosion before installation.

Discuss the steps for replacing the transformer.

Ensure that trainees understand when a rectifier stack can be repaired and when it should be replaced.

Laboratory

Set up stations with rectifier components—meters, breakers, fuses, transformers, and stacks. Have trainees identify the components, inspect them for damage, and explain how to repair or replace each.

Teaching Tip

See the Teaching Tip for Sections 4.0.0–4.5.0 at the end of this module.

Performance Testing

Have each trainee perform Performance Task 2 to your satisfaction. Fill out a Performance Profile Sheet for each trainee.

Homework

Have trainees review Sections 5.0.0–5.2.0 and complete the Review Questions for the next session.

Materials

Ensure that you have everything required for the demonstration and testing during this session.

Step 6 Disconnect the wires from the stack to the positive and negative DC output terminals.

> **NOTE**
> Mark the wire locations on the positive and negative DC output terminals.

Step 7 Remove the stack if the stack is selenium, or remove the defective diodes if the stack is silicon.

Step 8 Inspect the stack or diodes for signs of damage. Replace as required.

Step 9 Install the new stack or diodes

Step 10 Connect the wires from the stack to the positive and negative DC output terminals as they were removed and marked.

Step 11 Connect the wires from the fine and coarse tap panels to the stack as they were removed and marked.

> **NOTE**
> Ensure that the wires are positioned correctly on the stack and that all connections are secured.

Step 12 To avoid exceeding the rated limits of the rectifier components, adjust the tap settings before energizing the rectifier.

Step 13 Install the control panel cover to the rectifier housing.

Step 14 Place the internal power disconnect to the rectifier in the ON position or install the fuses.

Step 15 Switch the external disconnect to the ON position.

Step 16 Verify that the rectifier is operating at proper current and voltage output levels.

Step 17 Check the pipe-to-soil potential to ensure that the polarity on the pipe is correct and that the pipe-to-soil potential is at the desired range.

Step 18 Complete the company's required documentation and notifications.

5.0.0 ◆ ADJUST THE RECTIFIER

After a component repair or replacement, the rectifier needs to be checked and adjusted to the desired output. The adjustment of the rectifier is discussed in the following sections.

5.1.0 Adjustment of the Rectifier

Proper operation of the rectifier with the correct voltage and current is essential in maintaining the rectifier and the cathodic protection system that it is used to protect. Follow these steps to adjust the rectifier:

Step 1 Review the history of the rectifier operation before adjusting the rectifier.

Step 2 Increase the fine tap setting in progressive steps until desired settings have been achieved.

> **NOTE**
> The adjustment is complete when the required current output is obtained.

Step 3 If the fine tap setting reaches its highest setting, set the fine tap to the lowest setting and increase the coarse setting by 1.

Step 4 Determine the voltage by connecting a digital voltmeter across the output terminals of the rectifier.

> **NOTE**
> Obtaining accurate voltage and polarity is essential to maintaining a rectifier.

Step 5 Determine the current on a pre-installed shunt by reading the millivolt drop across the shunt and multiplying by the shunt ratio.

Step 6 Verify that the rectifier DC voltage and/or amperage output does not exceed the manufacturer's nameplate rating.

> **NOTE**
> Obtaining an accurate current is essential to maintaining a rectifier.

Step 7 Using a portable meter, adjust the rectifier **panel meters** to match the readings obtained.

Instructor's Notes:

Step 8 Measure the structure-to-electrolyte potential to ensure that the polarity on the pipe is correct and that the pipe-to-soil potential is in the desired range.

NOTE
Proper operation of the rectifier's meters provides a quick reference to the rectifier's output status.

Step 9 Inspect the rectifier for any defects or damage. Repair as required.

Step 10 Complete the required company documentation and notifications.

5.2.0 Setting Up Electronic Control Cards

SCRs serve the dual purpose of rectifying AC to DC and controlling the DC level from zero to full output. The SCR is a critical component of an electronic control card. Electronic control cards vary the rectifier's voltage output to maintain a constant-current output. The electronic control card or cards comprise automatic units. These units can alter the direction and amount of voltage supplied to the ground bed anodes.

Electronic control units do not have individual tap bars. The electronic control unit automatically adjusts the voltage into the stacks if operating in current-controlled mode. Feedback is provided by a pair of wires from the output to an electronic control circuit.

Current-controlled rectifiers usually sense output current by means of a shunt. The shunt develops a small voltage that is used for the feedback signal. The millivolt signal is called up by a precision amplifier and then compared to a preset reference or setpoint. To adjust the current setpoint, use the trim adjustment potentiometer located on the controller. Use the following steps to set up the electronic control card(s):

Step 1 Turn the unit to the OFF position by removing the negative lead.

Step 2 Turn the unit to the ON position.

Step 3 Set the voltage to 1 volt or less than the maximum rating, using the voltage adjustment potentiometer.

Step 4 Set the meter to read resistance.

Step 5 Check the resistance of each SCR.

NOTE
The resistance of each SCR should be about 1,000 ohms. If the resistance of each SCR is good, then reset the breaker.

Step 6 If the breaker does not hold with the AC leads disconnected, then check the transformer. Replace as required.

Step 7 Complete the required company documentation and notifications.

Summary

Pipeline operators install CP systems to protect pipelines from corrosion. Corrosion, a DC current phenomenon, is the result of the creation of a natural chemical by the installation of the metal pipeline in the ground, which generates a constant supply of DC current. DC current flowing away from the pipe at the point of exit extracts microscopic bits of metal, a condition called corrosion. Controlling the direction of DC current flow is the goal of CP systems.

Rectifiers are used to convert readily available AC current to DC current. A rectifier used for a CP system utilizes simple electrical circuits. Housed in a rectifier cabinet, most rectifiers contain a transformer, lightning arrester, circuit breakers, rectifying cells, bridge circuits, and meters. A control panel and connections for wires that run to the structure to be protected is located on the front of the cabinet.

Rectifier failure can often be prevented by a regular maintenance program. However, rectifiers can still fail. Trained and qualified corrosion technicians are responsible for troubleshooting and correcting the problems safely and efficiently. Many rectifier problems can be solved by technicians by simply looking at the components. Blown fuses, loose leads and terminals, damaged or detached wiring, and lightning damage are usually visually obvious. Various tests can be performed to isolate complex problems.

In this module, you have seen how rectifiers provide the DC current required for corrosion prevention and the simple system designed for its use on the pipeline. You have gained a strong knowledge of rectifier systems and their components so that you now understand the internal parts and their functions. With this knowledge, you may now be able to troubleshoot and repair a rectifier. You may use this training and knowledge to help advance to the level of a trained and qualified corrosion technician.

Classroom

Explain why obtaining the correct current and voltage is so important in maintaining the rectifier and cathodic protection system.

Demonstration

Demonstrate how to increase the fine tap setting until the proper setting has been achieved. Ensure that trainees understand how to determine the proper voltage.

Classroom

Discuss how electronic control cards work. Have a trainee explain how feedback is provided.

Discuss the purpose of an electronic control card. Explain how to set up an electronic control card.

Display several electronic control cards. Ask trainees to inspect the cards and identify the automatic units. Ask trainees to explain how the units alter the direction and amount of voltage.

Classroom

Go over the Review Questions for this module, and answer any questions trainees may have.

Examination

Administer the Module Examination. Be sure to record the results of the Exam on Craft Training Report Form 200 and submit the results to the Training Program Sponsor.

Performance Testing

Have each trainee perform Performance Task 3 to your satisfaction. Fill out a Performance Profile Sheet for each trainee.

Ensure that all Performance Tests have been completed and Performance Profile Sheets for each trainee are filled out. Be sure to record the results of the testing on Craft Training Report Form 200 and submit the results to the Training Program Sponsor.

Review Questions

1. Stepping voltage up or down is the purpose of a _____.
   a. rectifier
   b. bridge circuit
   c. transformer
   d. control meter

2. The ratio of the number of windings in the primary coil to the number of windings in the secondary coil determines the amount of change in the _____ power.
   a. output
   b. supply
   c. rectifier
   d. circuit

3. A simple lightning arrester design that works by providing two electrodes between which a high-voltage surge ionizes the air is a(n) _____ arrester.
   a. valve
   b. expulsion
   c. gas
   d. arc-gap

4. An arrester that has two electrodes and ground electrode is the _____ arrester.
   a. valve
   b. expulsion
   c. gas
   d. arc-gap

5. A metal oxide varistor (MOV) can be mounted across the _____ power output and the _____ power input to protect against voltage surges.
   a. AC; AC
   b. DC; AC
   c. DC; DC
   d. AC; DC

6. Rectifiers are turned off and on by means of _____.
   a. transformers
   b. SCRs
   c. bridge circuits
   d. circuit breakers

7. Three-phase equipment must have a(n) _____ on each supply line.
   a. transformer
   b. SCR
   c. bridge circuit
   d. circuit breaker

8. Within a rectifier, the component that passes current flow in one direction but blocks current flow in the other direction is a _____.
   a. rectifying cell
   b. lightning arrester
   c. transformer coil
   d. bimetallic element

9. An assembly of selenium plates or silicon diodes arranged to rectify AC current and produce DC output is called a(n) _____.
   a. selenium transformer
   b. rectifier stack
   c. bridge circuit
   d. SCR

10. To prevent full output current from passing through a meter coil, a(n) _____ is used.
    a. ammeter
    b. rectifier cell
    c. shunt resistor
    d. thyristor

11. The preferred method of ensuring that a rectifier is de-energized is with a(n) _____.
    a. light tap on the cabinet
    b. ground-fault-detection device
    c. digital multimeter
    d. ammeter

12. Most shunts are calibrated for _____ of drop.
    a. 50 millivolts
    b. 60 Hz
    c. 120 VAC
    d. 240 VAC

13. The positive and negative sides of a shunt are indicated _____.
    a. with an ammeter
    b. with a voltmeter
    c. by the output voltage
    d. by the schematic

14. A three-phase rectifier circuit has three poles on the _____.
    a. stack
    b. transformer
    c. lightning arrester
    d. circuit breaker

15. Electronic control cards vary the output of a(n) _________ to maintain a constant-current output.
    a. transformer
    b. rectifier
    c. SCR
    d. AC breaker

Instructor's Notes:

GLOSSARY

Glossary

This glossary will help you learn some terms specific to maintaining and repairing rectifiers. Trade Terms introduced in this module appear in italic text.

Bond: The electrical connection of two structures that usually have two different cathodic protection levels.

Circuit breaker: A device used to protect circuits from surges or lightning.

Closed circuit: One in which the path is complete, or made, so that current flows through it.

**Current:* 1) A flow of electric charge. 2) The amount of electric charge flowing past a specified circuit point per unit of time, measured in the direction of net transport of positive charges. (In a metallic conductor, this is the opposite direction of the electron flow.)

Diode: A device used to allow current to flow in only one direction.

Fuse: A device used to protect circuits from surges and lightning. It is installed in the circuit to protect other electrical devices.

Ganged: A method of mounting the controls of two or more circuits on a common shaft to provide simultaneous control through the use of a single device.

Open circuit: Discontinuity in the flow of current in an electrical circuit.

Panel meter: A device used to indicate voltage and/or current output of the rectifier.

Polarity: The condition of being electrically positive or negative. Power in AC. Power out (volts × amperes) DC. Polarity is used to describe the current flow, such as explaining the direction of current flow on a bond.

Rectification: The process of converting an alternating current to a direct current.

Rectifier stack: A dry-disk rectifier made up of layers or stacks of disks of individual rectifiers, as in a selenium rectifier or copper-oxide rectifier.

***Reference electrode:** An electrode whose open-circuit potential is constant under similar conditions of measurement, which is used for measuring the relative potentials of other electrodes.

Resistor: A device that controls the amount of current flow.

Shunt: A calibrated resistor used for determining current flow in the circuit.

Silicon diode: A crystal diode that uses silicon as a semiconductor; used as a detector in ultra-high and super-high frequency circuits.

Stack: An assembly of selenium rectifier plates or silicon diodes.

Varistor: A two-electrode semiconductor device having a voltage-dependent nonlinear resistance; its resistance drops as the applied voltage is increased. Also known as a voltage-dependent resistor.

Voltage: Electromotive force or difference of potential.

Watts: Unit of measurement of power.

REFERENCES AND ACKNOWLEDGMENTS

Additional Resources

This module is intended to present thorough resources for task training. The following reference works are suggested for further study. These are optional materials for continued education rather than for task training.

Cathodic Protection Rectifier Service Manual, March 1989. Fort Collins, CO: Good-All Electric, Inc.

Peabody's Control of Pipeline Corrosion, 2nd Edition, 2001. A.W. Peabody. Houston, TX: NACE International.

Corrosion Control and Cathodic Protection: Study Unit, 2000. Technical Training Solutions, Inc. Scranton, Pennsylvania: Amoco Pipeline Company.

Pipeline Corrosion and Cathodic Protection, 3rd Edition, June 1995. Marshall E. Parker, Edward G. Peattie. Houston, TX: Gulf Professional Publishing.

Figure Credits

CorrPro Companies, Inc.	202F01, 202F02, 202F03, 202F04 202F05, 202F06, 202F07, 202F08 202F10, 202F13, 202F14, 202F15, 202F16, 202F17, 202F18, 202F21, 202F22, 202F23
Universal Rectifiers	202F19, 202F20, 202F24

Instructor's Notes:

MODULE 61202-02 — TEACHING TIPS

The following are suggested activities or instructional methods to help you teach the material in this AIG.

General

When you call on someone to answer a question, the rest of the class relaxes or even tunes out because they expect that the question and answer will take place only between you and the trainee you called on. Instead, use this technique to involve more trainees in answering questions and to keep them on their toes.

1. Ask trainees to define a term or explain a concept.
2. After one trainee has answered, ask a trainee seated nearby if the answer is right. Then ask whether a trainee in the back of the room agrees.
3. Ask trainees to explain why they think an answer is right or wrong.
4. Use the session to clear up incorrect ideas and encourage trainees to learn from their mistakes.

Sections 2.0.0–2.5.4

Rectifier Systems

This exercise will reinforce trainees' understanding of the purpose of each component of a rectifier system. Trainees will need their Trainee Guides and pencils and paper to take notes for discussion. Allow 20 minutes for this exercise.

1. Distribute copies of *Figure 1* with the callouts covered. Near each component in the figure, such as transformer, have trainees identify the part.
2. Ask trainees to describe each component's function. For example, next to the circuit breaker a trainee might write, "turns the rectifier on and off."
3. Review trainees' work. Offer suggestions and corrections as needed.
4. Answer any questions trainees may have about rectifier systems.

Section 3.5.0

Troubleshooting Rectifier Problems

This exercise will strengthen trainees' understanding of the possible causes of several conditions that can be found on a rectifier system, as well as how to solve these common problems. Trainees will need their Trainee Guides and pencils and paper to take notes for discussion. Allow 30 minutes for this exercise.

1. Divide the class into small groups. Ask each group to review *Table 2.*
2. Ask one member of each group to present a rectifier condition. Ask other trainees in the group to suggest a possible cause of or solution for the condition.
3. Have trainees trade tasks so that each member of the group has an opportunity to present and diagnose common rectifier problems.
4. Offer suggestions and corrections as needed. Answer any questions the trainees have about troubleshooting rectifier problems.

Sections 4.0.0–4.5.0

Repair and Replace Defective Rectifier Components

This exercise will familiarize trainees with the procedures for repairing and replacing defective rectifier components. For this exercise, you will need an overhead projector, sheets of acetate, and transparency pens. Trainees will need their Trainee Guides and pencils and paper to take notes for discussion. Allow 30 minutes for this exercise.

1. Refer to the procedures in Sections 4.0.0–4.5.0. Explain that while the procedures can seem daunting, grouping the tasks into smaller parts can make them more manageable.
2. Write the first step for each procedure on the top of a separate sheet of acetate. Ask trainees to identify the subsequent steps until all of the steps are listed in order.
3. Review the procedures for each task. Ask trainees to divide the procedures into logical groups. For example, the first three steps for replacing a rectifier meter relate to disconnecting the meter's power, while the next three steps relate to the meter's removal.
4. Make suggestions, recommendations, and corrections as needed. Emphasize that understanding these steps will help trainees complete these tasks more efficiently.
5. Answer any questions the trainees have about repairing and replacing defective rectifier components.

MODULE 61202-02 — ANSWERS TO REVIEW QUESTIONS

Answer	Section
1. c	2.1.0
2. b	2.1.0
3. d	2.2.1
4. b	2.2.2
5. b	2.2.6
6. d	2.3.0
7. d	2.3.0
8. a	2.4.0
9. b	2.4.4
10. c	2.5.2
11. b	3.0.0
12. a	3.3.6
13. d	3.3.6
14. c	3.4.0
15. b	5.2.0

CONTREN™ LEARNING SERIES — USER UPDATES

The NCCER makes every effort to keep these textbooks up-to-date and free of technical errors. We appreciate your help in this process. If you have an idea for improving this textbook, or if you find an error, a typographical mistake, or an inaccuracy in NCCER's Contren™ textbooks, please write us, using this form or a photocopy. Be sure to include the exact module number, page number, a detailed description, and the correction, if applicable. Your input will be brought to the attention of the Technical Review Committee. Thank you for your assistance.

Instructors – If you found that additional materials were necessary in order to teach this module effectively, please let us know so that we may include them in the Equipment/Materials list in the Instructor's Guide.

Write: Curriculum Revision and Development Department
National Center for Construction Education and Research
P.O. Box 141104, Gainesville, FL 32614-1104

Fax: 352-334-0932

E-mail: curriculum@nccer.org

Craft ______________________ Module Name ______________________

Copyright Date ______________ Module Number ______________ Page Number(s) ______________

Description

____

____

____

____

(Optional) Correction

____

____

____

(Optional) Your Name and Address

____

____

____

Module 61203-02

Mitigate Interference

(CT 1.3 and 9.1)

Mitigate Interference
Annotated Instructor's Guide

Module 61203-02

MODULE OVERVIEW

This module explains how to identify, test for, and correct stray currents to prevent corrosion. Trainees will learn how to test for mitigation in a cathodic protection system and how to install bonds to mitigate DC interference.

PREREQUISITES

Please refer to the Course Map in the Trainee Module. Prior to training with this module, it is recommended that the trainee shall have successfully completed the following:

Core Curriculum; Pipeline Core, Modules 66101-02 and 66102-02; Pipeline Corrosion Control Level One; Pipeline Corrosion Control Level Two, Modules 61201-02 and 61202-02

OBJECTIVES

Upon completion of this module, the trainee will be able to:

1. Identify the sources of interference current in a cathodic protection system (CT 1.3).
2. Describe the techniques used to test for mitigation in a cathodic protection system (CT 9.1).

PERFORMANCE TASKS

Under the supervision of the instructor, the trainee should be able to:

1. Take structure-to-soil readings (CT 1.3).
2. Initiate proper remediation for readings that are outside of the desired range (CT 1.3).
3. Install bonds to mitigate DC interference (CT 9.1).

NCCER STANDARDIZED CRAFT TRAINING PROGRAM

The National Center for Construction Education and Research (NCCER) provides a standardized national program of accredited craft training. Key features of the program include instructor certification, competency-based training, and performance testing. The program provides trainees, instructors, and companies with a standard form of recognition through a National Craft Training Registry. The program is described in full in the *Guidelines for Accreditation*, published by the NCCER. For more information on standardized craft training, contact the NCCER by writing us at P.O. Box 141104, Gainesville, FL 32614-1104; calling 352-334-0911; or e-mailing info@nccer.org. More information may be found at our Web site, www.nccer.org.

HOW TO USE THIS ANNOTATED INSTRUCTOR'S GUIDE

Each page presents two sections of information. The larger section displays each page exactly as it appears in the Trainee Module. The narrow column ties suggested trainee and instructor actions to each page and provides icons to call your attention to material, safety, audiovisual, or testing requirements. The bottom of each page includes space for your notes.

If you see the Teaching Tip icon, that means there is a teaching tip associated with this section. Also refer to any suggested teaching tips at the end of the module.

SAFETY CONSIDERATIONS

Ensure that the trainees are equipped with appropriate personal protective equipment. Emphasize the importance of following all safety precautions and procedures when working with power tools.

PREPARATION

Before teaching this module, you should review the Module Outline, Objectives, Performance Tasks, and the Materials and Equipment List. Be sure to allow ample time to prepare your own training or lesson plan and gather all required equipment and materials.

MATERIALS AND EQUIPMENT LIST

- Overhead projector and screen
- Whiteboard/chalkboard
- Transparencies
- Markers/chalk
- Blank acetate sheets
- Transparency pens
- Pencils and scratch paper
- Module Examinations*
- Performance Profile Sheets*
- Copies of Performance Verifications for Covered Tasks 1 and 9**
- Appropriate personal protective equipment
- Copies of *49 CFR Part 192* (Gas) and/or *49 CFR Part 195* (Liquid)
- Copies of your company policy and procedures manual
- Automatic current interrupter
- Sample field data sheets
- Recording survey notes
- Voltmeter
- Shunts
- Test wires
- Test boxes
- Tools to attach and terminate test wires
- Tools to install shunts inside test boxes
- Sample strip maps

* Located in the Test Booklet.
**Available for download from www.nccer.org.

ADDITIONAL RESOURCES

This module is intended to present thorough resources for task training. The following reference works are suggested for both instructors and motivated trainees interested in further study. These are optional materials for continued education rather than for task training.

Peabody's Control of Pipeline Corrosion, 2nd Edition, 2001. A. W. Peabody. Houston, TX: NACE International.

Pipeline Corrosion and Cathodic Protection, 3rd Edition, June 1995. Marshall E. Parker, Edward G. Peattie. Houston, TX: Gulf Professional Publishing.

TEACHING TIME FOR THIS MODULE

An outline for use in developing your lesson plan is presented below. Note that each Roman numeral in the outline equates to one session of instruction. Each session has a suggested time period of 3 hours. This includes 10 minutes at the beginning of each session for administrative tasks and one 10-minute break during the session. Approximately 18 hours are suggested to cover *Mitigate Interference (CT 1.3 and 9.1).* You will need to adjust the time required for hands-on activity and testing based on your class size and resources.

Topic	Planned Time
Session I. Interference Current Sources, Part One	
A. Stray Current Causes	____________
B. Testing for Interference Current	____________
Session II. Interference Current Sources, Part Two	
A. Recording Survey Notes	____________
B. Performance Testing	____________
Session III. Mitigation/Reduction Techniques, Part One	
A. Reduction Techniques	____________
B. Bonds	____________
1. Laboratory	____________
Session IV. Mitigation/Reduction Techniques, Part Two	
A. Coating	____________
B. Galvanic Anodes	____________
Session V. Mitigation/Reduction Techniques, Part Three	
A. Electrical Shield	____________
B. Notification of Stray Interference Current	____________
Session VI. Mitigation Review	
A. Guest Speaker	____________
B. Summary	
1. Summarize module	____________
2. Answer questions	____________
C. Module Examination	____________
1. Trainees must score 70% or higher to receive recognition from the NCCER.	
2. Record the testing results on Craft Training Report Form 200 and submit the results to the Training Program Sponsor.	
D. Performance Testing	____________
1. Trainees must perform each task to the satisfaction of the instructor to receive recognition from the NCCER.	
2. Record the testing results on Craft Training Report Form 200 and submit the results to the Training Program Sponsor.	

Module 61203-02

Mitigate Interference
(CT 1.3 and 9.1)

Instructor's Notes:

Instructor's Notes:

Introduction to the Pipeline Operator Qualification Rule, API, and NCCER

It could be said that petroleum products keep America running. Petroleum products like oil and natural gas are the principal sources of energy in the United States and are used in many manufacturing and chemical processes. Petroleum products are primarily transported by pipeline. There are more than a million miles of liquid and natural gas pipeline in this country, ranging in diameter from two inches to four feet. These pipelines bring crude oil and natural gas from production fields to processing plants and refineries—and ultimately to the consumer.

Pipeline Operator Qualification

Petroleum products are considered hazardous, so pipeline personnel must take extra precautions and follow specific procedures when working with them. To help ensure the safe handling and transportation of petroleum products, the United States Department of Transportation (DOT) mandates safe practices as federal regulations.

One such regulation, the Operator Qualification (OQ) rule, requires pipeline operators to develop a written qualification plan to evaluate their personnel and their contract personnel (see *49 CFR Part 192 Subpart N—Gas*, and *49 CFR Part 195 Subpart G—Liquid*). The evaluation hinges on the operator's ability to perform certain covered tasks and to be able to recognize and respond to abnormal operating conditions that may be encountered while performing these covered tasks.

Covered Tasks

A pipeline performance task is deemed *covered* if the activity passes the following four-part test:

1. Is it performed on a pipeline facility?
2. Is it an operations or maintenance task?
3. Is it performed as a requirement of *49 CFR Part 195 (Liquid)* or *49 CFR Part 192 (Gas)*?
4. Does it affect the operation or integrity of the pipeline facility?

If the answer to all four questions is *yes*, then the activity is considered *covered* and must figure into the evaluation process of pipeline personnel.

American Petroleum Institute

The OQ rule and its related covered tasks provided the catalyst to develop a unified pipeline training and assessment program—so that pipeline operators and owners have ready access to the assessment and training tools they need. This effort was the direct result of the vision and leadership provided by The American Petroleum Institute (API) and API members.

Efforts to unify the industry are nothing new to API. Indeed, the impetus for forming the Institute in 1919 was the need to standardize engineering specifications for drilling and production equipment. Today, the most pressing issues revolve around public perceptions and government policies toward the industry. Speaking with one voice on these issues has become as essential as having interchangeable parts in the field. API's members determine the petroleum industry's positions on public policy as well as the standards that govern its day-to-day operations worldwide.

Consortium on Operator Qualification

API assembled a focus group whose primary purpose was to identify covered tasks and provide compliance recommendations to pipeline operators.

The Consortium on Operator Qualification (COOQ) became the guiding force for the identification and development of covered tasks for the liquid pipeline industry. Their work product, API publication 1161, *Guidance Document for the Qualification of Liquid Pipeline Personnel*, formed the basis for the resulting Pipeliner Training and Assessment Program.

The covered task list created in API publication 1161 has now evolved to include gas covered tasks and is maintained by NCCER. The most current covered task list can be downloaded at www.nccer.org.

National Center for Construction Education and Research

With the guidance and support of API, its members, and the Consortium on Operator Qualification, the National Center for Construction Education and Research (NCCER) was able to design and implement the Pipeliner Training and Assessment Program.

NCCER is a not-for-profit 501(c)(3) education foundation established in 1995 by the world's largest and most progressive construction and maintenance contractors and national contractor associations. NCCER was founded to address the

severe workforce shortage facing the construction and maintenance industries and to develop a standardized training process and curricula. Today, NCCER is supported by hundreds of leading construction and maintenance companies, manufacturers, and national associations and developed more than 35 construction and maintenance titles.

NCCER also maintains a National Registry, which contains the records of trainees and instructors who have received NCCER training. All completion information for every trainee is sent to the NCCER and filed within the Registry. The Registry can then confirm training and skills for workers as they move from company to company, state to state, or even within their own company.

Standardized Curriculum (NCCER's Contren™ Learning Series)

NCCER curricula are developed using the expertise of industry subject matter experts (SMEs). The use of SMEs as an integral part of every new development or revision ensures that the curriculum is current, industry-driven, and indicative of industry practices. The development and/or revision process is facilitated from start to finish by NCCER project managers, who work closely with the SMEs and a select team of technical writer(s) on the development and presentation of the curriculum.

Pipeline: Five Training Paths

The pipeline curriculum focuses on essential skills and competencies within five skill areas:

- Corrosion Control
- Electrical and Instrumentation
- Maintenance
- Mechanical
- Operations (Field and Control Center)

SMEs from the pipeline industry have determined the skills and competencies needed based on API's covered tasks and the consensus of the group/industry. Specific company information is generally not incorporated into the curriculum unless the SMEs determine that the information is the industry standard.

Training Modules and Covered Tasks

The smallest unit of instruction in the curriculum is a *module*. Each curriculum is composed of modules (the equivalent of stand-alone chapters) that are logically sequenced into complete *levels* of instruction. Every module is task-oriented and requires the successful completion of a written (knowledge-based) examination and a performance (hands-on) test. Modules may be individually selected and/or combined for efficient task training. Modules in the pipeline curricula, to the extent possible, will align with API-developed covered tasks. While not all modules will relate to covered tasks, all covered tasks will relate to at least one module.

Benefits of Training to the Pipeline Marketplace

Applying NCCER's standardized training process allows pipeline owners and contractors in the United States to establish the benchmark for acceptable skill and competency levels within the pipeline industry. This process will assist training organizations in developing and/or improving standardized pipeline training programs that effectively match their resources and operating conditions while responding to regulatory demands and to the needs of the industry. The process will also recognize training organizations that meet the minimum standards of competence, warranting regulatory, public, and professional confidence. This in turn will foster national unity within the pipeline industry for the mutual benefit of individual technicians, owners, contractors, and ultimately pipeline-users and society.

Additional Resources

To review the Pipeliner Training and Assessment Program Implementation and Resource Guide and the latest version of pipeline covered tasks, visit NCCER's Web site at www.nccer.org, or contact:

National Center for Construction Education and Research
PO Box 141104
Gainesville, FL 32614-1104
(888) 622-3720
Web site: www.nccer.org

For more information on API, please contact:

American Petroleum Institute
1220 L Street, Northwest
Washington, D.C. 20005
(202) 682-8000
Web site: www.api.org

To order *API Pub 1161*, contact Global Engineering Documents by calling 1-800-854-7149 or visiting their Web site at www.global.his.com.

Additional sources of information on the pipeline operator qualification regulation include:

DOT – Office of Pipeline Safety
www.ops.dot.gov

Transportation Safety Institute
www.tsi.dot.gov

Instructor's Notes:

COURSE MAP

This course map shows all of the modules in the second level of the Pipeline Corrosion Control curriculum. The suggested training order begins at the bottom and proceeds up. Skill levels increase as you advance on the course map. The local Training Program Sponsor may adjust the training order.

PIPELINE CORROSION CONTROL LEVEL TWO

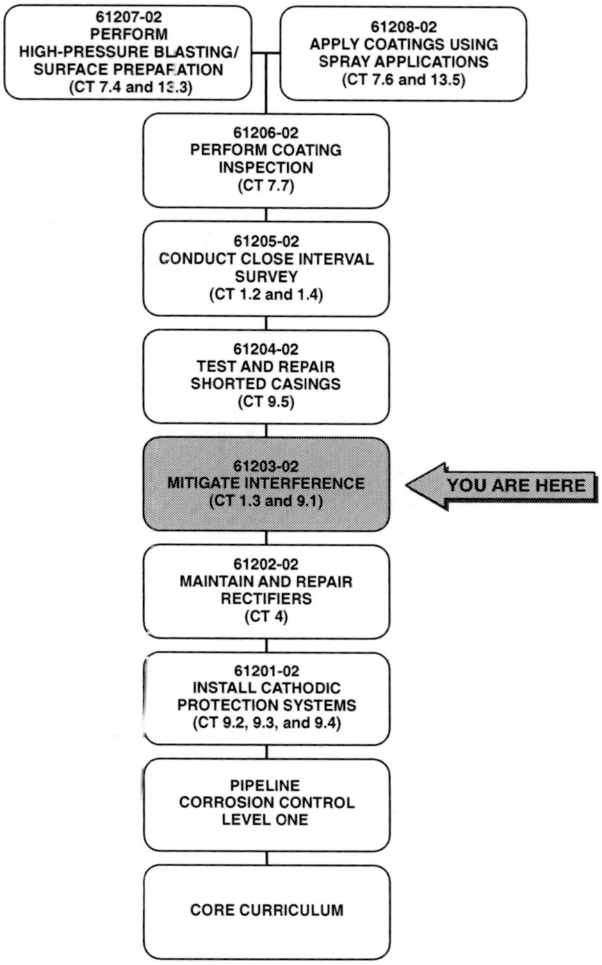

203CMAP.EPS

Homework

Assign reading of Module 61203-02.

Instructor's Notes:

MODULE 61203-02 CONTENTS

Figures

Instructor's Notes:

MODULE 61203-02

Mitigate Interference (CT 1.3 and 9.1)

Objectives

When you have completed this module, you will be able to do the following:

1. Identify the sources of interference current in a cathodic protection system (CT 1.3).
2. Describe the techniques used to test for mitigation in a cathodic protection system (CT 9.1).

Prerequisites

Before you begin this module, it is recommended that you successfully complete the following: Core Curriculum; Pipeline Core, Modules 66101-02 and 66102-02; Pipeline Corrosion Control, Level One; Pipeline Corrosion Control, Level Two, Modules 61201-02 and 61202-02.

Required Trainee Materials

1. Pencil and paper
2. Appropriate personal protective equipment

1.0.0 ◆ INTRODUCTION

According to *49 CFR 192.473*, "Each operator whose pipeline system is subjected to **stray currents** shall have in effect a continuing program to minimize the detrimental effects of such currents." Each impressed current protection system or galvanic anode system must be designed to help protect surrounding or adjacent underground metallic structures. Corrosion on a pipeline can be caused by stray currents. This is called **interference**. These stray currents are direct currents flowing in the earth that are from a source other than the corrosion protection source. Normally this outside current enters the pipeline at one point, exits the pipeline at another point, and returns to the earth. The area where the current returns to earth, completing the circuit, is the point at which the corrosion occurs.

Stray currents are either static or dynamic; sources of stray current include cathodic protection (CP) systems on other pipelines, DC transit systems, DC welding operations, DC mining operations, high-voltage DC transmission systems, or disturbances in the earth's magnetic field. If environmental conditions allow rectifying junctions such as copper oxide films, AC current flow can be rectified. This results in a direct current that can be mitigated.

This module explains identifying, testing for, and correcting stray currents to prevent corrosion. Corrosion individuals must be able to locate and correct these conditions. When the magnitude of stray currents exceeds the corrosion protection currents applied to the pipeline, the pipeline will develop leaks before its life expectancy is exceeded.

2.0.0 ◆ INTERFERENCE CURRENT SOURCES

There are many different sources of interference current in a cathodic protection system. Some of the sources and tests for interference current are discussed in the following sections.

WARNING!
Be aware that a release may be found when testing.

Materials

Ensure that you have everything required to teach the course. Check the Materials and Equipment List at the front of this module.

Classroom

Explain that terms shown in bold are defined in the Glossary at the back of this module.

Audiovisual

Show Transparency 1, Course Objectives.

Show Transparency 2, Performance Tasks.

Teaching Tip

See the general Teaching Tip at the end of this module.

Classroom

Explain how stray currents, called interference, cause corrosion.

Discuss problems that result when the magnitude of the stray current exceeds the corrosion protection current.

Safety

Emphasize that a release may be found when testing for interference currents.

Audiovisual

Show Transparency 3 (Figure 1). Review the causes of stray current interference. Answer any questions trainees may have.

Classroom

Explain how the location of a ground bed and the characteristics of a cathodic protection system affect stray current interference.

Refer to Figure 2. Explain what happens when a foreign pipeline travels through ground beds and picks up current.

2.1.0 Stray Current Causes

Stray current interference on adjacent pipelines can be caused by impressed current cathodic protection systems. This depends on the locations of the ground beds and the operating characteristics of the cathodic protection system. *Figure 1* shows stray current to a crossover foreign pipeline.

In this example, the foreign pipeline picks up current from the ground bed and transfers the current to the protected pipeline when it crosses under it. The potential for corrosion is greatly increased in the location of current flow between the foreign and the protected pipelines. *Figure 2* shows stray current to a parallel foreign pipeline.

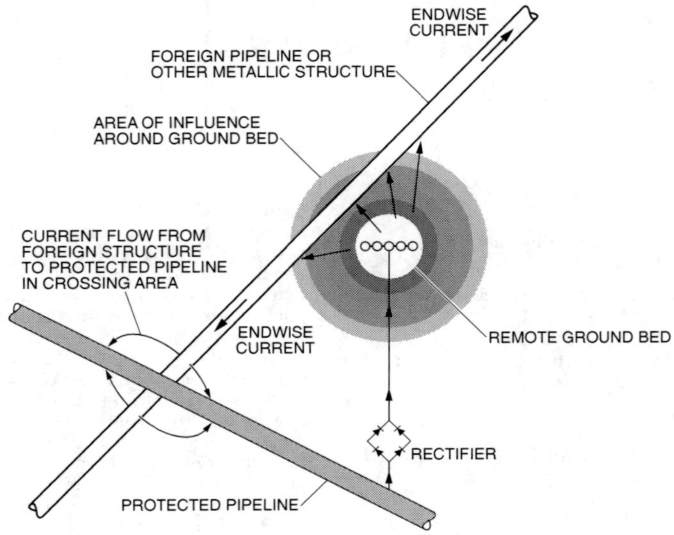

Figure 1 ◆ Stray current to a crossover foreign pipeline.

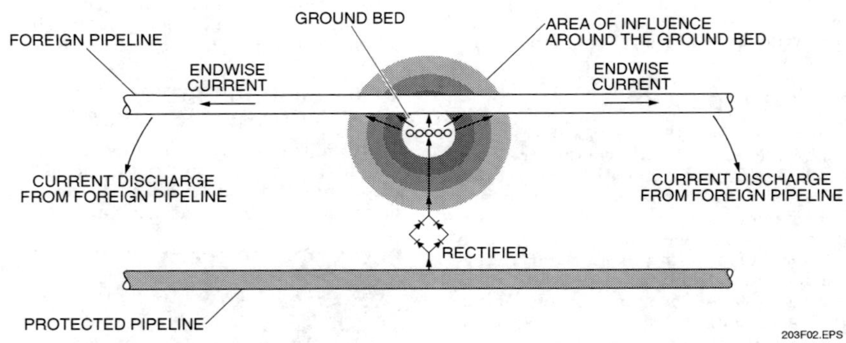

Figure 2 ◆ Stray current to a parallel foreign pipeline.

Instructor's Notes:

In this example, the parallel foreign pipeline is affected by the ground bed of the protected pipe. This is because the foreign pipeline travels through the ground bed and picks up current. When this occurs, the effectiveness of the CP system will be reduced in this area of the pipeline.

Potential gradients cause natural occurrences of stray current in the earth that are transferred to the protected pipeline. *Figure 3* shows potential gradients around a protected pipeline. *Figure 4* shows how a foreign pipeline is affected by passing through a protected pipeline potential gradient.

In the affected area of the foreign pipeline, it becomes positive to the soil and discharges current to the earth. This increases the potential for corrosion on the foreign pipeline in the affected area. The greatest chance of corrosion is at the crossing point of the pipelines.

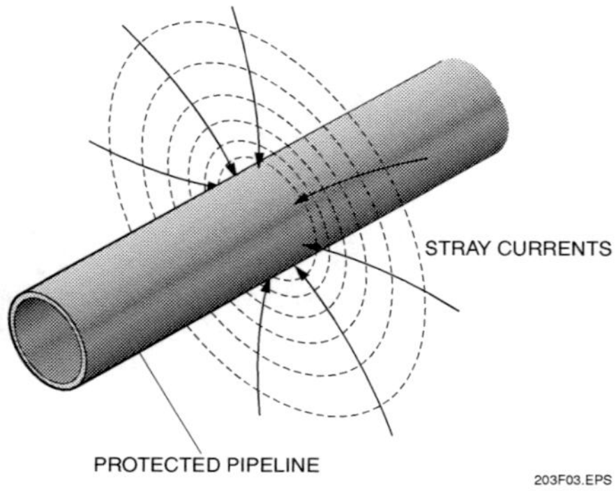

Figure 3 ◆ Potential gradients around a protected pipeline.

2.2.0 Testing for Interference Current

Static stray DC current interference is easy to test for as long as there is no presence of **dynamic stray currents.** Dynamic stray current can be caused by sources such as DC-powered transit systems. In *Figure 5* a section of coated pipeline has an impressed current CP system applied to it and has several foreign pipelines crossing it. In this example, "B" is a strong indicator of interference because it has the largest difference in voltage. Always look for large changes in voltage to identify interference. Follow company guidelines for determining the need to mitigate interference. To distinguish the current in the foreign pipeline from the current in the protected line, an automatic current interrupter is installed in the output of the rectifier. The current interrupter is a device that automatically opens and closes an electrical circuit at a preset time. With the current interrupter operating, measure the potential of the lines under both on and off conditions. *Figure 6* shows how to measure this potential at the test station.

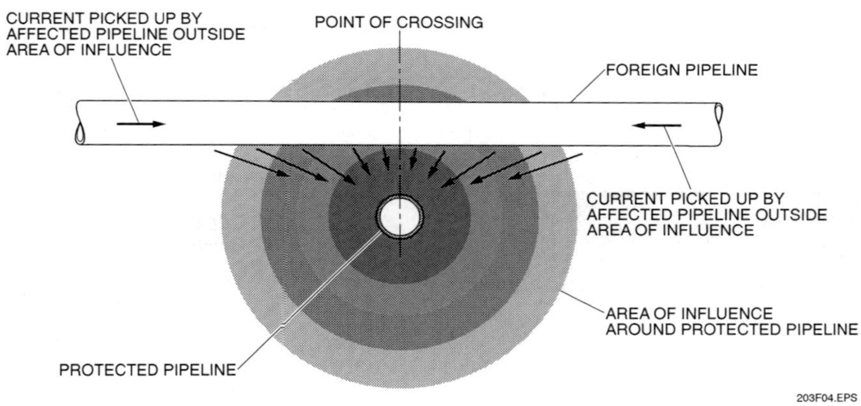

Figure 4 ◆ Foreign pipeline crossing protected pipeline potential gradient.

Classroom

Explain the point at which the greatest chance of corrosion exists.

Audiovisual

Show Transparency 4 (Figure 4). Review the sources that cause dynamic stray currents.

Classroom

Refer to Figure 5. Explain how to identify interference using change in voltage.

Review the role of an automatic current interrupter in a rectifier.

Demonstration

Bring in an automatic current interrupter. Demonstrate how to measure the potential of the lines under on and off conditions. Go over any questions trainees may have.

Safety

Discuss the consequences of failing to remain alert when testing for interference current.

Homework

Have trainees prepare for testing during the next session.

Materials

Ensure that you have everything required for testing during this session.

Teaching Tip

See the Teaching Tip for Sections 2.0.0–2.2.0 at the end of this module.

Performance Testing

Have each trainee perform Performance Task 1 to your satisfaction. Fill out a Performance Profile Sheet for each trainee.

Homework

Assign the reading of Sections 3.0.0–3.1.0 for the next session.

Materials

Ensure that you have everything required for the demonstration and laboratory during this session.

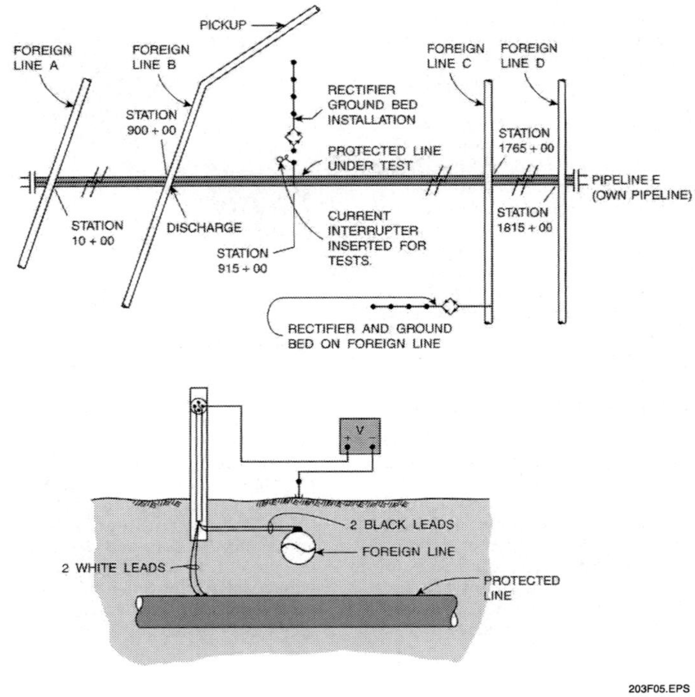

Figure 5 ◆ Testing foreign pipeline crossings for stray current interference from CP installations

These tests are performed using a copper sulfate electrode placed over the crossing point of the lines. Pipeline locators can be used to determine the exact location of the crossing.

> **WARNING!**
> Always be alert when testing for interference current. A release could be found while testing.

The data that is obtained at each pipeline crossing must be properly recorded. *Figure 7* shows interference testing potentials. Some of the types of stray current that may be encountered during a survey are also shown. These records are also called field data sheets. Field data sheets should also include any information about the protected line, the date, and the current output of the rectifier.

3.0.0 ◆ MITIGATION/REDUCTION TECHNIQUES

There are many different methods used to reduce or eliminate stray current interference from a CP system. Some of the most common methods include **bonds**, coatings, galvanic anodes, electrical shield, and notification of stray current, which are explained in the following sections.

Instructor's Notes:

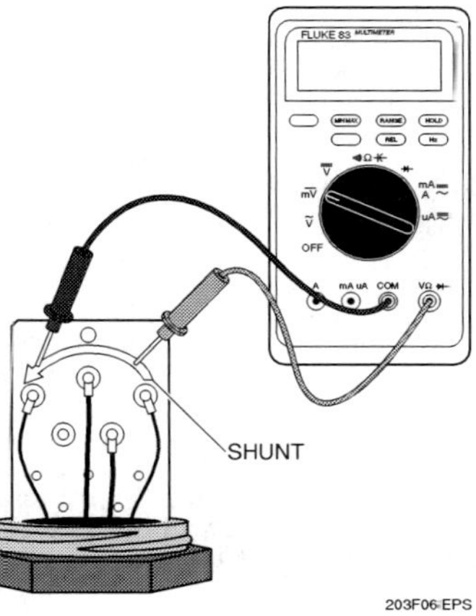

Figure 6 ◈ Measuring potential at the test station.

3.1.0 Bonds

Foreign pipelines can be subject to stray current damage from the impressed current cathodic protection system on another pipeline. To correct this problem, a resistance bond is connected between the two pipelines. The amount of resistance in the bond is adjusted to drain enough current from the affected line to eliminate the corrosion potential.

> **CAUTION**
> For liability reasons, always have the representative of the foreign pipeline company on site before proceeding to correct the problem on the affected line.

For a resistance bond, **diodes** are installed as basic components of a rectifier circuit. Three-phase rectifiers are probably the most common units used for cathodic protection. The diode is installed to allow current to flow in one direction only. This characteristic of the diode causes all the current waves to be more aligned when entering into the cables. Along with other components, the diodes reduce ripple effects in the power supply and filter out interference.

Once it is determined that a damaging stray current exists, the source of the current must be found. In order to locate the source of the stray current, follow these steps:

Step 1 Test for the presence of DC current interference.

> **NOTE**
> Potential sources include, but are not limited to, other pipeline cathodic protection facilities, DC traction systems (subways, light rail systems, trolley), high-voltage DC transmission, welding, or other facilities.

Step 2 Conduct joint cooperative interference tests with the representative of the foreign company to determine the location of current discharge and the magnitude of the interference current.

Step 3 Install test wires on both structures at the location of the current discharge.

Step 4 Attach test wires to the structure to yield a permanent, low-resistance connection.

Step 5 Terminate test wires inside of the appropriate test box that is accessible to both structures and representatives of both companies.

FOREIGN LINE		OWN PIPELINE (volts)			INTERFERING PIPELINE (volts)		
NAME	PIPELINE STATION	CURRENT ON	CURRENT OFF	Δ volts	CURRENT ON	CURRENT OFF	Δ volts
A	10 + 00	-0.88	-0.85	-0.03	-0.87	-0.89	+0.02
B	900 + 00	-1.98	-1.02	-0.96	-0.32	-0.68	+0.36
C	1765 + 00	-0.68	-0.64	-0.04	-0.78	-0.78	0
D	1815 + 00	-0.95	-0.91	-0.04	-0.68	-0.68	0

203F07.EPS

Figure 7 ◈ Sample of recording survey notes.

Classroom

Explain how to correct stray current damage using a resistance bond.

Safety

Emphasize the importance of having a foreign pipeline company representative on site before correcting problems on the line.

Classroom

Review the considerations for installing diodes to create a resistance bond.

Explain how to locate the source of a stray current. Ask whether trainees have any questions.

Demonstration

Bring in a voltmeter. Demonstrate how to set up the meter to measure the potential of the foreign line at the crossing point of a protected line.

Audiovisual

Show Transparency 5 (Figure 8). Explain how to restore the foreign pipeline potential to the original value.

Laboratory

Set up stations with shunts, test wires, and test boxes. Ask trainees to practice attaching and terminating test wires. Have trainees install shunts inside the test box.

Homework

Have trainees review Sections 3.2.0–3.3.0 for the next session.

Step 6 Install **shunts** for measurement of current flow inside of the test box.

Step 7 Install resistance inside of the test box as required in order to limit current interchange.

Step 8 Install blocking diodes as required.

Step 9 Conduct tests to determine the effectiveness of the installed interference bond.

Step 10 Install rectifiers or reverse current switches in unusual situations where a conventional metallic bond is not effective.

Step 11 Identify potential abnormal operating conditions and determine the appropriate actions to be taken.

Once all the test data has been collected and the appropriate action to eliminate the problem has been determined, technicians can begin repairs to bring the system to a proper efficiency.

Determining when the stray current interference has been eliminated is critical. To determine when the bond has been properly adjusted, set up a voltmeter to measure the potential of the foreign line to the protected line at the crossing point. The potential should be measured at the point of maximum exposure if it is different from the crossing point. The bond resistance is set so that the foreign line potential is the same with the affecting rectifier either on or off. This restores the foreign pipeline potential to the original value. *Figure 8* shows a typical foreign line test point with a bond in place. *Figure 8* also shows the instrument connections used for measuring the pipe-to-soil potentials on the affected pipeline.

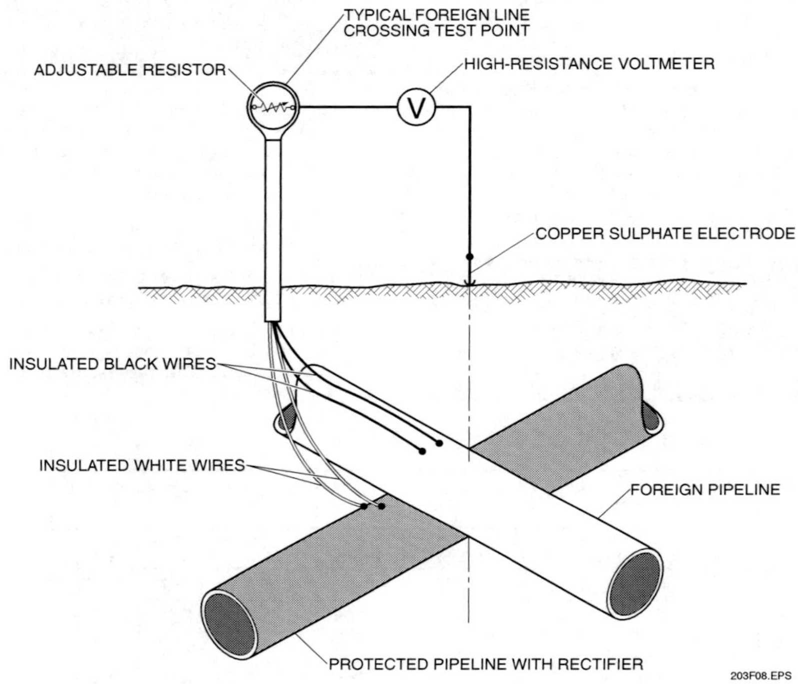

Figure 8 ◆ Foreign line test point with a bond in place.

Instructor's Notes:

3.2.0 Coating

If the testing of a foreign line indicates that the interference is caused by a foreign pipeline that is bare or poorly coated, the bond base may not be able to correct the condition. If a high current density is flowing through the soil and onto the bare foreign line, localized corrosion could occur. To correct this problem, reduce the density of the current flow to the crossing area of the foreign line.

The best way to accomplish this is to apply coating to the line. *Figure 9* shows the use of a coating to correct interference at foreign pipeline crossings.

With proper coating, the voltage drops in the earth around the foreign line are eliminated or at least reduced. Over-the-line potential profiles (*Figure 10*) determine the length of the pipeline to be coated.

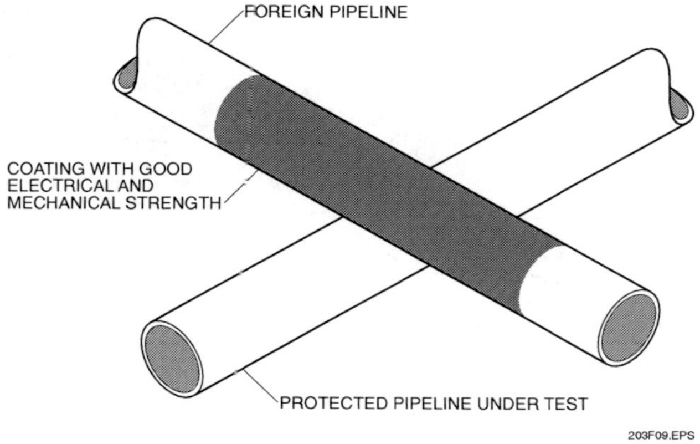

Figure 9 ◆ Using a coating to correct interference at foreign pipeline crossings.

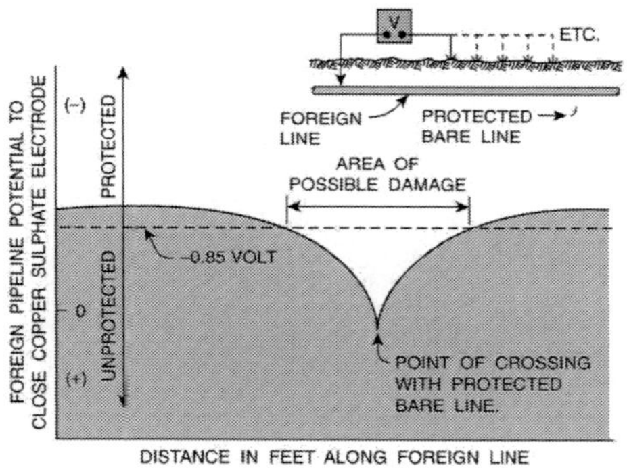

Figure 10 ◆ Pipe-to-earth potentials on foreign pipeline passing through area of influence around cathodically protected bare line.

Ensure that you have everything required for teaching this session.

Refer to Figure 9. Review the procedures for detecting and correcting localized corrosion.

Show Transparency 6 (Figure 10). Explain how to determine the length of the pipeline to be coated.

Safety

Ask a trainee to explain how to avoid holidays.

Classroom

Refer to Figure 11. Explain how anodic potential fields are used to offset the potential gradient field. Answer any questions trainees may have.

Homework

Assign the reading of Sections 3.4.0–3.5.0 for the next session.

Materials

Ensure that you have everything required for teaching this session.

Audiovisual

Show Transparency 7 (Figure 12). Explain how electrical shields reduce interference.

> **CAUTION**
> Coating the interfered line alone may not solve the problem, but if you do, take special precautions to avoid holidays.

Figure 10 identifies the length of line subject to interference. Use the profile of the interfered line to determine the length of the coating required on the interfering foreign line.

3.3.0 Galvanic Anodes

Galvanic anodes can also be used to correct interference. The anodes are attached to the affected line (see *Figure 11*). They are located in the area that is subject to interference from the foreign line.

This method uses the anodic potential fields that are around the galvanic anode to offset the potential gradient field around the protected pipeline. Usually a single line of anodes between the pipeline and the foreign pipeline causing the interference will mitigate the effects of the interference. The remaining current will discharge from the anodes instead of the pipeline. Several strings of anodes may be required when potential differences are very high. The anode life using this arrangement is short unless the surrounding soil has a high resistivity.

Magnesium anodes have a higher anodic potential gradient field than zinc anodes. This is why magnesium anodes are normally recommended. The length of the affected area of the pipe determines the length of the anode string. It is a good idea to have a test point to periodically measure potentials, and a permanent reference electrode should be installed. This installation will monitor the interference on the underside of the line at the maximum exposure point as shown in *Figure 11*.

> **NOTE**
> Although it is possible to fit the interfered pipeline with anode systems alone, it is not possible to achieve a correction by fitting the interfering pipeline only. Strapping the anodes together will reduce the consumption rate.

The heaviest anode is placed at the point of crossing. This is where the exposure is the most intense. The lighter anodes are used elsewhere on the line.

3.4.0 Electrical Shield

Electrical shields are defined as barriers to stray current flow. Electrical shielding can be provided by nonmetallic insulating barriers or by diverting the current from the affected pipeline to another structure. Shields are particularly useful for protecting pipeline in congested areas such as pumping stations and tank farms. *Figure 12* shows bare pipe used as electrical shields around a pipeline.

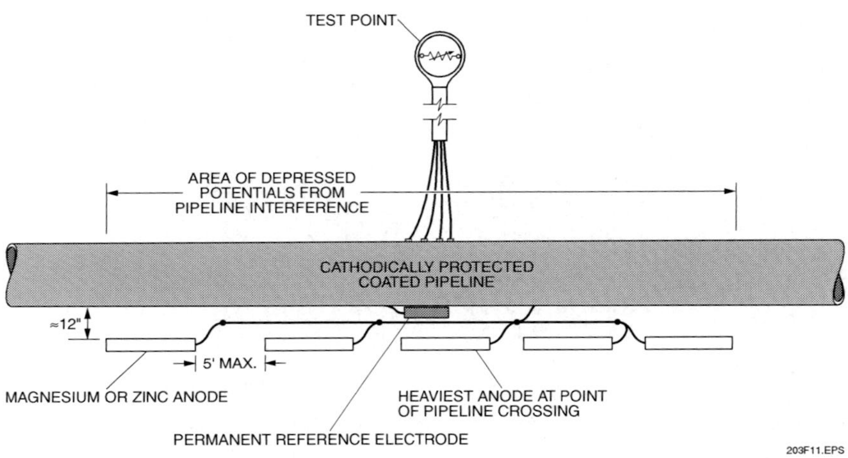

Figure 11 ◆ Galvanic anodes attached to a line under test.

Instructor's Notes:

Ground beds, both remote from the pipeline and close by, work to ensure that the electrical potential of the pipeline is negative with respect to the surrounding earth. Both approaches address the same problem, but in different ways. Remote ground beds can be used to protect large sections of pipeline, while close ground beds are more suited for protecting smaller, nearby areas of pipeline.

A buried pipeline extending across an area influenced by a ground bed may be subject to influence from stray electrical current surrounding the bed. Specific conditions in the area where this takes place determine how serious the effect will be to the foreign pipeline. In cases where the electrical potential of the stray current is high enough, protective measures must be put in place to prevent corrosion damage from current flowing away from anodic areas on the pipe.

One approach is to employ bare pipe shields on either side of the foreign pipeline where it lies within the potential gradient of the ground bed and picks up stray current. If the bare pipe shields are electrically connected to the rectifier for the ground bed, they will emit a negative cathodic field that opposes the positive field of the ground bed.

High installation and operating costs must be weighed against other alternatives. In the end it may be more cost effective to relocate the ground bed than to consider the use of a shield.

3.5.0 Notification of Stray Interference Current

Pipeline operators must notify owners of any nearby structures when planning any changes to existing CP systems or any new installations. The courtesy should be reciprocated. Stray currents may weaken or completely eliminate existing CP on structures within the same field of influence. Open communications between all owners and operators in an area should be encouraged. Owners should provide strip maps to each other, with each owner responsible for marking applicable structures. The exact location, size, and scope of any new projects can have serious effects on others. Test points in the same area may be shared. To avoid future problems, put all communications in writing.

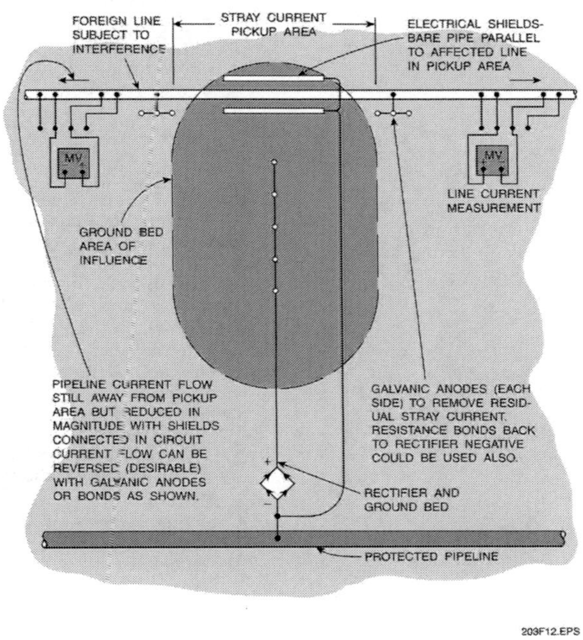

Figure 12 ◆ Use of electrical shields to reduce stray current interference.

Classroom

Compare the methods that are used to ensure that the electrical potential of the pipeline is negative.

Review the considerations for installing electrical shields. Go over any questions trainees may have.

Emphasize the importance of following proper procedures when notifying owners of stray interference currents.

Distribute sample strip maps. Ask trainees to review the maps and to locate test points and new structures that could affect existing structures. Discuss skills for communicating with owners/operators.

Teaching Tip

See the Teaching Tip for Sections 3.0.0–3.5.0 at the end of this module.

Homework

Have trainees brainstorm questions for the guest speaker, complete the Review Questions, and prepare for testing during the next session.

Materials

Ensure that you have everything required for testing during this session.

Teaching Tip

See the Summary Teaching Tip at the end of this module.

Classroom

Go over the Review Questions for this module, and answer any questions trainees may have.

Summary

Pipeline corrosion can be caused by stray currents from foreign sources. The source of the stray currents, or interference, should be identified if possible. Static interference can be tested with a current interrupter using copper sulfate electrodes at crossing points to measure the potential.

Among the techniques used to mitigate interference are bonds, coatings, galvanic anodes, and electrical shields. A resistance bond drains enough current to eliminate the corrosion potential. Coatings should be surveyed for poor condition, holidays, and other defects. When a stray current has been identified, galvanic anodes may be placed in the area to offset the potential gradient field. Electrical shielding provided by nonmetallic insulating barriers can divert the interference to another structure. Another aid is written notification of stray currents and CP systems to and by all affected companies, in accordance with *49 CFR 192.473*.

Review Questions

1. Corrosion on a pipeline can be caused by stray currents. This is called _____.
   a. stray corrosion
   b. incidental effects
   c. induction
   d. interference

2. If environmental conditions allow rectifying junctions, AC current flow can be _____.
   a. rectified
   b. increased
   c. decreased
   d. eliminated

3. _____ on adjacent pipelines can be caused by impressed current cathodic protection systems.
   a. Endwise current
   b. Stray current interference
   c. Current discharge
   d. Potential gradients

4. _____ cause natural occurrences of stray current in the earth that are transferred to the protected pipeline.
   a. Potential gradients
   b. Burial pressures
   c. Ground water changes
   d. Climate variations

5. _____ can be caused by DC-powered transit systems.
   a. Potential shifts
   b. Static DC interference
   c. Dynamic stray currents
   d. Impressed currents

6. To distinguish the current in the foreign pipeline from the current in the protected pipeline, a _____ is used.
   a. potentiometer
   b. current interrupter
   c. voltmeter
   d. pipeline locator

7. _____ rectifiers are probably the most common units used for cathodic protection units.
   a. Three-phase
   b. Single-phase
   c. Two-phase
   d. Delta-wye

Instructor's Notes:

8. To determine when the bond has been properly adjusted, set up a(n) _____ to measure the potential.
   a. ohmmeter
   b. potentiometer
   c. voltmeter
   d. current interrupter

9. If a high current density is flowing through the soil and onto the bare foreign line, _____ could occur.
   a. widespread corrosion
   b. CP system shorts
   c. resistance bonding
   d. localized corrosion

10. _____ can also be used to correct interference.
   a. Isolators
   b. Galvanic anodes
   c. Current interrupters
   d. Insulators

11. Magnesium anodes have a higher anodic _____ than zinc anodes.
   a. life
   b. resistivity
   c. current capacity
   d. potential gradient field

12. Electrical shielding can be provided by _____ or by diverting the current.
   a. resistance bonding
   b. nonmetallic barriers
   c. current interrupters
   d. galvanic anodes

13. Ground beds work to ensure that the electrical potential of the pipeline is _____ with respect to the surrounding earth.
   a. larger
   b. smaller
   c. positive
   d. negative

14. If bare pipe shields are electrically connected to the rectifier for the ground bed, they will emit a _____ field.
   a. positive anodic
   b. negative anodic
   c. negative cathodic
   d. positive cathodic

15. Pipeline operators must _____ any nearby structures when planning any changes to existing CP systems.
   a. notify owners of
   b. bypass
   c. remove
   d. provide CP for

Examination

Administer the Module Examination. Be sure to record the results of the Exam on Craft Training Report Form 200 and submit the results to the Training Program Sponsor.

Performance Testing

Have each trainee perform Performance Tasks 2 and 3 to your satisfaction. Fill out a Performance Profile Sheet for each trainee.

Ensure that all Performance Tests have been completed and Performance Profile Sheets for each trainee are filled out. Be sure to record the results of the testing on Craft Training Report Form 200 and submit the results to the Training Program Sponsor.

GLOSSARY

Glossary

This glossary will help you learn some terms specific to mitigating interference. Trade Terms introduced in this module appear in italic text.

Bond: The electrical connection of two structures that usually have two different cathodic protection levels.

Current flow through bond: Transfer of electrons between structures due to a difference in potential, either impressed or natural.

Diode: A device used to allow current to flow in only one direction.

Dynamic stray current: A portion of a current that flows over a path other than the intended path in an on/off or intermittent fashion and may cause electrochemical corrosion of metals in contact with electrolytes.

***Foreign structure:** Any metallic structure that is not intended as a part of a system under cathodic protection.

High input impedance voltage meter: An electronic devise used to measure the structure-to-soil potential.

Interference: The effect upon cathodic protection levels of a structure due to the interaction between its cathodic protection system and the cathodic protection system of another structure.

***IR drop:** The voltage across a resistance in accordance with Ohm's law.

Potential gradient: A change in the potential with respect to distance, expressed in millivolts per unit of distance.

Shunt: A calibrated resistor used for determining current flow in the circuit.

**Stray current:* Current through paths other than the intended circuit.

Static stray current: A portion of a current that flows over a path other than the intended path at a constant, unbroken rate and that may cause electrochemical corrosion of metals in contact with electrolytes.

***Structure-to-soil potential:** See structure-to-electrolyte potential.

***Structure-to-electrolyte potential:** The potential difference between the surface of a buried or submerged metallic structure and the electrolyte that is measured with reference to an electrode in contract with the electrolyte.

Test lead: A connection to the structure being tested, usually a wire in a supporting stand with an easy connection point for the meter.

Ultrasonic thickness meter: An instrument that determines the thickness of a structure using ultrasonic sound waves.

Voltage shift: Change in voltage as a result of a change in the application of cathodic protection.

Instructor's Notes:

Additional Resources

This module is intended to present thorough resources for task training. The following reference works are suggested for further study. These are optional materials for continued education rather than for task training.

Peabody's Control of Pipeline Corrosion, 2nd Edition, 2001. A. W. Peabody. Houston, TX: NACE International.

Pipeline Corrosion and Cathodic Protection, 3rd Edition, June 1995. Marshall E. Parker, Edward G. Peattie. Houston, TX: Gulf Professional Publishing.

Figure Credits

Figure 5 courtesy of NACE International

(Source: *Peabody's Control of Pipeline Corrosion, Second Edition*, Pg 214, Fig 11.5 Testing foreign pipeline crossings for stray current interference from CP installations) ©NACE International – The Corrosion Society, 1440 South Creek Drive, Houston, Texas 77084. Additional publications on corrosion control technology are available from www.nace.org or phone (281) 228-6200.

Figure 10 courtesy of NACE International

(Source: *Peabody's Control of Pipeline Corrosion, Second Edition*, Pg 222, Fig 11.9 Pipe-to-earth potentials on foreign pipeline passing through area of influence around cathodically protected bare line) ©NACE International – The Corrosion Society, 1440 South Creek Drive, Houston, Texas 77084. Additional publications on corrosion control technology are available from www.nace.org or phone (281) 228-6200.

Figure 12 courtesy of NACE International

(Source: *Peabody's Control of Pipeline Corrosion, Second Edition*, Pg 224, Fig 11.11 Use of electrical shields to reduce stray current interference) ©NACE International – The Corrosion Society, 1440 South Creek Drive, Houston, Texas 77084. Additional publications on corrosion control technology are available from www.nace.org or phone (281) 228-6200.

MODULE 61203-02 — TEACHING TIPS

The following are suggested activities or instructional methods to help you teach the material in this AIG.

General

When you call on someone to answer a question, the rest of the class relaxes or even tunes out because they expect that the question and answer will take place only between you and the trainee you called on. Instead, use this technique to involve more trainees in answering questions and to keep them on their toes.

1. Ask trainees to define a term or explain a concept.
2. After one trainee has answered, ask a trainee seated nearby if the answer is right. Then ask whether a trainee in the back of the room agrees.
3. Ask trainees to explain why they think an answer is right or wrong.
4. Use the session to clear up incorrect ideas and encourage trainees to learn from their mistakes.

Sections 2.0.0–2.2.0

Interference Current Sources

This exercise will give trainees the opportunity to review and record interference testing data. For this exercise, you will need sample field data sheets and recording survey notes. Trainees will need their Trainee Guides and pencils and paper to take notes for discussion. Allow 20 minutes for this exercise.

1. Distribute copies of the sample field data sheets and recording survey notes. Divide the class into small groups according to the number of samples.
2. Have trainees review and identify the types of stray current that are indicated in the survey notes.
3. Ask trainees to identify information about the line and the rectifier's output and potential sources of interference.
4. Ensure that trainees understand how to interpret and record interference data.
5. Answer any questions trainees may have about interference current sources.

Sections 3.0.0–3.5.0

Mitigation/Reduction Techniques

This exercise will familiarize trainees with the methods used to reduce or eliminate stray current interference. For this exercise, trainees will need their Trainee Guides and pencils and paper to take notes for discussion. Allow between 15 and 20 minutes for this exercise.

1. Present a variety of scenarios that could cause stray currents.
2. For each scenario, ask trainees to determine the mitigation/reduction technique that would be most useful for reducing or eliminating the interference.
3. Review and discuss each method. Provide examples from your own experience with mitigation/reduction techniques.
4. Review safety considerations associated with each method.
5. Answer any questions trainees may have about mitigation/reduction techniques.

Summary

Mitigation Review

This presentation will give trainees the opportunity to speak with a foreign pipeline company representative. For this presentation, invite a company representative to speak to the class about mitigation/reduction techniques. Trainees will need their Trainee Guides and pencils and paper to take notes for discussion. Allow between 45 and 60 minutes for this exercise.

1. Tell the trainees that a guest speaker will be presenting information on mitigation/reduction techniques.
2. Have trainees brainstorm questions for the guest speaker before the speaker arrives. Suggested questions could address the following:
   - Joint cooperative interference tests
   - Determination of the location of current discharge
   - Determination of the magnitude of the interference current
   - Notification of stray current
3. Introduce the guest speaker. Remember to include the speaker's name, title, company, and relevant credentials.
4. Allow 30 to 45 minutes for the speaker and 15 minutes for questions.
5. Following the presentation, ask trainees to share their questions with the guest speaker.

MODULE 61203-02 — ANSWERS TO REVIEW QUESTIONS

Answer	Section
1. d	1.0.0
2. a	1.0.0
3. b	2.1.0
4. a	2.1.0
5. c	2.2.0
6. b	2.2.0
7. a	3.1.0
8. c	3.1.0
9. d	3.2.0
10. b	3.3.0
11. d	3.3.0
12. b	3.4.0
13. d	3.4.0
14. c	3.4.0
15. a	3.5.0

CONTREN™ LEARNING SERIES — USER UPDATES

The NCCER makes every effort to keep these textbooks up-to-date and free of technical errors. We appreciate your help in this process. If you have an idea for improving this textbook, or if you find an error, a typographical mistake, or an inaccuracy in NCCER's Contren™ textbooks, please write us, using this form or a photocopy. Be sure to include the exact module number, page number, a detailed description, and the correction, if applicable. Your input will be brought to the attention of the Technical Review Committee. Thank you for your assistance.

Instructors – If you found that additional materials were necessary in order to teach this module effectively, please let us know so that we may include them in the Equipment/Materials list in the Instructor's Guide.

Write: Curriculum Revision and Development Department
National Center for Construction Education and Research
P.O. Box 141104, Gainesville, FL 32614-1104

Fax: 352-334-0932

E-mail: curriculum@nccer.org

Craft ______ Module Name ______

Copyright Date ______ Module Number ______ Page Number(s) ______

Description ______

______

______

______

(Optional) Correction ______

______

______

(Optional) Your Name and Address ______

______

______

Module 61204-02

Test and Repair Shorted Casings

(CT 9.5)

Test and Repair Shorted Casings
Annotated Instructor's Guide

Module 61204-02

MODULE OVERVIEW

This module reviews the types and causes of shorted casings. Trainees will learn how to test for and repair shorted casings, as well as how to use insulated fittings.

PREREQUISITES

Please refer to the Course Map in the Trainee Module. Prior to training with this module, it is recommended that the trainee shall have successfully completed the following:

Core Curriculum; Pipeline Core, Modules 66101-02 and 66102-02; Pipeline Corrosion Control Level One; Pipeline Corrosion Control Level Two, Modules 61201-02 through 61203-02

OBJECTIVES

Upon completion of this module, the trainee will be able to:

1. Identify causes of shorted casings.
2. Identify a shorted casing.
3. Test for shorted casings.
4. Identify methods for repairing shorted casings.
5. Describe the use of insulated fittings.

PERFORMANCE TASK

Under the supervision of the instructor, the trainee should be able to:

1. Identify and properly repair shorted casings (CT 9.5).

NCCER STANDARDIZED CRAFT TRAINING PROGRAM

The National Center for Construction Education and Research (NCCER) provides a standardized national program of accredited craft training. Key features of the program include instructor certification, competency-based training, and performance testing. The program provides trainees, instructors, and companies with a standard form of recognition through a National Craft Training Registry. The program is described in full in the *Guidelines for Accreditation*, published by the NCCER. For more information on standardized craft training, contact the NCCER by writing us at P.O. Box 141104, Gainesville, FL 32614-1104; calling 352-334-0911; or e-mailing info@nccer.org. More information may be found at our Web site, www.nccer.org.

HOW TO USE THIS ANNOTATED INSTRUCTOR'S GUIDE

Each page presents two sections of information. The larger section displays each page exactly as it appears in the Trainee Module. The narrow column ties suggested trainee and instructor actions to each page and provides icons to call your attention to material, safety, audiovisual, or testing requirements. The bottom of each page includes space for your notes.

Teaching Tip

If you see the Teaching Tip icon, that means there is a teaching tip associated with this section. Also refer to any suggested teaching tips at the end of the module.

SAFETY CONSIDERATIONS

Ensure that the trainees are equipped with appropriate personal protective equipment. Emphasize the importance of following all safety precautions and procedures when working with power tools.

PREPARATION

Before teaching this module, you should review the Module Outline, Objectives, Performance Task, and the Materials and Equipment List. Be sure to allow ample time to prepare your own training or lesson plan and gather all required equipment and materials.

MATERIALS AND EQUIPMENT LIST

Overhead projector and screen

Whiteboard/chalkboard

Transparencies

Markers/chalk

Blank acetate sheets

Transparency pens

Pencils and scratch paper

Module Examinations*

Performance Profile Sheets*

Copies of Performance Verifications for Covered Task 9**

Appropriate personal protective equipment

Copies of *49 CFR Part 192* (Gas) and/or *49 CFR Part 195* (Liquid)

Copies of your company policy and procedures manual

Calculators

Tools and materials to fill shorted casings, including the following
- Casings
- High dielectric materials
- Casing end seals
- Vents

Materials to flush the casing annulus

Casing insulators

Center cradles

Tools to install casing insulators and center cradles into a shorted casing

Shorted casings

Tools and materials to repair shorted casings

* Located in the Test Booklet.
**Available for download from www.nccer.org.

ADDITIONAL RESOURCES

This module is intended to present thorough resources for task training. The following reference works are suggested for both instructors and motivated trainees interested in further study. These are optional materials for continued education rather than for task training.

Peabody's Control of Pipeline Corrosion, 2nd Edition, 2001. A. W. Peabody. Houston, TX, NACE International.

Pipeline Corrosion and Cathodic Protection, 3rd Edition, June 1995. Marshall E. Parker, Edward G. Peattie. Houston, TX, Gulf Professional Publishing.

TEACHING TIME FOR THIS MODULE

An outline for use in developing your lesson plan is presented below. Note that each Roman numeral in the outline equates to one session of instruction. Each session has a suggested time period of 3 hours. This includes 10 minutes at the beginning of each session for administrative tasks and one 10-minute break during the session. Approximately 18 hours are suggested to cover *Test and Repair Shorted Casings (CT 9.5).* You will need to adjust the time required for hands-on activity and testing based on your class size and resources.

Topic	Planned Time
Session I. Test Casings	
A. Causes of Shorted Casings	____________
B. Recognizing Shorted Casings	____________
Session II. Testing for Shorted Casings, Part One	
A. Potential Shift Test	____________
B. Internal Resistance Test	____________
C. Casing Depolarization Test	____________
Session III. Testing for Shorted Casings, Part Two	
A. Four-Wire IR Drop Test	____________
B. Analyzing Test Results	____________
Session IV. Repair Shorted Casings, Part One	
A. Clearing the Shorted Casing	____________
B. Filling the Shorted Casing	____________
1. Laboratory	____________
C. Repairing the Shorted Casing	____________
Session V. Repair Shorted Casings, Part Two	
A. Replace a Shorted Casing	____________
B. Replace the Carrier Pipe	____________
Session VI. Shorted Casing Review	
A. Site Visit	____________
B. Summary	
1. Summarize module	____________
2. Answer questions	____________
C. Module Examination	____________
1. Trainees must score 70% or higher to receive recognition from the NCCER.	
2. Record the testing results on Craft Training Report Form 200 and submit the results to the Training Program Sponsor.	
D. Performance Testing	____________
1. Trainees must perform each task to the satisfaction of the instructor to receive recognition from the NCCER.	
2. Record the testing results on Craft Training Report Form 200 and submit the results to the Training Program Sponsor.	

Module 61204-02

Test and Repair Shorted Casings

(CT 9.5)

Instructor's Notes:

Instructor's Notes:

Introduction to the Pipeline Operator Qualification Rule, API, and NCCER

It could be said that petroleum products keep America running. Petroleum products like oil and natural gas are the principal sources of energy in the United States and are used in many manufacturing and chemical processes. Petroleum products are primarily transported by pipeline. There are more than a million miles of liquid and natural gas pipeline in this country, ranging in diameter from two inches to four feet. These pipelines bring crude oil and natural gas from production fields to processing plants and refineries—and ultimately to the consumer.

Pipeline Operator Qualification

Petroleum products are considered hazardous, so pipeline personnel must take extra precautions and follow specific procedures when working with them. To help ensure the safe handling and transportation of petroleum products, the United States Department of Transportation (DOT) mandates safe practices as federal regulations.

One such regulation, the Operator Qualification (OQ) rule, requires pipeline operators to develop a written qualification plan to evaluate their personnel and their contract personnel (see *49 CFR Part 192 Subpart N—Gas,* and *49 CFR Part 195 Subpart G—Liquid*). The evaluation hinges on the operator's ability to perform certain covered tasks and to be able to recognize and respond to abnormal operating conditions that may be encountered while performing these covered tasks.

Covered Tasks

A pipeline performance task is deemed *covered* if the activity passes the following four-part test:

1. Is it performed on a pipeline facility?
2. Is it an operations or maintenance task?
3. Is it performed as a requirement of *49 CFR Part 195 (Liquid)* or *49 CFR Part 192 (Gas)*?
4. Does it affect the operation or integrity of the pipeline facility?

If the answer to all four questions is *yes*, then the activity is considered *covered* and must figure into the evaluation process of pipeline personnel.

American Petroleum Institute

The OQ rule and its related covered tasks provided the catalyst to develop a unified pipeline training and assessment program—so that pipeline operators and owners have ready access to the assessment and training tools they need. This effort was the direct result of the vision and leadership provided by The American Petroleum Institute (API) and API members.

Efforts to unify the industry are nothing new to API. Indeed, the impetus for forming the Institute in 1919 was the need to standardize engineering specifications for drilling and production equipment. Today, the most pressing issues revolve around public perceptions and government policies toward the industry. Speaking with one voice on these issues has become as essential as having interchangeable parts in the field. API's members determine the petroleum industry's positions on public policy as well as the standards that govern its day-to-day operations worldwide.

Consortium on Operator Qualification

API assembled a focus group whose primary purpose was to identify covered tasks and provide compliance recommendations to pipeline operators.

The Consortium on Operator Qualification (COOQ) became the guiding force for the identification and development of covered tasks for the liquid pipeline industry. Their work product, API publication 1161, *Guidance Document for the Qualification of Liquid Pipeline Personnel,* formed the basis for the resulting Pipeliner Training and Assessment Program.

The covered task list created in API publication 1161 has now evolved to include gas covered tasks and is maintained by NCCER. The most current covered task list can be downloaded at www.nccer.org.

National Center for Construction Education and Research

With the guidance and support of API, its members, and the Consortium on Operator Qualification, the National Center for Construction Education and Research (NCCER) was able to design and implement the Pipeliner Training and Assessment Program.

NCCER is a not-for-profit 501(c)(3) education foundation established in 1995 by the world's largest and most progressive construction and maintenance contractors and national contractor associations. NCCER was founded to address the

severe workforce shortage facing the construction and maintenance industries and to develop a standardized training process and curricula. Today, NCCER is supported by hundreds of leading construction and maintenance companies, manufacturers, and national associations and developed more than 35 construction and maintenance titles.

NCCER also maintains a National Registry, which contains the records of trainees and instructors who have received NCCER training. All completion information for every trainee is sent to the NCCER and filed within the Registry. The Registry can then confirm training and skills for workers as they move from company to company, state to state, or even within their own company.

Standardized Curriculum (NCCER's Contren™ Learning Series)

NCCER curricula are developed using the expertise of industry subject matter experts (SMEs). The use of SMEs as an integral part of every new development or revision ensures that the curriculum is current, industry-driven, and indicative of industry practices. The development and/or revision process is facilitated from start to finish by NCCER project managers, who work closely with the SMEs and a select team of technical writer(s) on the development and presentation of the curriculum.

Pipeline: Five Training Paths

The pipeline curriculum focuses on essential skills and competencies within five skill areas:

- Corrosion Control
- Electrical and Instrumentation
- Maintenance
- Mechanical
- Operations (Field and Control Center)

SMEs from the pipeline industry have determined the skills and competencies needed based on API's covered tasks and the consensus of the group/industry. Specific company information is generally not incorporated into the curriculum unless the SMEs determine that the information is the industry standard.

Training Modules and Covered Tasks

The smallest unit of instruction in the curriculum is a *module*. Each curriculum is composed of modules (the equivalent of stand-alone chapters) that are logically sequenced into complete *levels* of instruction. Every module is task-oriented and requires the successful completion of a written (knowledge-based) examination and a performance (hands-on) test. Modules may be individually selected and/or combined for efficient task training. Modules in the pipeline curricula, to the extent possible, will align with API-developed covered tasks. While not all modules will relate to covered tasks, all covered tasks will relate to at least one module.

Benefits of Training to the Pipeline Marketplace

Applying NCCER's standardized training process allows pipeline owners and contractors in the United States to establish the benchmark for acceptable skill and competency levels within the pipeline industry. This process will assist training organizations in developing and/or improving standardized pipeline training programs that effectively match their resources and operating conditions while responding to regulatory demands and to the needs of the industry. The process will also recognize training organizations that meet the minimum standards of competence, warranting regulatory, public, and professional confidence. This in turn will foster national unity within the pipeline industry for the mutual benefit of individual technicians, owners, contractors, and ultimately pipeline-users and society.

Additional Resources

To review the Pipeliner Training and Assessment Program Implementation and Resource Guide and the latest version of pipeline covered tasks, visit NCCER's Web site at www.nccer.org, or contact:

National Center for Construction Education and Research
PO Box 141104
Gainesville, FL 32614-1104
(888) 622-3720
Web site: www.nccer.org

For more information on API, please contact:

American Petroleum Institute
1220 L Street, Northwest
Washington, D.C. 20005
(202) 682-8000
Web site: www.api.org

To order *API Pub 1161*, contact Global Engineering Documents by calling 1-800-854-7149 or visiting their Web site at www.global.his.com.

Additional sources of information on the pipeline operator qualification regulation include:

DOT – Office of Pipeline Safety
www.ops.dot.gov

Transportation Safety Institute
www.tsi.dot.gov

Instructor's Notes:

COURSE MAP

This course map shows all of the modules in the second level of the Pipeline Corrosion Control curriculum. The suggested training order begins at the bottom and proceeds up. Skill levels increase as you advance on the course map. The local Training Program Sponsor may adjust the training order.

PIPELINE CORROSION CONTROL LEVEL TWO

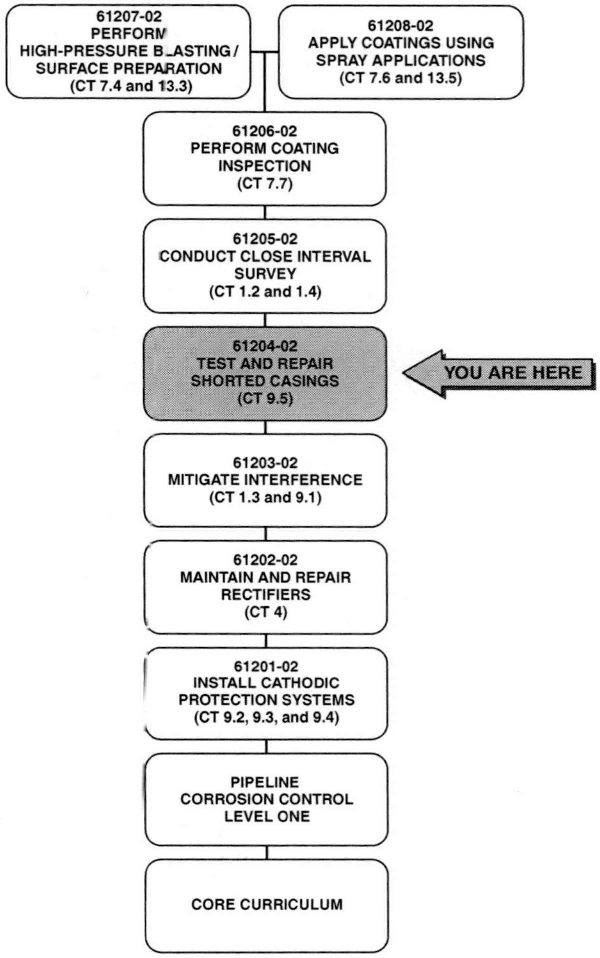

204CMAP.EPS

Homework

Assign reading of Module 61204-02.

Instructor's Notes:

MODULE 61204-02 CONTENTS

Figures

Instructor's Notes:

MODULE 61204-02

Test and Repair Shorted Casings (CT 9.5)

Objectives

When you have completed this module, you will be able to do the following:

1. Identify causes of shorted casings.
2. Identify a shorted casing.
3. Test for shorted casings.
4. Identify methods for repairing shorted casings.
5. Describe the use of insulated fittings.

Prerequisites

Before you begin this module, it is recommended that you successfully complete the following: Core Curriculum; Pipeline Core, Modules 66101-02 and 66102-02; Pipeline Corrosion Control, Level One; Pipeline Corrosion Control, Level Two, Modules 61201-02 through 61203-02.

Required Trainee Materials

1. Pencil and paper
2. Appropriate personal protective equipment

1.0.0 ◆ INTRODUCTION

The effectiveness of a cathodic protection (CP) system can be compromised if the pipeline comes into contact with the **casing**. Pipelines inside casings must have spacers that will prevent any electrical contact between the **carrier pipe** and the casing. Casing seals that prevent water from entering the area between the pipe and the casing should be used. Water in the area between the casing and the carrier pipe can cause corrosion damage. This module describes some techniques used to maintain insulated fittings. *Figure 1* shows a typical casing installation.

2.0.0 ◆ TEST CASINGS

On an in-service pipeline, it is important to be able to perform testing to identify the existence of a cased crossing short. Quickly and accurately locating shorts saves time and expense. Testing for **shorted casings** is discussed in the following sections.

2.1.0 Causes of Shorted Casings

The following is a list of the most common causes of contact between the carrier pipe and the casing:

- Settling of the soil around the casing or carrier pipe, usually occurring at each end of the casing where the carrier pipe enters or leaves the casing
- Test station wire with damaged insulation touching the casing
- Carrier pipe movement caused by internal stresses, expansion, and/or contraction
- Stresses remaining from jacking the pipe for lineup during new construction
- Improper use of casing insulation spacers
- Misaligned casing pipes
- Mechanical contacts installed during construction
- Foreign objects left inside the casing during installation

Materials

Ensure that you have everything required to teach the course. Check the Materials and Equipment List at the front of this module.

Classroom

Explain that terms shown in bold are defined in the Glossary at the back of this module.

Audiovisual

Show Transparency 1, Course Objectives.

Show Transparency 2, Performance Task.

Teaching Tip

See the general Teaching Tip at the end of this module.

Classroom

Discuss the role of spacers and casing seals in preventing corrosion.

Review the common causes of shorted casings.

Classroom

Explain how to compare the casing with the carrier potential readings.

Discuss the procedure for testing for shorted casings. Answer any questions trainees may have.

Assign values for pipe-to-soil potential and casing-to-soil potential. Ask trainees to calculate absolute difference and determine whether the calculations indicate that the casing is shorted.

Homework

Have trainees review Sections 2.3.0–2.3.3 for the next session.

Materials

Ensure that you have everything required for the demonstrations during this session.

Classroom

Compare the methods of testing for shorted casings. Go over any questions trainees may have.

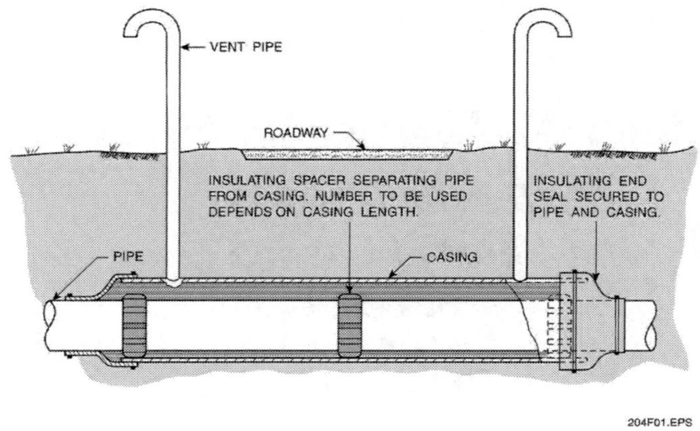

Figure 1 ◆ Casing installation requirements.

2.2.0 Recognizing Shorted Casings

The most common method used for identifying possible shorted casings is the comparison of casing and carrier potential readings. This is normally done while conducting the annual survey.

The following procedure describes a typical recognition test for possible shorted casings.

> **NOTE**
> All casings should be included in the annual survey, and their electrical isolation status verified.

Step 1 Measure the carrier **pipe-to-soil potential** (P) with respect to a suitable reference electrode (for example, copper-copper sulfate).

Step 2 With the reference electrode placed at the same location as in Step 1, measure the **casing-to-soil potential** (C).

Step 3 Subtract the casing-to-soil reading from the carrier pipe-to-soil reading (P – C).

Step 4 If the absolute difference (P – C) is equal to or greater than 100 millivolts (0.100 volt), the casing is considered electrically isolated (non-shorted).

If the absolute difference (P – C) is less than 100 millivolts (0.100 volt), the casing may be shorted and further testing is required to confirm the possible short.

2.3.0 Testing for Shorted Casings Condition

When the absolute difference is less than 100 millivolts, perform one or a combination of the following test(s) to determine if a casing short is present:

- Potential shift test
- Casing **depolarization** test
- Internal resistance test
- Four-wire IR drop test

2.3.1 Potential Shift Test

The potential shift test is easy to perform and requires minimal equipment for pipelines equipped with impressed current CP systems. *Figure 2* shows a potential shift test schematic. Perform the potential shift test using the following procedure:

Step 1 With the reference electrode placed at the same location, measure the cathodic protection potential for the casing (C1) and the carrier pipe (P1).

Step 2 Increase the cathodic protection current on the system by adjusting the rectifier having the most cathodic protection influence on the carrier pipe.

Step 3 Measure and record the new casing potential (C2) and the pipe potential (P2) due to the increase in cathodic protection current.

Instructor's Notes:

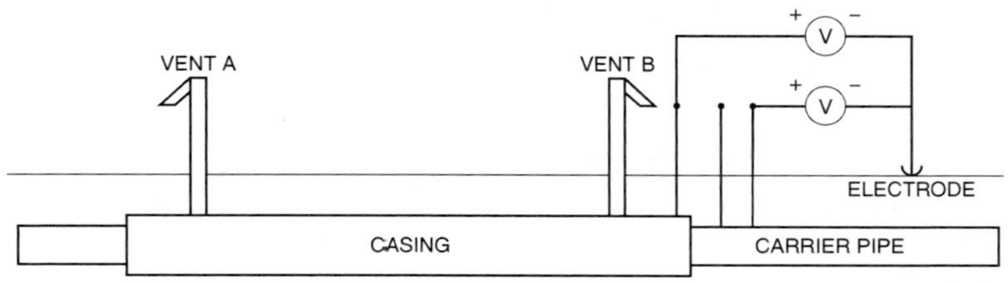

Figure 2 ◆ Potential shift test schematic.

NOTE
The rectifier selected must be sufficiently remote from the casing under test so that the ground bed voltage gradients do not influence the potential measurements. Distances greater than a mile are preferable.

Step 4 The casing is considered electrically isolated (non-shorted) if one or both of the following conditions exist:

- The pipe potential increases (goes more negative), while the casing potential remains constant.
- The potential difference between the pipe and casing potential (P2 – C2) becomes more than 100 millivolts (0.100 volt) after the increase in cathodic protection current.

Step 5 If both the pipe and casing increase in the same direction (both go more negative) and the potential difference is less than 100 millivolts, the casing may be shorted and further testing is required to verify the effectiveness of the electrical isolation.

2.3.2 Internal Resistance Test

The internal resistance test is used to determine if a direct metal-to-metal contact exists between the casing and the carrier pipe. This is done by determining the electrical resistance between the casing and the carrier pipe. *Figure 3* shows an internal resistance test schematic.

WARNING!
When conducting this test, large current flows may be generated by small voltages if a short exists between the carrier and the casing. Use dry cell batteries when conducting this test. Because a direct short can exist, wet cell batteries should not be used for this testing due to the potential explosive condition.

Perform the internal resistance test using the following procedure (refer to *Figure 3*):

Step 1 Connect the positive terminal of a DC source (6- or 12-volt battery) to the casing vent, and the negative terminal to the carrier pipe.

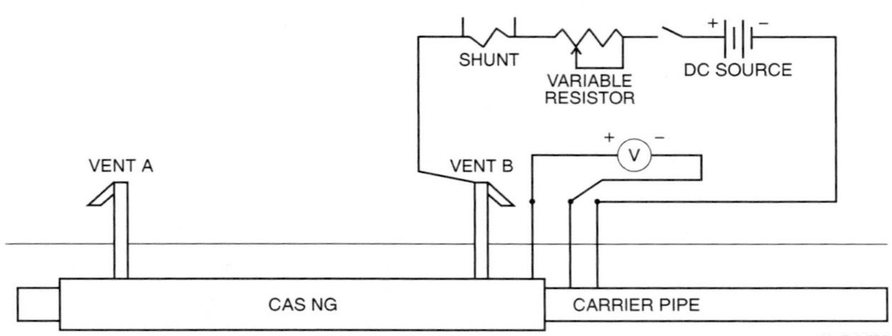

Figure 3 ◆ Internal resistance test schematic.

Audiovisual

Show Transparency 3 (Figure 2). Review the procedure for performing the potential shift test.

Safety

Emphasize the importance of using dry cell batteries when conducting internal resistance tests.

Classroom

Refer to Figure 3. Explain how to perform an internal resistance test. Ensure that trainees understand how to connect the terminal to the casing.

Demonstration

On the whiteboard/ chalkboard, demonstrate how to calculate internal casing-to-pipe resistance.

Safety

Discuss the consequences of arcing.

Ask a trainee to explain how to minimize arcing.

Classroom

Discuss procedures for performing a casing depolarization test. Explain how to construct a temporary ground bed.

> **NOTE**
> It is recommended that you use a 4½-digit voltmeter for making the voltage drop measurements.

Step 2 Insert in the positive lead:

- An ON/OFF switch to interrupt the circuit
- A variable resistor to limit and adjust the test current
- A shunt to measure the circuit current

Step 3 With the circuit switch in the OFF position (no current), measure and record the voltage between the casing and the carrier pipe (E1) in volts.

> **NOTE**
> To avoid introducing an error into the measurement for the casing-to-carrier voltage, do not use the current-carrying leads.

Step 4 Use an insulated probe bar to make casing contact for the voltage measurement if there is no separate casing lead available.

Step 5 Use the carrier pipe test lead to connect the negative battery terminal, and use an insulated probe bar to make pipe contact for the voltage measurement.

> **CAUTION**
> To avoid arcing, do not use the insulated probe bar as a contact on a current-carrying lead.

Step 6 Use the probe bars for voltage measurements, ensuring good metallic contact.

Step 7 With the circuit switch in the ON position, simultaneously measure and record the voltage between the casing and the carrier pipe (E2) in volts, and the circuit current (I2) in amps.

Step 8 Calculate the internal casing-to-pipe resistance (R) by dividing the change in voltage (E2 – E1) by the circuit current applied (I2).

$$R = \frac{E2 - E1}{I2}$$

> **NOTE**
> If the internal casing-to-pipe resistance is typically less than or equal to 0.01 ohm, a metal-to-metal contact is indicated and the casing is shorted. If the resistance is greater than 0.01 ohm, the casing may be shorted and further testing is required.

Step 9 Repeat the previous steps to obtain a minimum of three casing-to-pipe resistance calculations. Increase the circuit current at least 0.5 amp for each test. If the casing status results repeat for all tests, the test is considered valid.

> **CAUTION**
> To minimize arcing between the carrier pipe and casing, do not exceed 10 amps circuit current.

2.3.3 Casing Depolarization Test

The casing depolarization test is used to determine if a casing short exists. In this test, an external DC current is applied to the casing and the potential shift between the casing and the carrier pipe is analyzed. *Figure 4* shows a casing depolarization test schematic.

Perform the casing depolarization test using the following procedure (refer to *Figure 4*):

Step 1 Construct a temporary ground bed a minimum of 50' from, and perpendicular to, the carrier pipe. Bare steel bars driven into the ground and electrically connected may be used as the temporary ground bed.

Step 2 Connect the positive terminal of a DC source (6- or 12-volt battery) to the casing, and the negative terminal to the temporary ground bed.

Step 3 Insert in the positive lead:

- An ON/OFF switch to interrupt the circuit
- A variable resistor to limit and adjust the test current
- A shunt to measure the circuit current

> **NOTE**
> In this circuit, the casing is made the anode and the temporary ground bed is made the cathode.

Instructor's Notes:

____

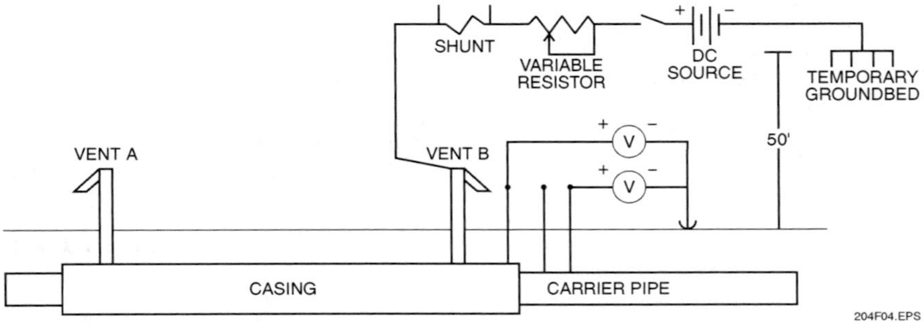

Figure 4 ◆ Casing depolarization test schematic.

Step 4 With the circuit switch in the OFF position (no current), measure and record the potential (in volts) of the casing (C1) and of the carrier pipe (P1), with the reference electrode placed at the same location.

CAUTION

Do not use the current carrying casing vent in measuring the casing potentials because this introduces an error in the measurement. If a separate casing test lead is not available, connect the positive of the DC source to the far casing vent (vent A), or use an insulated probe bar to make contact with the casing.

Step 5 With the circuit switch in the ON position, and the current increased approximately 0.5 amp, simultaneously measure and record the new casing (C2) and carrier pipe (P2) potentials in volts and the applied current (I2) in amps.

NOTE

To minimize **polarization** caused by the increase in current, do not energize the circuit for more than 5 seconds at a time.

Step 6 Obtain the casing potential shift by subtracting the new casing potential (C2) from the initial potential (C1). Obtain the carrier pipe potential shift by subtracting the new carrier potential (P2) from the initial potential (P1).

Step 7 Calculate the effective casing-to-pipe resistance by subtracting the shift in the carrier pipe potential (P2 – P1) from the shift in the casing potential (C2 – C1), and dividing the result by the current applied (I2).

$$R = \frac{(C2 - C1) - (P2 - P1)}{I2}$$

NOTE

If the effective casing-to-pipe resistance is typically less than or equal to 0.080 ohm, a short is indicated. If the resistance is greater than 0.080 ohm, the casing has effective electrical isolation.

Step 8 Repeat the previous steps to obtain a minimum of three casing-to-pipe resistance calculations. Increase the circuit current at least 0.5 amp for each test. If the casing status results repeat for all tests, the test is considered valid.

NOTE

It is usually not necessary to exceed 10 amps from the DC source to the temporary ground bed to develop significant potential shift.

2.3.4 Four-Wire IR Drop Test

The four-wire IR drop test is used to indicate if a casing **metallic short** exists and the location of the short. The four-wire IR drop test is divided into four parts.

Safety

Explain problems that could result if the current carrying casing vent is used to measure the casing potential.

Demonstration

On the whiteboard/chalkboard, demonstrate how to calculate effective casing-to-pipe resistance. Ask whether trainees have any questions.

Provide the values needed to calculate internal casing-to-pipe and effective casing-to-pipe resistance. Ask trainees to obtain the resistance calculations and explain how the formulas are applied.

Homework

Assign the reading of Section 2.3.4 for the next session.

Materials

Ensure that you have everything required for teaching this session.

Audiovisual

Show Transparency 4 (Figure 5). Explain how to perform the four-wire IR drop test. Emphasize the importance of properly labeling the casing contact points.

Classroom

Refer to Figure 6. Review the second part of the four-wire IR drop test. Explain how to move the DC source negative terminal.

> **NOTE**
> It is recommended that you use a 4½-digit voltmeter for making the voltage drop measurements. Casing vents may be doglegged and not necessarily located at the casing ends.

Perform the four-wire IR drop test using the following procedure. Part 1 of the four-wire IR drop test includes the following eight steps. *Figure 5* shows the four-wire IR drop test, Part 1 schematic.

Step 1 Connect the positive terminal from a DC source (6- or 12-volt battery) to casing vent A and the negative terminal to casing vent B.

Step 2 Insert in the positive lead the following:
- An ON/OFF switch to interrupt the circuit
- A variable resistor to limit and adjust the test current
- A shunt to measure the circuit current

Step 3 With the circuit switch in the OFF position (no current), measure and record the voltage drop (E1) in millivolts, and the polarity across the casing length.

Step 4 Contact to the casing may be made with the existing casing test lead and/or insulated probe bar.

> **NOTE**
> Do not use the current carrying casing vents as contacts in making this voltage drop measurement. If insulated probe bars are used, the contact should be made near and inside the casing vents.

Step 5 Label the casing contact point nearest vent A as *casing contact A,* and the casing contact point nearest vent B as *casing contact B.*

> **NOTE**
> When using insulated probe bars, good metallic contact is vital for accurate measurement interpretation.

Step 6 With the circuit switch in the ON position, measure and record the voltage drop (E2) in millivolts, and polarity across the same casing length contact points as in previous steps.

Step 7 Simultaneously measure and record the circuit current (I2) in amps.

Step 8 A casing voltage drop shift (E2 – E1) less than 5 millivolts (5 mV) may indicate poor contact or insufficient source current. Make the appropriate adjustments before proceeding to Part 2 of the four-wire IR drop test.

Part 2 of the four-wire IR drop test includes the following five steps. *Figure 6* shows the four-wire IR drop test, Part 2 schematic.

Step 1 Move the DC source negative terminal from vent B and connect to the carrier pipe while maintaining the positive lead on vent A.

Step 2 With the circuit switch in the OFF position (no current), measure and record the voltage drop (E3) in millivolts, and the polarity across the casing length.

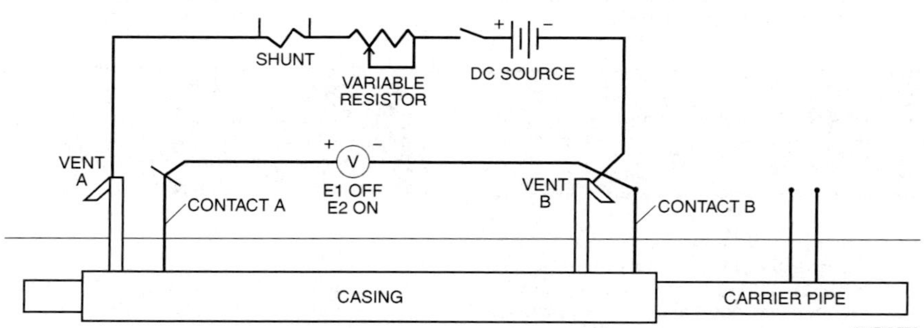

Figure 5 ◆ Four-wire IR drop test, Part 1 schematic.

Instructor's Notes:

Step 3 With the circuit switch in the ON position, measure and record the voltage drop (E4) in millivolts and polarity across the same casing length as in Step 2.

Step 4 Calculate and record the voltage shift (E4 – E3).

Step 5 Simultaneously measure the circuit current (I4) in amps and record.

Part 3 of the four-wire IR drop test includes the following six steps. *Figure 7* shows the four-wire IR drop-test, Part 3 schematic.

Step 1 Move the positive lead from vent A to vent B.

Step 2 With the circuit switch in the OFF position (no current), measure and record the voltage drop (E5) in millivolts and the polarity across the casing length.

NOTE
Use the same connecting wire length as in Part 2.

Step 3 With the circuit switch in the ON position, adjust the circuit current (if necessary) to the same value as I4 in Part 2.

Step 4 Measure and record the voltage drop (E6) in millivolts across the same casing length as in Step 3.

Step 5 Calculate and record the voltage shift (E6 – E5).

Step 6 Simultaneously measure the circuit current (I6) in amps and record.

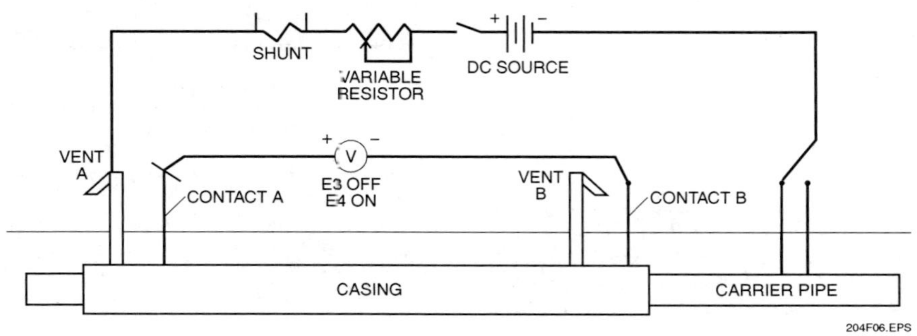

Figure 6 ◆ Four-wire IR drop test, Part 2 schematic.

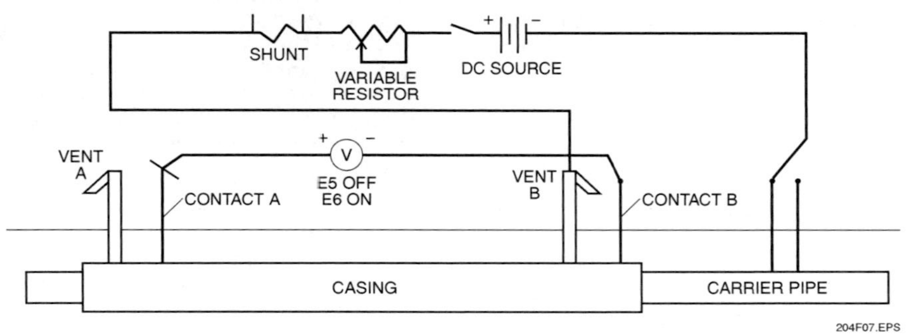

Figure 7 ◆ Four-wire IR drop test, Part 3 schematic.

Audiovisual

Show Transparency 5 (Figure 7). Discuss the steps in the third part of the four-wire IR drop test. Explain how to calculate the voltage shift.

Classroom

Ensure that trainees understand how to record data for the first three parts of the drop test. Answer any questions trainees may have.

Using hypothetical data for the first three parts of the four-wire IR drop test, have trainees practice analyzing test results. Ask trainees to calculate approximate distance.

Teaching Tip

See the Teaching Tip for Sections 2.0.0–2.3.4 at the end of this module.

Homework

Have trainees review Sections 3.0.0–3.1.4 for the next session.

Materials

Ensure that you have everything required for the laboratory and demonstration during this session.

Classroom

Review factors that affect shorted casing repair. Discuss the steps in clearing a shorted casing.

Part 4 is the analysis of the test results. Use the data collected in each of the preceding tests for the following analysis:

- If E3 = E4 and E5 = E6 (no change in voltage shift), the casing is not shorted.
- If the casing voltage shift (E4 – E3) is greater than zero in Part 2 and/or (E6 – E5) is greater than zero in Part 3, the casing is shorted. If the casing is shorted, the sum of the voltage shift in Part 2 and Part 3 is approximately the voltage shift determined in Part 1. (E2 – E1).
- If the casing is determined shorted and (E4 – E3) is zero and (E6 – E5) is greater than zero, the short is located near vent A.
- If the casing is determined shorted and (E6 – E5) is zero and (E4 – E3) is greater than zero, the short is located near vent B.
- If the casing is determined shorted and (E4 – E3) and (E6 – E5) are both greater than zero, the short is located between the two casing voltage drop contacts. An approximate distance of the metallic short from casing contact A is:

$$\text{Distance} = \frac{(E4 - E3) + (E6 - E5)\ L}{E4 - E3}$$

L is the distance in feet between the two casing contact points, and both (E4 – E3) and (E6 – E5) are positive numbers.

> **NOTE**
> If more than one casing short exists, this test may not accurately identify the location of the shorts.

3.0.0 ◆ REPAIR SHORTED CASINGS

The repair of shorted casings consists of:

- Listing all known nonrepaired **shorted casings** in the cathodic protection system under test
- Establishing a repair priority to the cathodic protection system under test
- Taking corrective action for the shorted casing, according to the established repair priority
- Documenting test results according to company requirements
- Notifying appropriate individuals of shorted casings

3.1.0 Corrective Actions to Take Upon Verification of Shorted Casing

A plan of action and the implementation of the plan should be initiated according to company specifications. The corrective actions that can be taken for shorted casings are discussed in the following sections.

3.1.1 Clearing the Shorted Casing

It is important to have a through understanding of shorted casings before attempting this procedure. When clearing a casing short, the following steps can be performed:

Step 1 Conduct tests to determine the approximate location of the short.

> **NOTE**
> If tests indicate the short is located somewhere other than at the ends of the casing, it may not be practical to clear the short.

Step 2 If tests indicate the short is located at one end of the casing, excavate that end and expose the pipe and casing.

> **NOTE**
> Follow current OSHA standards and company policy for excavation and trenching when excavating a ditch or bellhole.

Step 3 Check for test lead wire meeting the casing or casing vent.

> **NOTE**
> If the wire is the cause of the short, repair the wire insulation or replace the wire. Position the wire so backfilling does not cause a shorted condition.

Step 4 Once exposed, visually inspect the carrier pipe and casing.

Step 5 If the short is caused by the carrier pipe contacting the casing, realign the carrier pipe within the casing using company-approved lifting equipment.

Instructor's Notes:

CAUTION
Take care not to damage or cause any unusual stresses on the pipe during this process. Any coating damage to the carrier pipe should be repaired.

Step 6 Maintain clearance by installing new or additional insulators or end seals after clearing the contact.

NOTE
Use sandbags under the end of the casing and under the pipe for support, as the backfill will normally settle.

Step 7 Check for bonding cables or straps connecting the casing to the carrier. If possible, remove the bonding device.

Step 8 Retest the casing before backfilling to determine if the short has been cleared and that there are no other shorts present.

3.1.2 Filling the Shorted Casing

Follow these steps when filling a shorted casing:

Step 1 It may be necessary to fill the casing with a **high dielectric material**, such as inhibited petroleum-based compounds, if it is impractical to clear a short.

NOTE
While casing fillers may not clear the short, they can mitigate possible corrosion of the carrier pipe by minimizing the corrosive environment of water and oxygen. Follow the manufacturer's recommendations when installing casing fillers.

Step 2 Ensure that the casing end seals are adequate to contain the filler material during installation and while in service.

NOTE
Rubber link-type end seals or metal end seals that are welded to the casing but insulated from the carrier pipe provide an adequate casing seal, especially during the filling process when high pumping pressure is expected.

Step 3 Install vents at each end of the casing to permit filling of the casing.

NOTE
The opening where the vent is connected to the casing should be adequate in size to permit the filler to flow without requiring excessive pressure. If existing vents are used, the opening to the casing should be checked for adequate size. On the low casing end, consider connecting the vent to the bottom of the casing.

Step 4 Flush the casing annulus with clean water to remove trapped mud, dirt, and other corrosive products.

NOTE
This also ensures that a free flow exists before pumping the casing filler.

Step 5 Test the casing for sealability, as the casing itself may not be watertight.

Step 6 Pump the filler material from the vent on the low end if the casing filler material is heavier than water. Pump from the high end if the casing filler material is lighter than water.

NOTE
Always follow the casing filler manufacturer's recommendations.

Step 7 Calculate the theoretical gallons of filler required to fill the casing annulus.

Step 8 Measure and record the actual gallons of filler pumped.

Step 9 Compare the two amounts to approximate the effectiveness of the filler in that casing.

NOTE
Always follow manufacturer's recommendations when determining if casing is properly filled.

Safety

Stress the importance of preventing damage to the pipe when clearing a shorted casing.

Classroom

Review the procedure for filling a shorted casing.

Explain how to calculate theoretical gallons and how to measure and record actual gallons of filler pumped.

Laboratory

Set up stations with casings, high dielectric materials, casing end seals, vents, and materials to flush the casing annulus. Ask trainees to practice filling shorted casings.

Demonstration

Bring in casing insulators and center cradles. Demonstrate how to install casing insulators and center cradles into a shorted casing. Go over any questions trainees may have.

Classroom

Emphasize the importance of monitoring shorted casings at the appropriate intervals.

Homework

Assign the reading of Sections 3.2.0–3.3.0 for the next session.

Materials

Ensure that you have everything required for teaching this session.

Teaching Tip

See the Teaching Tip for Sections 3.0.0–3.3.0 at the end of this module.

Homework

Have trainees prepare for testing during the next session and complete the Review Questions.

3.1.3 Repairing the Shorted Casing

Follow these steps when repairing a shorted casing:

Step 1 Remove the end seal and check for any points of physical contact between the pipe and casing.

Step 2 Center the carrier pipe within the casing following company policy and procedures.

Step 3 Ensure that all coating is intact.

Step 4 Install **casing insulators** and center cradles while providing adequate support.

Step 5 Replace the end seal.

Step 6 Install test wires as required.

Step 7 Backfill supported pipe and casing ends in such a manner that stresses will not be added.

Step 8 Conduct the following testing for a shorted casing to ensure that the casing is repaired:

- Resistance test
- Potential shift test
- Internal resistance test
- Casing depolarization test
- Four-wire IR drop test

Step 9 Conduct pipe-to-casing potential difference tests to determine that pipe and casing are still clear after backfilling.

3.1.4 Monitor the Shorted Casing

The procedure for monitoring a shorted casing is described in the following steps:

Step 1 Monitor the casing vents with leak-detection equipment.

Step 2 Perform the leak-detection test at least twice each calendar year at an interval not exceeding 7½ months until the corrective action is performed.

> **NOTE**
> For gas pipelines located in Class 3 and Class 4 population density areas, perform the leak-detection test at least four times each calendar year at an interval not exceeding 4½ months.

Step 3 If a leak is detected during the monitoring, follow company procedures to repair the leak.

Step 4 If an in-line inspection tool survey indicates no corrosion activity on the carrier pipe within the shorted casing, an evaluation for appropriate corrective action should be taken and a time frame for corrective action established.

3.2.0 Replace a Shorted Casing with Uncased Heavy-Wall Steel Pipe

Whenever it is possible and appropriate, replace the cased crossing with an uncased heavy-wall steel pipe or a concrete-coated steel pipe. These types of pipes provide added mechanical strength without having the problems that may be caused by pipelines with casings. Always follow any state and local guidelines when replacing shorted casings.

3.3.0 Replace the Carrier Pipe Within the Casing

Casing insulating spacers have an insulated skid and are strapped around the pipe. There are many types of casing insulators available. These insulators can be all plastic or can have insulated blocks that are attached to steel bands. Some steel bands are lined with rubber or plastic and are called skids. Skids are placed evenly along the pipeline at a maximum of 10' apart. Insulating spacers are also located at the ends of the casing. Always follow manufacturer recommendations for the number, size, and spacing of insulated spacers.

Stresses exerted on spacers during installation are very great, especially when working with large pipe and long-cased crossings. If insulators are snagged and damaged during installation, the insulator should be replaced before continuing the installation. When the installation is complete, make sure that the end insulators are close to the end of the casing. If they are not, install another insulator into the casing and position it near the casing end to provide proper electrical separation at the ends of the casing.

End seals are designed to be flexible and to provide a tight seal between the pipe and the casing. Most seals are watertight. Proper casing preparation is required before the carrier pipe is pulled into it. Check the carrier pipe for the following to make sure it has been prepared properly:

- All welds must be smooth.
- The casing must be straight.
- All debris must be removed.
- All welding and hole burning must be performed before the carrier pipe is installed.
- All vent pipes must be set before the carrier pipe is installed.

Instructor's Notes:

Summary

Pipelines are often run through casings, especially at crossings. These pipelines are subject to corrosion from several sources: where the piping comes in contact with the casing or casing vent, where soil settles around the pipe entry into the casing, or where water enters the casing. To protect the pipeline, the casings are tested for shorts.

Shorted casings are identified by comparing potential readings taken from the carrier and the casing. Tests performed for shorted casings are the potential shift test, the casing potential depolarization test, the internal resistance test, and the four-wire IR drop test.

Repairing shorted casings consists of forming and implementing a plan of corrective action. The short must be cleared and then filled with a high dielectric material. If feasible, the shorted casing should be replaced.

Review Questions

1. The effectiveness of a cathodic protection system can be compromised if the pipeline comes into contact with the _____.
   a. carrier pipe
   b. insulation
   c. casing
   d. spacers

2. Misaligned casing pipes can cause _____.
   a. shorted casings
   b. bad potential readings
   c. pipeline movement
   d. pipeline damage

3. The most common method used for identifying possible shorted casings is the comparison of casing and carrier _____.
   a. alignments
   b. resistance
   c. depolarization
   d. potential readings

4. When performing a recognition test for possible shorted casings, measure the carrier pipe-to-soil potential with respect to a suitable _____.
   a. nearby measurement
   b. landmark
   c. reference electrode
   d. test station

5. If the absolute difference, P – C, is less than _____, the casing may be shorted.
   a. 100V
   b. 100mV
   c. 1V
   d. 200mV

6. With the potential shift test, the casing is considered electrically isolated if the pipe potential _____ while the casing potential _____.
   a. decreases; remains constant
   b. increases; remains constant
   c. remains constant; decreases
   d. remains constant; increases

7. The _____ test is used to determine if a direct metal-to-metal contact exists between the casing and the carrier pipe.
   a. internal resistance
   b. potential shift
   c. depolarization
   d. IR drop

8. If an internal casing-to-pipe resistance test reading is less than or equal to _____, a metal-to-metal contact is indicated.
   a. 1 ohm
   b. 0.001 ohm
   c. 0.5 ohm
   d. 0.01 ohm

9. In the casing depolarization test, a(n) _____ is applied to the _____.
   a. AC current; carrier pipe
   b. DC current; carrier pipe
   c. DC current; casing
   d. AC current; casing

10. The four-wire IR test is divided into _____ parts.
    a. two
    b. three
    c. four
    d. five

Materials

Ensure that you have everything required for testing during this session.

Teaching Tip

See the Summary Teaching Tip at the end of this module.

Classroom

Go over the Review Questions for this module, and answer any questions trainees may have.

Examination

Administer the Module Examination. Be sure to record the results of the Exam on Craft Training Report Form 200 and submit the results to the Training Program Sponsor.

Performance Testing

Have each trainee perform Performance Task 1 to your satisfaction. Fill out a Performance Profile Sheet for each trainee.

Ensure that all Performance Tests have been completed and Performance Profile Sheets for each trainee are filled out. Be sure to record the results of the testing on Craft Training Report Form 200 and submit the results to the Training Program Sponsor.

11. When using insulated probe bars, good _____ is vital for accurate measurement.
    a. pipe contact
    b. polarity
    c. test lead connection
    d. metallic contact

12. One step in repairing a shorted casing is conducting tests to determine the _____.
    a. location of the short
    b. magnitude of the short
    c. repair priority
    d. number of shorts

13. It may be necessary to fill the casing with a(n) _____ if it is impractical to clear a short.
    a. foam
    b. low dielectric material
    c. high dielectric material
    d. insulating fluid

14. Casing insulated spacers have a(n) _____.
    a. insulated bushing
    b. insulated skid
    c. electrical fitting
    d. insulated flange

15. Skids are placed evenly along a pipeline at a maximum of _____ apart.
    a. 4'
    b. 6'
    c. 8'
    d. 10'

Instructor's Notes:

GLOSSARY

Glossary

This glossary will help you learn some terms specific to testing and repairing shorted casings. Trade Terms introduced in this module appear in italic text.

Carrier pipe: A pipe that carries the product. The carrier pipe is installed through the casing.

Casing: A sleeve installed under a roadway, railroad tracks, etc., by boring, which will allow another pipe to be slid through.

Casing insulator: A device attached to the carrier pipe that aids in the insertion into the casing and provides electrical isolation from the casing.

Casing-to-soil potential: The voltage potential of the casing in the soil or electrolyte as related to a reference electrode.

**Depolarization:* The removal of factors resisting the current in an electrochemical cell.

Electrolytic short: Indirect or direct contact between two metallic structures via an electrolyte.

High dielectric material: A material having high electrical insulation properties. A non-conductor of electric current.

Holiday testing: Electrical testing of a coating to locate anomalies or imperfections (holidays).

Metallic short: Indirect or direct metallic contact between two metallic structures.

**Pipe-to-soil potential:* See structure-to-electrolyte potential.

**Polarization:* The change from the open-circuit potential as a result of current across the electrode/electrolyte interface.

Shorted casing: A casing in which a metallic contact exists between the casing and the carrier pipe. The metallic contact provides a low electrical resistance path that could cause the casing to shield cathodic protection from the portion of carrier pipe within the casing.

***Structure-to-electrolyte potential:** The potential difference between the surface of a buried or submerged metallic structure and the electrolyte that is measured with reference to an electrode in contact with the electrolyte.

REFERENCES & ACKNOWLEDGMENTS

Additional Resources

This module is intended to present thorough resources for task training. The following reference works are suggested for further study. These are optional materials for continued education rather than for task training.

Peabody's Control of Pipeline Corrosion, 2nd Edition, 2001. A. W. Peabody. Houston, TX, NACE International.

Pipeline Corrosion and Cathodic Protection, 3rd Edition, June 1995. Marshall E. Parker, Edward G. Peattie. Houston, TX, Gulf Professional Publishing.

Figure Credits

Figure 1 courtesy of NACE International

(Source: *Peabody's Control of Pipeline Corrosion, Second Edition*, Pg 245, Fig 12.6, Casing Installation Requirements) ©NACE International – The Corrosion Society, 1440 South Creek Drive, Houston, Texas 77084. Additional publications on corrosion control technology are available from www.nace.org or phone (281) 228-6200.

Instructor's Notes:

MODULE 61204-02 — TEACHING TIPS

The following are suggested activities or instructional methods to help you teach the material in this AIG.

General

When you call on someone to answer a question, the rest of the class relaxes or even tunes out because they expect that the question and answer will take place only between you and the trainee you called on. Instead, use this technique to involve more trainees in answering questions and to keep them on their toes.

1. Ask trainees to define a term or explain a concept.
2. After one trainee has answered, ask a trainee seated nearby if the answer is right. Then ask whether a trainee in the back of the room agrees.
3. Ask trainees to explain why they think an answer is right or wrong.
4. Use the session to clear up incorrect ideas and encourage trainees to learn from their mistakes.

Sections 2.0.0–2.3.4

Test Casings

This exercise will familiarize trainees with the various tests for shorted casings. For this exercise, trainees will need their Trainee Guides and pencils and paper to take notes for discussion. Allow 20 minutes for this exercise.

1. On the whiteboard/chalkboard, write a list of scenarios for shorted casings.
2. Ask trainees to determine which test would be most appropriate to test for the shorted casing condition described in each scenario.
3. Have trainees review the procedures for performing each test. Ask trainees to explain how to connect the terminal to the casing when performing each test.
4. Answer any questions trainees may have about testing shorted casings.

Sections 3.0.0–3.3.0

Repair Shorted Casings

This exercise will give trainees the opportunity to practice techniques for repairing shorted casings. For this exercise, you will need shorted casings and the tools and materials to repair and test shorted casings. Trainees will need their Trainee Guides and pencils and paper to take notes for discussion. Allow between 30 and 45 minutes for this exercise.

1. Set up stations with the materials to repair and test shorted casings. Divide the class into small groups according to the number of stations.
2. Ask trainees to practice the procedures for repairing the shorted casings.
3. If time permits, ask trainees to test a shorted casing to ensure that it has been properly repaired. Practice one of the following appropriate tests:
   - Resistance
   - Potential shift
   - Internal resistance
   - Casing depolarization
   - Four-wire IR drop
4. Explain when it is appropriate to replace a shorted casing and when it is necessary to replace the carrier pipe.
5. Observe and assess trainees' work. Answer any questions trainees may have about repairing shorted casings.

Summary

Test and Repair Shorted Casings

This exercise will give trainees the opportunity to observe the replacement of a carrier pipe. For this exercise, you will need to locate a nearby site where a replacement is scheduled. Trainees will need their Trainee Guides and pencils and paper to take notes for discussion. Allow at least 60 minutes for this exercise.

1. Schedule a visit to a local site where a carrier pipe is scheduled for replacement.
2. Ask the operations personnel to explain the scheduled procedure to the trainees.
3. Ask trainees to observe the replacement procedure and to assist with the procedures, as appropriate. Ask trainees to complete the following tasks:
   - Observe proper spacing of skids and insulators
   - Inspect end seals for water resistance and damage
   - Check the carrier pipe to ensure that it has been prepared properly
4. Following the site visit, review trainees' observations.
5. Answer any questions trainees may have about testing and repairing shorted casings.

MODULE 61204-02 — ANSWERS TO REVIEW QUESTIONS

Answer	Section
1. c	1.0.0
2. a	2.1.0
3. d	2.2.0
4. c	2.2.0
5. b	2.2.0
6. b	2.3.1
7. a	2.3.2
8. d	2.3.2
9. c	2.3.3
10. c	2.3.4
11. d	2.3.4
12. a	3.1.1
13. c	3.1.2
14. b	3.3.0
15. d	3.3.0

CONTREN™ LEARNING SERIES — USER UPDATES

The NCCER makes every effort to keep these textbooks up-to-date and free of technical errors. We appreciate your help in this process. If you have an idea for improving this textbook, or if you find an error, a typographical mistake, or an inaccuracy in NCCER's Contren™ textbooks, please write us, using this form or a photocopy. Be sure to include the exact module number, page number, a detailed description, and the correction, if applicable. Your input will be brought to the attention of the Technical Review Committee. Thank you for your assistance.

Instructors – If you found that additional materials were necessary in order to teach this module effectively, please let us know so that we may include them in the Equipment/Materials list in the Instructor's Guide.

Write: Curriculum Revision and Development Department
National Center for Construction Education and Research
P.O. Box 141104, Gainesville, FL 32614-1104

Fax: 352-334-0932

E-mail: curriculum@nccer.org

Craft ______________________ Module Name ______________________

Copyright Date ____________ Module Number ____________ Page Number(s) ____________

Description

____

____

____

____

(Optional) Correction

____

____

____

(Optional) Your Name and Address

____

____

____

Module 61205-02

Conduct Close Interval Survey

(CT 1.2 and 1.4)

Conduct Close Interval Survey
Annotated Instructor's Guide

Module 61205-02

MODULE OVERVIEW

This module describes the close interval survey. Trainees will learn how to select the proper equipment and practice the procedure itself.

PREREQUISITES

Please refer to the Course Map in the Trainee Module. Prior to training with this module, it is recommended that the trainee shall have successfully completed the following:

Core Curriculum; Pipeline Core, Module 66101-02 and 66102-02; Pipeline Corrosion Control Level One; Pipeline Corrosion Control Level Two, Modules 61201-02 through 61204-02

OBJECTIVES

Upon completion of this module, the trainee will be able to:

1. Identify common close interval survey equipment (CT 1.2).
2. Describe the common methods to perform a close interval survey (CT 1.2).
3. Describe the major steps in performing a close interval survey (CT 1.2).
4. Describe inspecting and testing bonds and protective devices (CT 1.4).

PERFORMANCE TASKS

Under the supervision of the instructor, the trainee should be able to:

1. Select the proper instrumentation, test leads, and reference electrode for a given close interval survey (CT 1.2).
2. Correctly perform a close interval survey (CT 1.2).
3. Visually inspect and perform electrical tests on bonds connecting multiple structures and isolating devices (CT 1.4).

NCCER STANDARDIZED CRAFT TRAINING PROGRAM

The National Center for Construction Education and Research (NCCER) provides a standardized national program of accredited craft training. Key features of the program include instructor certification, competency-based training, and performance testing. The program provides trainees, instructors, and companies with a standard form of recognition through a National Craft Training Registry. The program is described in full in the *Guidelines for Accreditation*, published by the NCCER. For more information on standardized craft training, contact the NCCER by writing us at P.O. Box 141104, Gainesville, FL 32614-1104; calling 352-334-0911; or e-mailing info@nccer.org. More information may be found at our Web site, www.nccer.org.

HOW TO USE THIS ANNOTATED INSTRUCTOR'S GUIDE

Each page presents two sections of information. The larger section displays each page exactly as it appears in the Trainee Module. The narrow column ties suggested trainee and instructor actions to each page and provides icons to call your attention to material, safety, audiovisual, or testing requirements. The bottom of each page includes space for your notes.

Teaching Tip

If you see the Teaching Tip icon, that means there is a teaching tip associated with this section. Also refer to any suggested teaching tips at the end of the module.

SAFETY CONSIDERATIONS

Ensure that the trainees are equipped with appropriate personal protective equipment. Emphasize the importance of following all safety precautions and procedures when working with power tools.

PREPARATION

Before teaching this module, you should review the Module Outline, Objectives, Performance Tasks, and the Materials and Equipment List. Be sure to allow ample time to prepare your own training or lesson plan and gather all required equipment and materials.

MATERIALS AND EQUIPMENT LIST

Overhead projector and screen

Whiteboard/chalkboard

Transparencies

Markers/chalk

Blank acetate sheets

Transparency pens

Pencils and scratch paper

Module Examinations*

Performance Profile Sheets*

Copies of the Performance Verifications for Covered Task 1**

Appropriate personal protective equipment

Copies of your local code

Wire reels

Current interrupter

Materials and equipment for selecting and maintaining survey test equipment, including manufacturer's operation manual

Materials and equipment for performing a close interval survey

Materials and equipment for inspecting and testing bonds and protective devices, including manufacturer's guidelines

* Located in the Test Booklet.

**Available for download from www.nccer.org.

TEACHING TIME FOR THIS MODULE

An outline for use in developing your lesson plan is presented below. Note that each Roman numeral in the outline equates to one session of instruction. Each session has a suggested time period of 3 hours. This includes 10 minutes at the beginning of each session for administrative tasks and one 10-minute break during the session. Approximately 18 hours are suggested to cover *Conduct Close Interval Survey (CT 1.2 and 1.4)*. You will need to adjust the time required for hands-on activity and testing based on your class size and resources.

Topic	Planned Time
Session I. Survey Test Equipment, Part One	
A. Introduction	____________
B. GPS	____________
Session II. Survey Test Equipment, Part Two	
A. Test Lead Reels	____________
1. Laboratory	____________
B. Current Interrupter	____________
C. Data Logger and Reference Electrode	____________
D. Test Equipment Care	____________
1. Performance Testing	____________
Session III. Close Interval Survey Methods	
A. Close Interval Survey, Continuous Method	____________
B. Close Interval Survey, Interrupted Method	____________
Session IV. Close Interval Survey Procedure	
A. Preparation	____________
B. Implementation	____________
C. Documentation	____________
1. Performance Testing	____________
Session V. Inspecting and Testing Bonds and Protective Devices, Part One	
A. Bonding Requirements	____________
B. Grounding System	____________
Session VI. Inspecting and Testing Bonds and Protective Devices, Part Two	
A. Bonding Method	____________
B. Earth System	____________
C. Summary	
1. Summarize module	____________
2. Answer questions	____________
D. Module Examination	____________
1. Trainees must score 70% or higher to receive recognition from the NCCER.	
2. Record the testing results on Craft Training Report Form 200 and submit the results to the Training Program Sponsor.	
E. Performance Testing	____________
1. Trainees must perform each task to the satisfaction of the instructor to receive recognition from the NCCER.	
2. Record the testing results on Craft Training Report Form 200 and submit the results to the Training Program Sponsor.	

Module 61205-02

Conduct Close Interval Survey

(CT 1.2 and 1.4)

Instructor's Notes:

Instructor's Notes:

Introduction to the Pipeline Operator Qualification Rule, API, and NCCER

It could be said that petroleum products keep America running. Petroleum products like oil and natural gas are the principal sources of energy in the United States and are used in many manufacturing and chemical processes. Petroleum products are primarily transported by pipeline. There are more than a million miles of liquid and natural gas pipeline in this country, ranging in diameter from two inches to four feet. These pipelines bring crude oil and natural gas from production fields to processing plants and refineries—and ultimately to the consumer.

Pipeline Operator Qualification

Petroleum products are considered hazardous, so pipeline personnel must take extra precautions and follow specific procedures when working with them. To help ensure the safe handling and transportation of petroleum products, the United States Department of Transportation (DOT) mandates safe practices as federal regulations.

One such regulation, the Operator Qualification (OQ) rule, requires pipeline operators to develop a written qualification plan to evaluate their personnel and their contract personnel (see *49 CFR Part 192 Subpart N—Gas,* and *49 CFR Part 195 Subpart G—Liquid*). The evaluation hinges on the operator's ability to perform certain covered tasks and to be able to recognize and respond to abnormal operating conditions that may be encountered while performing these covered tasks.

Covered Tasks

A pipeline performance task is deemed *covered* if the activity passes the following four-part test:

1. Is it performed on a pipeline facility?
2. Is it an operations or maintenance task?
3. Is it performed as a requirement of *49 CFR Part 195 (Liquid)* or *49 CFR Part 192 (Gas)*?
4. Does it affect the operation or integrity of the pipeline facility?

If the answer to all four questions is *yes,* then the activity is considered *covered* and must figure into the evaluation process of pipeline personnel.

American Petroleum Institute

The OQ rule and its related covered tasks provided the catalyst to develop a unified pipeline training and assessment program—so that pipeline operators and owners have ready access to the assessment and training tools they need. This effort was the direct result of the vision and leadership provided by The American Petroleum Institute (API) and API members.

Efforts to unify the industry are nothing new to API. Indeed, the impetus for forming the Institute in 1919 was the need to standardize engineering specifications for drilling and production equipment. Today, the most pressing issues revolve around public perceptions and government policies toward the industry. Speaking with one voice on these issues has become as essential as having interchangeable parts in the field. API's members determine the petroleum industry's positions on public policy as well as the standards that govern its day-to-day operations worldwide.

Consortium on Operator Qualification

API assembled a focus group whose primary purpose was to identify covered tasks and provide compliance recommendations to pipeline operators.

The Consortium on Operator Qualification (COOQ) became the guiding force for the identification and development of covered tasks for the liquid pipeline industry. Their work product, API publication 1161, *Guidance Document for the Qualification of Liquid Pipeline Personnel,* formed the basis for the resulting Pipeliner Training and Assessment Program.

The covered task list created in API publication 1161 has now evolved to include gas covered tasks and is maintained by NCCER. The most current covered task list can be downloaded at www.nccer.org.

National Center for Construction Education and Research

With the guidance and support of API, its members, and the Consortium on Operator Qualification, the National Center for Construction Education and Research (NCCER) was able to design and implement the Pipeliner Training and Assessment Program.

NCCER is a not-for-profit 501(c)(3) education foundation established in 1995 by the world's largest and most progressive construction and maintenance contractors and national contractor associations. NCCER was founded to address the

severe workforce shortage facing the construction and maintenance industries and to develop a standardized training process and curricula. Today, NCCER is supported by hundreds of leading construction and maintenance companies, manufacturers, and national associations and developed more than 35 construction and maintenance titles.

NCCER also maintains a National Registry, which contains the records of trainees and instructors who have received NCCER training. All completion information for every trainee is sent to the NCCER and filed within the Registry. The Registry can then confirm training and skills for workers as they move from company to company, state to state, or even within their own company.

Standardized Curriculum (NCCER's Contren™ Learning Series)

NCCER curricula are developed using the expertise of industry subject matter experts (SMEs). The use of SMEs as an integral part of every new development or revision ensures that the curriculum is current, industry-driven, and indicative of industry practices. The development and/or revision process is facilitated from start to finish by NCCER project managers, who work closely with the SMEs and a select team of technical writer(s) on the development and presentation of the curriculum.

Pipeline: Five Training Paths

The pipeline curriculum focuses on essential skills and competencies within five skill areas:

- Corrosion Control
- Electrical and Instrumentation
- Maintenance
- Mechanical
- Operations (Field and Control Center)

SMEs from the pipeline industry have determined the skills and competencies needed based on API's covered tasks and the consensus of the group/industry. Specific company information is generally not incorporated into the curriculum unless the SMEs determine that the information is the industry standard.

Training Modules and Covered Tasks

The smallest unit of instruction in the curriculum is a *module*. Each curriculum is composed of modules (the equivalent of stand-alone chapters) that are logically sequenced into complete *levels* of instruction. Every module is task-oriented and requires the successful completion of a written (knowledge-based) examination and a performance (hands-on) test. Modules may be individually selected and/or combined for efficient task training. Modules in the pipeline curricula, to the extent possible, will align with API-developed covered tasks. While not all modules will relate to covered tasks, all covered tasks will relate to at least one module.

Benefits of Training to the Pipeline Marketplace

Applying NCCER's standardized training process allows pipeline owners and contractors in the United States to establish the benchmark for acceptable skill and competency levels within the pipeline industry. This process will assist training organizations in developing and/or improving standardized pipeline training programs that effectively match their resources and operating conditions while responding to regulatory demands and to the needs of the industry. The process will also recognize training organizations that meet the minimum standards of competence, warranting regulatory, public, and professional confidence. This in turn will foster national unity within the pipeline industry for the mutual benefit of individual technicians, owners, contractors, and ultimately pipeline-users and society.

Additional Resources

To review the Pipeliner Training and Assessment Program Implementation and Resource Guide and the latest version of pipeline covered tasks, visit NCCER's Web site at www.nccer.org, or contact:

National Center for Construction Education and Research
PO Box 141104
Gainesville, FL 32614-1104
(888) 622-3720
Web site: www.nccer.org

For more information on API, please contact:

American Petroleum Institute
1220 L Street, Northwest
Washington, D.C. 20005
(202) 682-8000
Web site: www.api.org

To order *API Pub 1161*, contact Global Engineering Documents by calling 1-800-854-7149 or visiting their Web site at www.global.his.com.

Additional sources of information on the pipeline operator qualification regulation include:

DOT – Office of Pipeline Safety
www.ops.dot.gov

Transportation Safety Institute
www.tsi.dot.gov

Instructor's Notes:

COURSE MAP

This course map shows all of the modules in the second level of the Pipeline Corrosion Control curriculum. The suggested training order begins at the bottom and proceeds up. Skill levels increase as you advance on the course map. The local Training Program Sponsor may adjust the training order.

PIPELINE CORROSION CONTROL LEVEL TWO

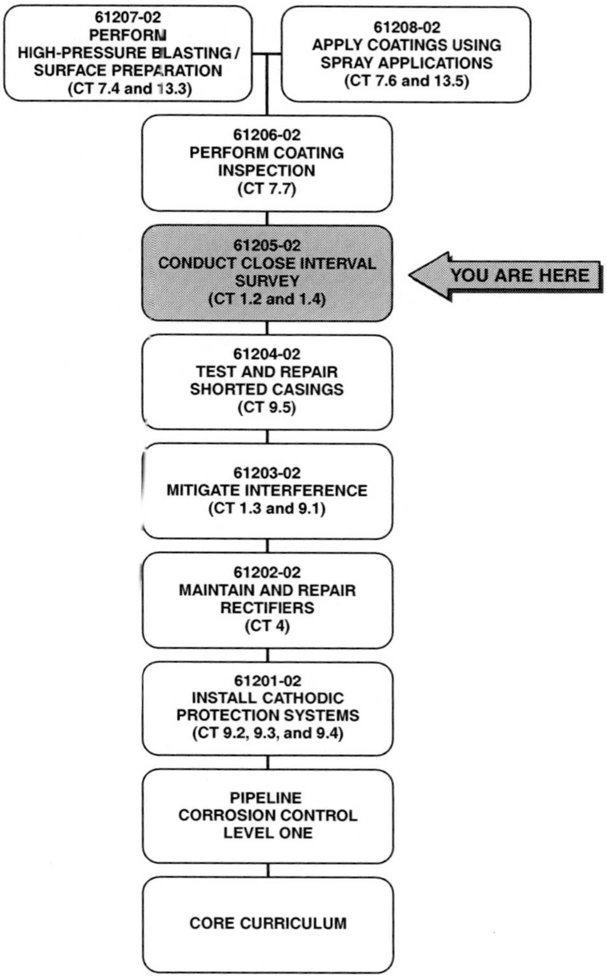

205CMAP.EPS

Homework

Assign reading of Module 61205-02.

Instructor's Notes:

MODULE 61205-02 CONTENTS

Figures

Instructor's Notes:

MODULE 61205-02

Conduct Close Interval Survey (CT 1.2 and 1.4)

Objectives

When you have completed this module, you will be able to do the following:

1. Identify common close interval survey equipment (CT 1.2).
2. Describe the common methods to perform a close interval survey (CT 1.2).
3. Describe the major steps in performing a close interval survey (CT 1.2).
4. Describe inspecting and testing bonds and protective devices (CT 1.4).

Prerequisites

Before you begin this module, it is recommended that you successfully complete the following: Core Curriculum; Pipeline Core, Modules 66101-02 and 66102-02; Pipeline Corrosion Control, Level One; Pipeline Corrosion Control, Level Two, Modules 61201-02 through 61204-02.

Required Trainee Materials

1. Pencil and paper
2. Appropriate personal protective equipment

1.0.0 ◆ INTRODUCTION

Close interval surveys are performed to establish proper cathodic protection and evaluate the coating. It is important to stay at the proper pipeline location and not move to the side.

Cathodically protected pipelines have permanent **test stations** attached to them. Electronic **test leads** that are attached to the pipeline are used to measure the **pipe-to-soil potential**. The potential should be in a range that will provide pipeline protection without being excessively cathodic or deficient. Protection that is too cathodic may cause coating damage and hydrogen embrittlement.

Only a small fraction of the pipeline length provides the potentials that are measured at permanent test stations. The area of measured potential on the pipeline is approximately two times the depth of pipeline burial. To assess the effectiveness of the cathodic protection over the entire length of the pipe, a close interval potential survey is performed. A drop in potential may be an indication of coating damage or holidays.

To perform a close interval survey, an electrical connection must be made to the pipeline using a trailing wire. The trailing wire unwinds from a spool as the pipeline is walked by the equipment operator. This wire is normally coated copper. A set of electrodes is positioned directly over the pipeline at approximately 3' intervals. These **reference electrodes** measure pipeline potential at ground level.

The potential readings provided from the cathodic protection current flow contain IR error. *Figure 1* shows some typical examples of bad coating. To get a more accurate reading of pipe-to-soil potential, the IR error must be minimized. To minimize the IR error, the current flow for the cathodic protection must be interrupted for an instant-off **potential measurement**. All sources of cathodic protection current must be interrupted at the same time. To accomplish this synchronous current interruption, **rectifiers** and other foreign sources of current are shut off one at a time. **Global positioning system (GPS)** timing devices are becoming more common as synchronous switching devices. *Figure 2* shows the use of GPS as a switching device.

Materials

Ensure that you have everything required to teach the course. Check the Materials and Equipment List at the front of this module.

Classroom

Explain that terms shown in bold are defined in the Glossary at the back of this module.

Audiovisual

Show Transparency 1, Course Objectives.

Show Transparency 2, Performance Tasks.

Teaching Tip

See the general Teaching Tip at the end of this module.

Classroom

Ask a trainee to name the purpose of a close interval survey. Have a trainee explain the problems that result from protection that is too cathodic.

Have a trainee explain how to minimize infrared (IR) error.

Audiovisual

Show Transparency 3 (Figure 2). Ask a trainee to explain how GPS is used as a switching device.

Audiovisual

Show Transparency 4 (Figure 3). Have a trainee explain why an interrupter is used.

Classroom

Ask a trainee to explain how to ensure the accuracy of the close interval survey.

Classroom

Have trainees name the three components of GPS. Ask a trainee to explain the types of information GPS can provide.

Have trainees name the uses of GPS in pipeline technology.

Figure 1 ◆ Examples of disbonding and rusting coating.

The interruption cycles provided by the switching device vary. The potential measurement must take place at the proper time within the interruption cycle. The cycle must be sufficient to obtain the readings and not to allow **depolarization** to occur. *Figure 3* shows an **interrupter** with GPS timing installed at a rectifier.

Although this survey technique is based on simple potential measurements, it still requires a high degree of operator skill and experience. Extensive support logistics are required for physical measurements and data management.

This module describes the methods and techniques used to perform a close interval survey.

2.0.0 ◆ SURVEY TEST EQUIPMENT

The accuracy of the close interval survey depends on the proper operation of the survey test equipment and recording the information correctly. Always consult the test equipment manufacturer's operation manual for proper operation procedures. The following sections describe some of the common survey test equipment and tools used in the performance of a close interval survey.

2.1.0 GPS

The GPS is a navigational system developed and maintained by the U.S. Department of Defense (DOD). This system can compute position and time anywhere on Earth.

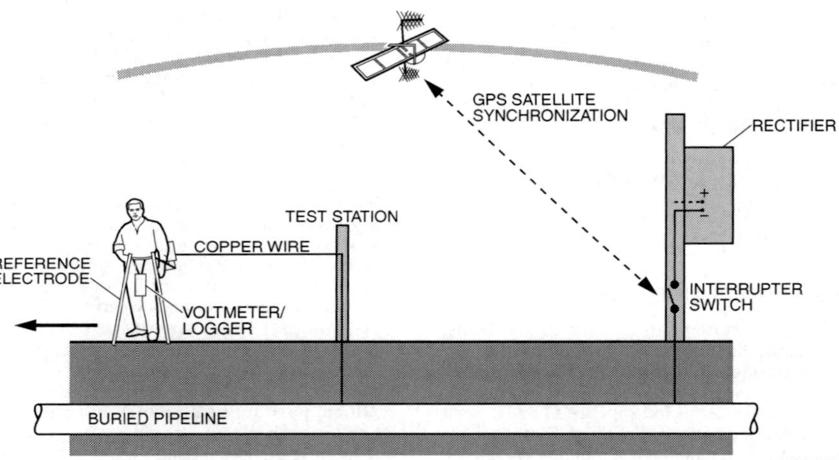

Figure 2 ◆ Use of GPS synchronization.

Instructor's Notes:

The GPS consists of three components:

- Orbiting satellites (space segment)
- Ground level monitoring and satellite tracking (control segment)
- User segment

GPS receivers at ground level are employed for the user segment. Three-dimensional position coordinates and timing are converted by the receivers from the radio signals transmitted from the satellites. The position and time signals can also identify speed and direction of travel. Usually the three-parameter position location and the time parameter are computed from the signals of at least four satellites. GPS technology has many uses in the process of pipeline monitoring, including:

- Finding the location of corrosion control equipment and monitoring points in the field
- Providing positional coordinates and mapping for geographical information systems
- Linking position to coating defect locations and corrosion data
- Synchronous timing of devices such as interrupting a series of rectifiers for a close internal survey

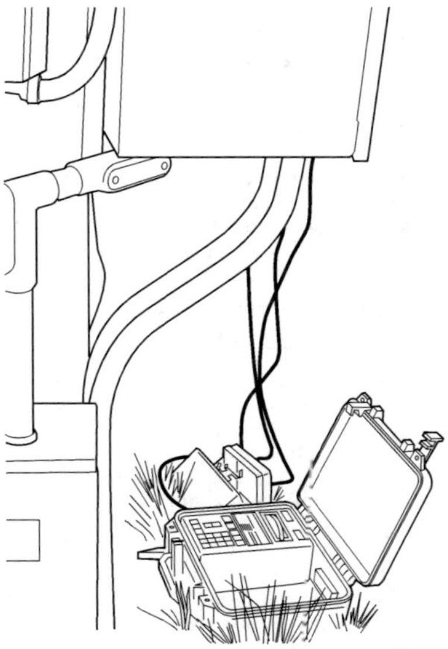

205F03.EPS

Figure 3 ◆ Interrupter with GPS timing installed at a rectifier.

2.2.0 Test Lead Reels

Test lead reels are required to perform a close interval survey. The test lead reel is usually carried on a backpack. The backpack is used to feed the wire during the survey and provides continuous operation by using a slip ring hub. The integrity of the insulation must be maintained to get accurate readings. Bare wire can come into contact with other objects or earth and give inaccurate measurements. *Figure 4* shows types of wire reels.

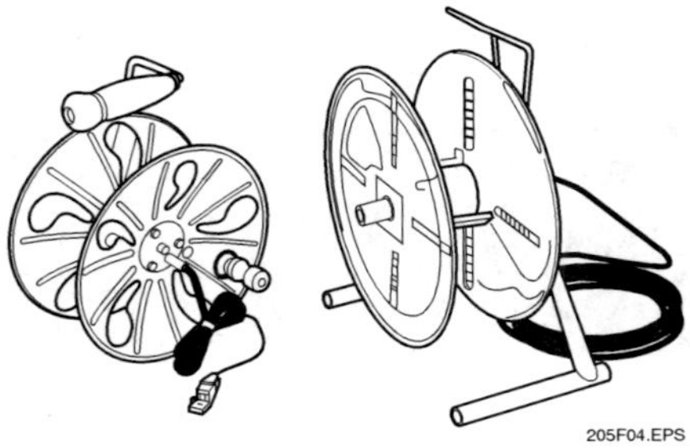

205F04.EPS

Figure 4 ◆ Wire reels.

2.3.0 Current Interrupter

Current interrupters (see *Figure 5*) are usually compact, lightweight, and portable. This makes them easier to transport into the field and use. Current interrupters are used to turn the current supplied to a CP system on and off. The timing for both the ON and OFF cycles of the current interrupter is adjustable.

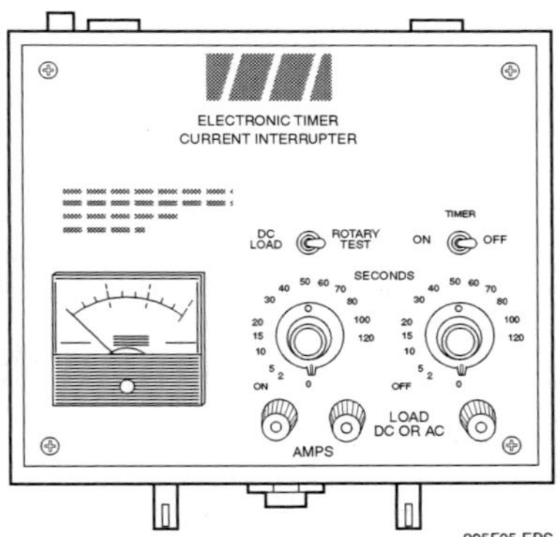

205F05.EPS

Figure 5 ◆ Typical current interrupter.

Homework

Have trainees review Sections 2.2.0–2.5.0 for the next session.

Materials

Ensure that you have everything required for the laboratory, demonstration, and testing during this session.

Audiovisual

Show Transparency 5 (Figure 4). Ask a trainee to explain how each wire reel is used.

Laboratory

Set up stations with a few varieties of wire reels. Have trainees practice using wire reels.

Audiovisual

Show Transparency 6 (Figure 5). Ask a trainee to explain the purpose of a current interrupter.

Demonstration

Bring in a current interrupter. Demonstrate how the current interrupter works. Have trainees practice using the current interrupter.

Safety

Stress the importance of not allowing the survey equipment or wire to come into contact with electricity. Ask a trainee to simulate the proper steps to take if a worker is caught in an electrical storm while performing a close interval survey.

Audiovisual

Show Transparency 7 (Figure 6). Have a trainee identify each piece of close interval survey equipment in the figure.

Safety

Ask trainees why metallic sandpaper, a pocket knife, or steel wool should not be used on copper rods.

Teaching Tip

See the Teaching Tip for Sections 2.1.0–2.5.0 at the end of this module.

Performance Testing

Have each trainee perform Performance Task 1 to your satisfaction. Fill out a Performance Profile Sheet for each trainee.

Homework

Have trainees review Section 3.0.0 for the next session.

2.4.0 Data Logger and Reference Electrode

The data logger and reference electrode are very important pieces of equipment for performing close interval surveys. The data logger takes a reading with reference to the buried pipe and the reference electrode by using a built-in voltmeter. The reading can then be downloaded into a computer from the data logger memory.

Always follow these safety tips when performing a close interval survey:

- Never perform a close interval survey during an electrical storm.

> **WARNING!**
> If caught in an electrical storm while performing a close interval survey, break the wire, move away from the pipeline, and crouch down as close to the ground as possible.

- Do not allow the survey equipment (*Figure 6*) or wire to come into contact with an electric fence or any other power supply.

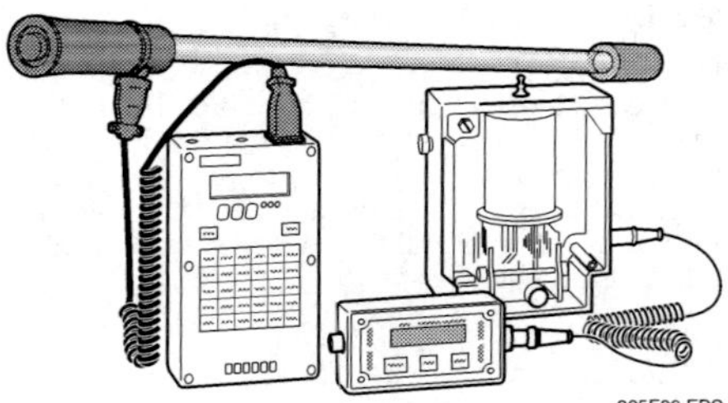

205F06.EPS

Figure 6 ◆ Typical close interval survey equipment.

2.5.0 Test Equipment Care

To ensure the continued reliability of the instruments and the accuracy of the test data collected, the corrosion test equipment must be properly maintained and handled. Inadequate protection during transportation is the most common cause of damage to the test equipment. Sensitive instruments should be secured against movement and insulated from shock while being transported in a car or truck. Moisture and road-dust protection is also required when transporting electronic equipment.

The following guidelines will also help with proper test equipment care:

- Always keep the instruction manual with the test equipment.
- Keep calibration records up to date and properly documented.
- Always carry spare batteries for each instrument that uses batteries.
- Fill reference cells with antifreeze if they will be stored where they are subject to freezing.
- Change the electrode periodically.
- Keep copper sulfate reference electrodes clean and in good condition.
- Always make sure that the copper sulfate solution is a saturated solution. (Crystals are visible at ambient temperatures.)
- Clean the copper rod with fine, nonmetallic sandpaper.

> **CAUTION**
> Never use metallic sandpaper, a pocket knife, or steel wool on copper rods. The metal particles can imbed in the copper and cause inaccurate readings.

3.0.0 ◆ CLOSE INTERVAL SURVEY METHODS

Possible problem areas on underground pipelines may be identified through the use of continuous pipe-to-soil potential profiles. The profiles are usually developed through close interval surveys. The commonly used types of close interval surveys are the continuous ON method and the interrupted ON/OFF method.

3.1.0 Close Interval Survey, Continuous Method

This is a survey conducted with the current either on or off. Current is not interrupted at intervals during this test.

To perform the continuous method of close interval survey, readings are taken continuously at predetermined intervals along the pipeline. This is performed with the current sources either all on or all off. *Figure 7* shows the continuous close interval survey method.

A negligible voltage drop can be expected between any two points tested when surveying an all-welded steel pipeline. A data logger and a measuring device are connected to the pipeline with a wire. The wire provides continuous contact

Instructor's Notes:

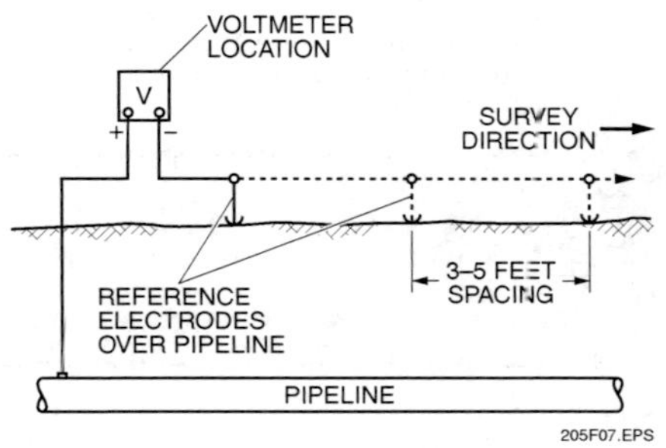

Figure 7 ◆ Close interval survey, continuous method.

with the pipeline. The data logger is used to store the information gathered during the survey. The measured potential at each electrode location is recorded.

The distance of a pipeline survey is limited only by the length of the test conductor in areas where obstructions do not interfere. As long as the spools of light-wire test conductors are properly maintained, they will provide an accurate reading over long distances. This is because the resistance in the light-wire test lead is not enough to affect the measurements taken.

3.2.0 Close Interval Survey, Interrupted Method

Typical close interval survey equipment (refer to *Figure 7*) is fitted with interrupters to turn the current off and on. Technicians must record an ON reading and an OFF reading on the documentation sheet or logbook.

4.0.0 ◆ CLOSE INTERVAL SURVEY PROCEDURE

This close interval survey procedure lists the common steps involved in conducting a survey and is not intended to cover all requirements. When conducting a survey, always refer to your company's policy and procedures. Your company may have special requirements that need to be addressed during the survey or special safety considerations for the survey location.

The following preparation is recommended before performing a close interval survey.

Step 1 Verify the availability and serviceability of the following tools, materials, and equipment. *Figure 8* shows some common equipment used in a close interval survey.

Figure 8 ◆ Common equipment used in performing a close interval survey.

- Company procedures, applicable forms, and maps
- Data logger
- Portable computer with applicable close interval survey software
- Printer
- Footage counter
- Pipeline locator
- Company vehicle and/or all-terrain vehicle (boat when required to cross water)
- Personal protective equipment/clothing
- Spools of adequately insulated wire

> **NOTE**
> Short surveys performed for troubleshooting low potentials may not require long lengths of wire. Ensure that you have enough wire to perform the survey.

Materials

Ensure that you have everything required for teaching this session.

Audiovisual

Show Transparency 8 (Figure 7). Ask a trainee to explain the continuous method.

Classroom

Have a trainee compare and contrast the continuous and interrupted methods. Discuss the reasons for using one method instead of the other.

Homework

Have trainees review Section 4.0.0 for the next session.

Materials

Ensure that you have everything required for teaching and testing during this session.

Classroom

Review the steps for conducting a close interval survey.

Safety

Discuss why a gas detector/LEL monitor may be required if the pipeline passes through a casing.

- Portable reference electrodes (Copper/Copper Sulfate)
- Current interrupters (for **interrupted surveys**)
- Mobile telephone/company radio and/or other communication equipment

> **CAUTION**
> A gas detector/LEL monitor may be required if the pipeline passes through a casing.

Step 2 Proceed to right-of-entrance for the beginning of the survey.

Step 3 Visually observe and inspect the right-of-way for the survey using all of your senses to check for any indications of a leak such as:

- Eroding or stained soil
- Unusual odor
- Sheen on the water
- Bubbling in the water
- Blowing **casing vents**
- Washouts
- Dead vegetation
- Damaged equipment

Step 4 If a pipeline leak is found, immediately move to a safe distance and notify your control center.

Step 5 When you notify the control center, provide the following information, if available at the time of your call:

- Location of the leak
- Product that is leaking
- Possible hazards to the spill response team personnel
- Identify approach routes and danger zones
- Any other information you may have available that will assist the spill response team personnel

Step 6 Restrict access to the potentially hazardous area if possible.

Step 7 Take any action requested by the control center that you are qualified to perform.

> **NOTE**
> If a leak is confirmed by the spill response team, follow your company spill procedures and complete your company-required forms.

Step 8 Electronically coordinate the ON/OFF cycles of all the current interrupters to be synchronized to each other and to the data logger.

> **NOTE**
> Place current interrupters on all potential sources of current such as your company systems, foreign systems, and bonds if the pipeline is protected by an impressed current system.

Step 9 Measure the voltage and current output of all company rectifiers and foreign rectifiers with a meter and document the results on the survey form.

Step 10 Check and record the soil conditions as dry, moist, or wet for each day of the survey.

Step 11 Locate and mark the location of the pipeline. *Figure 9* shows a close interval survey being performed in a right-of-way.

Step 12 Perform a safety meeting each morning before starting the close interval survey.

Step 13 Attach appropriate leads to the high resistance voltmeter or data logger and the reference electrode.

> **NOTE**
> Ensure that good contact has been made with the test leads.

Step 14 Take a pipe-to-soil potential reading according to company requirements.

205F09.EPS

Figure 9 ◆ Close interval survey being performed in a right-of-way.

Instructor's Notes:

Step 15 Record the following information along with the pipe-to-soil potentials:

- Date and time
- Location of potential reading
- Rectifier ON/OFF reading
- Test point locations
- Soil conditions if different from what was listed as the daily condition
- Prominent features such as fences, roads, pipeline markers, creeks, and any other obstruction during the course of the survey
- Exposed pipe and washouts

Step 16 Record the number and description of all rectifiers and test points used on the survey.

Step 17 Measure the potential drop in the pipe between the adjacent test point location each time a new connection to a test point is made.

> **NOTE**
> For interrupted surveys, both the ON and OFF cycles are measured. The potential drop will vary for the appropriate line size and the distance between the test points. For interrupted surveys, verify that the potential drop in the pipe is close to zero as possible during the OFF cycle.

Step 18 Perform the following actions if the potential between the test points does not meet the conditions listed in the previous steps:

- Investigate the source of the unexpected current
- Check the operation of the interrupters (for interrupted survey only)
- Resurvey as necessary
- Note the location and nature of the current flow in the survey report, if a diligent effort is made and the unexpected current cannot be identified

Step 19 Consider drilling holes in paved areas and collecting data every 50 or less in high-risk areas.

Step 20 Record potentials for all water crossings under 3' deep, if dangerous conditions are not present.

> **WARNING!**
> Failure to use amphibious transportation in swamp and marsh areas where walking is prohibited could result in personnel injury or death.

Step 21 Survey river crossings over 3' deep or where dangerous conditions exist with a trailing wire and weighted electrode survey method, using a boat that is traveling at a speed of approximately 1 knot.

Step 22 Record both rectifier potentials for all foreign lines and company lines (interrupted surveys only).

Step 23 Measure the current in all **bonds** during the survey.

> **NOTE**
> Report the polarity and magnitude for all bonds.

Step 24 Measure and record AC voltage at test leads whenever the pipeline is adjacent to high-voltage lines.

> **NOTE**
> It may be necessary to use a filter to mitigate the effects of AC on the recorded survey.

Step 25 If necessary, the effects of **telluric current** and other extraneous currents should be considered and evaluated.

Step 26 At the end of each day, make sure that all rectifiers are on. Synchronize the interrupters again prior to placement the following day.

Step 27 Generate a survey plot each day and review the results with company officials.

> **NOTE**
> For interrupted surveys, ensure that plots of the ON and instant OFF potentials are placed.

Step 28 Ensure the cover page includes the following information:

- Survey dates
- Inspector's name
- Pipeline name
- Crew chief name
- Contractor's name
- Total number of books

Step 29 Ensure that there is a tabular listing of all locations that do not meet the predetermined company specifications.

Step 30 Ensure that there is a tabular listing of the operating condition of all rectifiers interrupted and dates measured.

Safety

Stress the importance of using amphibious transportation in swamp and marsh areas where walking is prohibited.

Teaching Tip

See the Teaching Tip for Section 4.0.0 at the end of this module.

Performance Testing

Have each trainee perform Performance Task 2 to your satisfaction. Fill out a Performance Profile Sheet for each trainee.

Homework

Have trainees review Sections 5.0.0–5.1.0 for the next session.

Materials

Ensure that you have everything required for teaching this session.

Classroom

Ask trainees to explain each of the two requirements for bonding gas piping.

Have trainees name the components that must be tested in an equipment grounding test. Ask a trainee to explain the importance of proper grounding.

Homework

Have trainees review Sections 5.2.0–5.3.0 for the next session.

Materials

Ensure that you have everything required for teaching and testing during this session.

Classroom

Ask a trainee to explain which equipment requires equal potential bonding and why.

Have a trainee explain what earth testing involves. Ask trainees if they have any questions.

Step 31 Ensure that there is a tabular listing of all information obtained at all bonds, including bond polarity.

Step 32 Ensure that there is a tabular listing of any unusual areas observed on the survey plot such as:

- OFF potential higher than ON potential
- Potentials that are lower than company specifications
- Sudden changes in potential over a short distance
- Poor correlation between near ground, far ground, and potential (IR) drop in pipe
- Greater pipe potential (IR) than permitted in specifications

Step 33 Ensure that the survey plots show the following information in clean areas of the plots:

- Potentials on **foreign structures**
- Potentials on current bonds
- Rectifier outputs
- Pipe potential (IR) drops
- Other potentials measured or useful information

Step 34 Ensure that all equipment is secured after completing the survey and any damaged equipment is reported. Ensure that the system is in proper operating order and is set up the same way as before the survey was done.

5.0.0 ◆ INSPECTING AND TESTING BONDS AND PROTECTIVE DEVICES

Gas piping has special bonding requirements. The National Fuel Gas Code Committee is the lead committee responsible for interpreting requirements for bonding of gas piping. In the 1999 NEC, the use of the equipment grounding conductor, while applicable to all other metal piping systems, is not applicable to fuel gas piping.

NFPA 70 - 1999 National Electric Code, 250-104 (b) Metal Gas Piping: Each aboveground portion of a gas piping system upstream from the equipment shutoff valve shall be electrically continuous and bonded to the grounding electrode system.

ANSI Z223.1/NFPA 54 – 1999 National Fuel Gas Code, 3.14 Electrical Bonding and Grounding, (a): Each aboveground portion of a gas piping system upstream from the equipment shutoff valve shall be electrically continuous and bonded to any grounding electrode, as defined by *NFPA 70*, National Electrical Code.

Testing and inspecting includes three parts: the equipment grounding system, the bonding method, and the earth system.

5.1.0 Grounding System

The equipment grounding system testing includes inspecting and testing the grounding methods of the electrical system equipment to ensure compliance with the National Electrical Code (NEC), and other industry electrical standards such as the Institute of Electrical and Electronics Engineers (IEEE). Personal safety requirements, as defined by the Occupational Safety and Health Administration (OSHA), manufacturers' safety recommendations, and company safety rules must also be met. Testing of the equipment grounding includes the electrical service, the distribution system, and branch circuits to ensure that the grounding system is adequate for the electrical system and all connected equipment.

To prevent excessive voltage caused by lightning, line surges, or unintentional contact with higher voltage lines during normal operation, electrical distribution systems must be properly grounded to limit the voltage to ground during normal operation, and the grounding system to be compatible with the equipment requirements.

5.2.0 Bonding Method

Inspecting and testing the bonding covers the electrical system, from the electrical service entrance and to the equipment, to determine whether the bonding is adequate. Bonding must meet the requirements of the NEC, IEEE, OSHA, and any other applicable regulations, as well as company standards. The bonding for both the electrical service and the electrical distribution system bonding are included in the testing.

Structures, equipment, and systems with more than one ground reference require equal potential bonding. Without proper and appropriate equal potential bonding, even the most sophisticated surge protection devices and systems cannot protect networked equipment from damage; and without low-impedance grounding and bonding, a facility cannot be protected reliably from lightning.

5.3.0 Earth System

Earth system testing consists of inspecting and testing the electrical system connection to earth to ensure compliance with the NEC and other industry electrical standards. Personal safety requirements, as defined by OSHA, manufacturers, and company safety rules must also be met. The testing process covers the earth system and the methods and connection of the electrical service to ensure that it is adequate for the electrical system.

Instructor's Notes:

Summary

The detection of pipeline anomalies is usually the result of continuous potential measurement and recording. These are several methods of taking soil potential measurements, but they often contain an **IR drop** error. An IR drop error occurs when the CP current interacts with the resistance of the soil. To overcome this, the CP current is briefly switched off, or interrupted, and the potential is then measured. This is called instant-off potential measurement. This method is used with impressed current systems; the current cannot be switched off in a galvanic **anode** system.

The accuracy of the results depends on the use of the correct tools. GPS is often used to provide synchronous off/on switching of the CP devices along a pipeline. Test lead reels are used to feed the wire during the survey. Portable current interrupters are used in the field to turn the current off and on. A data logger takes the readings from a reference electrode and the buried pipeline. To obtain accurate results, this equipment must be kept in good condition.

Close interval surveys are taken every few feet along the length of a pipeline to detect active corrosion in localized areas. One method is to take continuous readings, and the other is the interrupted current on/off method. These types of surveys require careful attention to specified procedures.

Review Questions

1. A drop in potential in the cathodic protection system may be an indication of _____.
   a. soil irregularities
   b. pipe leaks
   c. bad connections
   d. coating damage

2. To perform a close interval survey, an electrical connection must be made to the pipeline using _____.
   a. battery leads
   b. trailing wire
   c. jumper cables
   d. rectifiers

3. The accuracy of the close interval survey depends on _____ and _____.
   a. ambient conditions; electrode spacing
   b. calibration; data interpretation
   c. operator skill; test duration
   d. equipment operation; correct recording

4. The GPS system can compute _____ and _____ anywhere on earth.
   a. mass; acceleration
   b. position; temperature
   c. position; time
   d. time; depth

5. The test lead reel is usually carried on a _____.
   a. backpack
   b. winch drum
   c. cart
   d. truck rack

6. Current interrupters are usually compact, lightweight, and _____.
   a. stationary
   b. portable
   c. disposable
   d. easily assembled

7. The _____ takes a reading with reference to the buried pipe and the reference electrode by using a built-in voltmeter.
   a. data logger
   b. interrupter
   c. GPS
   d. test station

8. _____ is the most common cause of damage to the test equipment.
   a. Humidity
   b. High voltage
   c. Temperature
   d. Transportation

9. Possible problem areas on underground pipelines may be identified through the use of continuous _____.
   a. monitoring
   b. pipe-to-soil profiles
   c. current profiles
   d. internal surveys

10. The distance of a pipeline survey is limited by the _____.
    a. visibility
    b. electrode spacing
    c. test conductor length
    d. power source

Teaching Tip

See the Teaching Tip for Sections 5.0.0–5.3.0 at the end of this module.

Classroom

Go over the Review Questions for this module. Answer any questions trainees may have.

Examination

Administer the Module Examination. Be sure to record the results of the Exam on Craft Training Report Form 200 and submit the results to the Training Program Sponsor.

Performance Testing

Have each trainee perform Performance Task 3 to your satisfaction. Fill out a Performance Profile Sheet for each trainee.

Ensure that all Performance Tests have been completed and Performance Profile Sheets for each trainee are filled out. Be sure to record the results of the testing on Craft Training Report Form 200 and submit the results to the Training Program Sponsor.

Instructor's Notes:

GLOSSARY

Glossary

This glossary will help you learn some terms specific to conducting close interval surveys. Trade Terms introduced in this module appear in italic text.

Ammeter: A meter that measures amperes. Usually comes as direct-connect type and clamp-on type.

**Anode:* The electrode of an electrochemical cell at which oxidation occurs. Electrons flow away from the anode in the external circuit. Corrosion usually occurs and metal ions enter the solution at the anode.

Bond: The electrical connection of two structures, which usually have two different cathodic protection levels.

***Cathode:** The electrode of an electrochemical cell at which reduction is the principal reaction. Electrons flow toward the cathode in the external cell.

Casing-to-soil potential: The voltage potential of the casing in the soil as related to a reference electrode.

Casing vent: A metal pipe aboveground connected to the buried casing, which provides access to and venting of the atmosphere within the casing.

Current interrupter: Mechanism connected into a current-carrying line to periodically interrupt current flow to allow no-current tests of system components.

Data logger: Instrument used for conversion of electrical impulses from process instrument into digital data to be recorded, stored, and periodically tabulated.

**Depolarization:* The removal of factors resisting the current in an electrochemical cell.

***Electrical isolation:** The condition of being electrically separated from other metallic structures or the environment.

**Foreign structure:* Any metallic structure that is not intended as a part of a system under cathodic protection.

Global positioning system (GPS): A positioning or navigation system designed to use 18 to 24 satellites, each carrying atomic clocks, to provide a receiver anywhere on earth with extremely accurate measurements of its three-dimensional position.

Interference: The effect upon cathodic protection levels of a structure due to the interaction between its cathodic protection systems and the cathodic protection system of other structures.

Interrupter: An electronic device used to activate and deactivate the cathodic protection system at a specified frequency.

Interrupted survey, synchronous (also known as an *"on-off" survey*): A procedure in which structure-to-soil potentials are taken with the cathodic protection system alternating between the activated or "on" mode, and the interrupted or "instant off" mode.

**IR drop:* The voltage across a resistance in accordance with Ohm's law.

Multimeter: A meter that allows measurement of different quantities and the functions are made available through a selector switch.

* Term published in NACE *Glossary of Corrosion-Related Terms.* Reproduced with permission from NACE International, Houston, TX. For a copy of the complete *Glossary of Corrosion-Related Terms,* contact NACE International, 1440 South Creek Dr., Houston, TX 77084. The Glossary is revised periodically, so information in an outdated version may not be accurate.

"On" survey: A procedure in which structure-to-soil potentials are taken with the protective current applied using approved procedures to account for measurement of IR drop.

"Off" survey: A procedure in which structure-to-soil potentials are taken with the structure depolarized. Also known as a *native* or *static* survey.

******Pipe-to-soil potential:*** See *structure-to-electrolyte potential.*

***Polarization:** The change from the open-circuit potential as a result of current across the electrode/electrolyte interface.

Potential measurement: The activity of measuring the structure-to-soil potential.

Rectifier: A device used to convert AC current to DC current. In an impressed cathodic protection system, it supplies current to the structure in the soil.

******Reference electrode:*** An electrode whose open-circuit potential is constant under similar conditions of measurement, which is used for measuring the relative potentials of other electrodes.

Shorted casing: A metallic short is a metal casing that is in contact with the metal carrier pipe. It is usually caused by improper installation or settling of the casing or pipeline. An electrolytic short is where the metal casing is not in metallic contact with the carrier pipe; however, there is current flow between the casing and the carrier pipe due to an electrolyte in the casing.

***Structure-to-electrolyte potential:** The potential difference between the surface of a buried or submerged metallic structure and the electrolyte that is measured with reference to an electrode in contact with the electrolyte.

***Structure-to-soil potential:** See *structure-to-electrolyte potential.*

Shunt: A calibrated resistor used for determining current flow in the circuit.

Tank-to-soil potential: The voltage potential of the tank in the soil as related to a reference electrode.

Telluric current: Naturally occurring changes in the earth's magnetism.

Test lead: A connection to the structure being tested, usually a wire in a supporting stand with an easy connection point for the meter.

Test station or test point: Specific area or location where a connection to the structure is made to allow a structure-to-soil potential to be measured.

* Term published in NACE *Glossary of Corrosion-Related Terms*. Reproduced with permission from NACE International, Houston, TX. All rights reserved. © NACE International 2002. For a copy of the complete *Glossary of Corrosion-Related Terms,* contact NACE International, 1440 South Creek Dr., Houston, TX 77084. The Glossary is revised periodically, so information in an outdated version may not be accurate.

Instructor's Notes:

REFERENCES AND ACKNOWLEDGMENTS

Additional Resources

This module is intended to present thorough resources for task training. The following reference works are suggested for further study. These are optional materials for continued education rather than for task training.

Peabody's Control of Pipeline Corrosion, 2nd Edition, 2001. A. W. Peabody. Houston, TX: NACE International.

Cathodic Protection Level 1 Training Manual, September 2001. NACE International.

Figure Credits

Corrosion-club.com 205F02

Figure 7 courtesy of NACE International

(Source: *Peabody's Control of Pipeline Corrosion, Second Edition,* Pg 74, Fig 5.6 Close Interval Survey Method 1) © NACE International – The Corrosion Society, 1440 South Creek Drive, Houston, Texas 77084. Additional publications on corrosion control technology are available from www.nace.org or phone (281) 228-6200

MODULE 61205-02 — TEACHING TIPS

The following are suggested activities or instructional methods to help you teach the material in this AIG.

General

When you call on someone to answer a question, the rest of the class relaxes or even tunes out because they expect that the question and answer will take place only between you and the trainee you called on. Instead, use this technique to involve more trainees in answering questions and to keep them on their toes.

1. Ask trainees to define a term or explain a concept.
2. After one trainee has answered, ask a trainee seated nearby if the answer is right. Then ask whether a trainee in the back of the room agrees.
3. Ask trainees to explain why they think an answer is right or wrong.
4. Use the session to clear up incorrect ideas and encourage trainees to learn from their mistakes.

Sections 2.1.0–2.5.0

Survey Test Equipment

In this exercise, trainees will learn how to select the proper instrumentation, test leads, and reference electrode for a given close interval survey, as well as how to maintain the test equipment. Trainees will need appropriate personal protective equipment, pencils, and paper. You will need to provide necessary materials and equipment and the test equipment manufacturer's operation manual. Allow 1 to 2 hours for this exercise.

1. Divide the class into small groups.
2. Describe a close interval survey scenario. Ask each group of trainees to list the equipment that will be needed to perform this survey.
3. Ask each group of trainees to explain in writing how each piece of equipment would be used in this scenario. Have trainees note any safety considerations.
4. Have each group of trainees take turns either demonstrating or explaining maintenance procedures for each piece of test equipment.
5. Review trainees' written work. Ask trainees if they have any questions.

Section 4.0.0

Close Interval Survey Procedure

In this exercise, trainees will practice performing a close interval survey. Trainees will need appropriate personal protective equipment, pencils, paper, and their Trainee Guides. You will need to arrange access to a field site and provide necessary materials and equipment. Alternatively, you may have trainees simulate the procedure. Allow 2 to 3 hours for this exercise.

1. Discuss safety considerations and any special company requirements.
2. Have trainees check that they have the equipment needed for the survey.
3. Supervise trainees as they perform the close interval survey, following company procedures or the Trainee Guide.
4. After the survey is completed, check that all equipment is secured and the system is in proper operating order. Ensure that trainees document their findings.
5. Discuss the results. Ask trainees if they have any questions.

Sections 5.0.0–5.3.0

Inspecting and Testing Bonds and Protective Devices

This exercise will give trainees practice inspecting and testing bonds and protective devices. Trainees will need appropriate personal protective equipment, paper, pencils, and their Trainee Guides. You will need to provide access to a grounding system, electrical system, and earth system. You will also need to provide any necessary equipment and manufacturer's guidelines. Alternatively, have trainees simulate the procedures for inspecting and testing bonds and protective devices. Allow 2 to 3 hours for this exercise.

1. Have trainees review the NEC requirements for bonding and grounding gas piping.
2. Present the procedures for inspecting the grounding system and bonding method and testing the earth system. Ask trainees to explain why each of these inspections/tests is important.
3. Have trainees inspect the grounding system and bonding method. Then have them test the earth system.
4. Ask trainees to write a brief summary of their findings, pointing out any problems they encountered. Ask trainees if they have any questions.

MODULE 61205-02 — ANSWERS TO REVIEW QUESTIONS

Answer	Section
1. d	1.0.0
2. b	1.0.0
3. d	2.0.0
4. c	2.1.0
5. a	2.2.0
6. b	2.3.0
7. a	2.4.0
8. d	2.5.0
9. b	3.0.0
10. c	3.1.0

CONTREN™ LEARNING SERIES — USER UPDATES

The NCCER makes every effort to keep these textbooks up-to-date and free of technical errors. We appreciate your help in this process. If you have an idea for improving this textbook, or if you find an error, a typographical mistake, or an inaccuracy in NCCER's Contren™ textbooks, please write us, using this form or a photocopy. Be sure to include the exact module number, page number, a detailed description, and the correction, if applicable. Your input will be brought to the attention of the Technical Review Committee. Thank you for your assistance.

Instructors – If you found that additional materials were necessary in order to teach this module effectively, please let us know so that we may include them in the Equipment/Materials list in the Instructor's Guide.

Write: Curriculum Revision and Development Department
National Center for Construction Education and Research
P.O. Box 141104, Gainesville, FL 32614-1104

Fax: 352-334-0932

E-mail: curriculum@nccer.org

Craft ______________________ Module Name ______________________

Copyright Date __________ Module Number __________ Page Number(s) __________

Description ____

____

____

____

(Optional) Correction ____

____

____

(Optional) Your Name and Address ____

____

____

Module 61206-02

Perform Coating Inspection

(CT 7.7)

Perform Coating Inspection
Annotated Instructor's Guide

Module 61206-02

MODULE OVERVIEW

This module introduces the procedures for verifying the quality of professional coating jobs. Trainees will learn to take the measurements and make the visual inspections that are part of the quality control process.

PREREQUISITES

Please refer to the Course Map in the Trainee Module. Prior to training with this module, it is recommended that the trainee shall have successfully completed the following:

> Core Curriculum; Pipeline Core, Module 66101-02 and 66102-02; Pipeline Corrosion Control Level One; Pipeline Corrosion Control Level Two, Modules 61201-02 through 61205-02

OBJECTIVES

Upon completion of this module, the trainee will be able to:

1. Describe the required pre-inspection activities.
2. Perform quality inspections (CT 7.7).
3. Identify and explain causes of coating failures (CT 7.7).

PERFORMANCE TASK

Under the supervision of the instructor, the trainee should be able to:

1. Properly conduct pre-inspection activities.
2. Correctly conduct and record results from inspections performed using various inspection methods (CT 7.7).

NCCER STANDARDIZED CRAFT TRAINING PROGRAM

The National Center for Construction Education and Research (NCCER) provides a standardized national program of accredited craft training. Key features of the program include instructor certification, competency-based training, and performance testing. The program provides trainees, instructors, and companies with a standard form of recognition through a National Craft Training Registry. The program is described in full in the *Guidelines for Accreditation*, published by the NCCER. For more information on standardized craft training, contact the NCCER by writing us at P.O. Box 141104, Gainesville, FL 32614-1104; calling 352-334-0911; or e-mailing info@nccer.org. More information may be found at our Web site, www.nccer.org.

HOW TO USE THIS ANNOTATED INSTRUCTOR'S GUIDE

Each page presents two sections of information. The larger section displays each page exactly as it appears in the Trainee Module. The narrow column ties suggested trainee and instructor actions to each page and provides icons to call your attention to material, safety, audiovisual, or testing requirements. The bottom of each page includes space for your notes.

If you see the Teaching Tip icon, that means there is a teaching tip associated with this section. Also refer to any suggested teaching tips at the end of the module.

SAFETY CONSIDERATIONS

Ensure that the trainees are equipped with appropriate personal protective equipment. Emphasize the importance of following all safety precautions and procedures when working with power tools.

PREPARATION

Before teaching this module, you should review the Module Outline, Objectives, Performance Task, and the Materials and Equipment List. Be sure to allow ample time to prepare your own training or lesson plan and gather all required equipment and materials.

MATERIALS AND EQUIPMENT LIST

- Overhead projector and screen
- Whiteboard/chalkboard
- Transparencies
- Markers/chalk
- Blank acetate sheets
- Transparency pens
- Pencils and scratch paper
- Module Examinations*
- Performance Profile Sheets*
- Copies of the Performance Verifications for Covered Task 7**
- Appropriate personal protective equipment
- Copies of your company's policies and procedures manual
- Sling-type and battery-operated psychrometers, water, and a psychrometric chart
- Electronic thermometer/hygrometer and manufacturer's calibration instructions
- Electronic and surface temperature thermometers
- Wind meter
- Materials for checking a compressed air supply
- Copy of *SSPC-SP1* and *SSPC-SP13*
- Equipment and materials for measuring surface profile
- Equipment and materials for measuring coating viscosity, including Zahn cup
- Testing equipment for wet-film and destructive and nondestructive dry-film testing
- Coated materials
- Pencils of varying hardnesses
- Coated section of pipe
- Manufacturer's data sheets and application bulletins
- Blank or completed forms for quality control/assurance
- Examples of coating failures
- Necessary equipment for diagnosing coating failure, including manufacturer's data sheet/application bulletins

* Located in the Test Booklet.
** Available for download from www.nccer.org.

TEACHING TIME FOR THIS MODULE

An outline for use in developing your lesson plan is presented below. Note that each Roman numeral in the outline equates to one session of instruction. Each session has a suggested time period of 3 hours. This includes 10 minutes at the beginning of each session for administrative tasks and one 10-minute break during the session. Approximately 18 hours are suggested to cover *Perform Coating Inspection (CT 7.7).* You will need to adjust the time required for hands-on activity and testing based on your class size and resources.

Topic	Planned Time
Session I. Quality Inspections	
A. Inspection	____________
B. Pre-Surface Preparation	____________
C. Ambient Conditions	____________
1. Laboratories	____________
Session II. Compressor Air Cleanliness and Surface Preparation Equipment	
A. Compressed Air Cleanliness	____________
1. Laboratory	____________
B. Evaluation of the Blast Cleaning System	____________
C. Checking Blast System Abrasives	____________
Session III. Preparation of Surface and Coating Materials	
A. Surface Preparation	____________
B. Surface Preparation Standards	____________
C. Evaluating Surface Profile	____________
D. Measuring the Temperature of Coating	____________
E. Measuring Viscosity of Coating Materials	____________
1. Laboratory	____________
F. Evaluation of Application Equipment	____________
Session IV. Inspection and Testing	
A. Inspecting Coating Applications and Cleanliness Between Coats	____________
B. Determination of Wet-Film and Dry-Film Thickness	____________
C. Pinhole and Holiday Testing	____________
D. Evaluating Coating Adhesion/Cure	____________
E. Quality Control/Assurance Documents	____________
1. Performance Testing	____________
Session V. Causes of Coating Failures	
A. Application Problems	____________
B. Specification-Related Problems	____________
C. Coating-Related Problems	____________
D. Coating Failures on Steel	____________
Session VI. Diagnosing Coating Failures	
A. Guidelines for Analysis Data Gathering	____________
B. Guidelines for Sample Taking	____________
C. Summary	
1. Summarize module	____________
2. Answer questions	____________

D. Module Examination ____________

1. Trainees must score 70% or higher to receive recognition from the NCCER.
2. Record the testing results on Craft Training Report Form 200 and submit the results to the Training Program Sponsor.

E. Performance Testing ____________

1. Trainees must perform each task to the satisfaction of the instructor to receive recognition from the NCCER.
2. Record the testing results on Craft Training Report Form 200 and submit the results to the Training Program Sponsor.

Module 61206-02

Perform Coating Inspection

(CT 7.7)

Instructor's Notes:

Instructor's Notes:

Introduction to the Pipeline Operator Qualification Rule, API, and NCCER

It could be said that petroleum products keep America running. Petroleum products like oil and natural gas are the principal sources of energy in the United States and are used in many manufacturing and chemical processes. Petroleum products are primarily transported by pipeline. There are more than a million miles of liquid and natural gas pipeline in this country, ranging in diameter from two inches to four feet. These pipelines bring crude oil and natural gas from production fields to processing plants and refineries—and ultimately to the consumer.

Pipeline Operator Qualification

Petroleum products are considered hazardous, so pipeline personnel must take extra precautions and follow specific procedures when working with them. To help ensure the safe handling and transportation of petroleum products, the United States Department of Transportation (DOT) mandates safe practices as federal regulations.

One such regulation, the Operator Qualification (OQ) rule, requires pipeline operators to develop a written qualification plan to evaluate their personnel and their contract personnel (see *49 CFR Part 192 Subpart N—Gas*, and *49 CFR Part 195 Subpart G—Liquid*). The evaluation hinges on the operator's ability to perform certain covered tasks and to be able to recognize and respond to abnormal operating conditions that may be encountered while performing these covered tasks.

Covered Tasks

A pipeline performance task is deemed *covered* if the activity passes the following four-part test:

1. Is it performed on a pipeline facility?
2. Is it an operations or maintenance task?
3. Is it performed as a requirement of *49 CFR Part 195 (Liquid)* or *49 CFR Part 192 (Gas)*?
4. Does it affect the operation or integrity of the pipeline facility?

If the answer to all four questions is *yes*, then the activity is considered *covered* and must figure into the evaluation process of pipeline personnel.

American Petroleum Institute

The OQ rule and its related covered tasks provided the catalyst to develop a unified pipeline training and assessment program—so that pipeline operators and owners have ready access to the assessment and training tools they need. This effort was the direct result of the vision and leadership provided by The American Petroleum Institute (API) and API members.

Efforts to unify the industry are nothing new to API. Indeed, the impetus for forming the Institute in 1919 was the need to standardize engineering specifications for drilling and production equipment. Today, the most pressing issues revolve around public perceptions and government policies toward the industry. Speaking with one voice on these issues has become as essential as having interchangeable parts in the field. API's members determine the petroleum industry's positions on public policy as well as the standards that govern its day-to-day operations worldwide.

Consortium on Operator Qualification

API assembled a focus group whose primary purpose was to identify covered tasks and provide compliance recommendations to pipeline operators.

The Consortium on Operator Qualification (COOQ) became the guiding force for the identification and development of covered tasks for the liquid pipeline industry. Their work product, API publication 1161, *Guidance Document for the Qualification of Liquid Pipeline Personnel*, formed the basis for the resulting Pipeliner Training and Assessment Program.

The covered task list created in API publication 1161 has now evolved to include gas covered tasks and is maintained by NCCER. The most current covered task list can be downloaded at www.nccer.org.

National Center for Construction Education and Research

With the guidance and support of API, its members, and the Consortium on Operator Qualification, the National Center for Construction Education and Research (NCCER) was able to design and implement the Pipeliner Training and Assessment Program.

NCCER is a not-for-profit 501(c)(3) education foundation established in 1995 by the world's largest and most progressive construction and maintenance contractors and national contractor associations. NCCER was founded to address the

severe workforce shortage facing the construction and maintenance industries and to develop a standardized training process and curricula. Today, NCCER is supported by hundreds of leading construction and maintenance companies, manufacturers, and national associations and developed more than 35 construction and maintenance titles.

NCCER also maintains a National Registry, which contains the records of trainees and instructors who have received NCCER training. All completion information for every trainee is sent to the NCCER and filed within the Registry. The Registry can then confirm training and skills for workers as they move from company to company, state to state, or even within their own company.

Standardized Curriculum (NCCER's Contren™ Learning Series)

NCCER curricula are developed using the expertise of industry subject matter experts (SMEs). The use of SMEs as an integral part of every new development or revision ensures that the curriculum is current, industry-driven, and indicative of industry practices. The development and/or revision process is facilitated from start to finish by NCCER project managers, who work closely with the SMEs and a select team of technical writer(s) on the development and presentation of the curriculum.

Pipeline: Five Training Paths

The pipeline curriculum focuses on essential skills and competencies within five skill areas:

- Corrosion Control
- Electrical and Instrumentation
- Maintenance
- Mechanical
- Operations (Field and Control Center)

SMEs from the pipeline industry have determined the skills and competencies needed based on API's covered tasks and the consensus of the group/industry. Specific company information is generally not incorporated into the curriculum unless the SMEs determine that the information is the industry standard.

Training Modules and Covered Tasks

The smallest unit of instruction in the curriculum is a *module*. Each curriculum is composed of modules (the equivalent of stand-alone chapters) that are logically sequenced into complete *levels* of instruction. Every module is task-oriented and requires the successful completion of a written (knowledge-based) examination and a performance (hands-on) test. Modules may be individually selected and/or combined for efficient task training. Modules in the pipeline curricula, to the extent possible, will align with API-developed covered tasks. While not all modules will relate to covered tasks, all covered tasks will relate to at least one module.

Benefits of Training to the Pipeline Marketplace

Applying NCCER's standardized training process allows pipeline owners and contractors in the United States to establish the benchmark for acceptable skill and competency levels within the pipeline industry. This process will assist training organizations in developing and/or improving standardized pipeline training programs that effectively match their resources and operating conditions while responding to regulatory demands and to the needs of the industry. The process will also recognize training organizations that meet the minimum standards of competence, warranting regulatory, public, and professional confidence. This in turn will foster national unity within the pipeline industry for the mutual benefit of individual technicians, owners, contractors, and ultimately pipeline-users and society.

Additional Resources

To review the Pipeliner Training and Assessment Program Implementation and Resource Guide and the latest version of pipeline covered tasks, visit NCCER's Web site at www.nccer.org, or contact:

National Center for Construction Education and Research
PO Box 141104
Gainesville, FL 32614-1104
(888) 622-3720
Web site: www.nccer.org

For more information on API, please contact:

American Petroleum Institute
1220 L Street, Northwest
Washington, D.C. 20005
(202) 682-8000
Web site: www.api.org

To order *API Pub 1161*, contact Global Engineering Documents by calling 1-800-854-7149 or visiting their Web site at www.global.his.com.

Additional sources of information on the pipeline operator qualification regulation include:

DOT – Office of Pipeline Safety
www.ops.dot.gov

Transportation Safety Institute
www.tsi.dot.gov

Instructor's Notes:

COURSE MAP

This course map shows all of the modules in the second level of the Pipeline Corrosion Control curriculum. The suggested training order begins at the bottom and proceeds up. Skill levels increase as you advance on the course map. The local Training Program Sponsor may adjust the training order.

PIPELINE CORROSION CONTROL LEVEL TWO

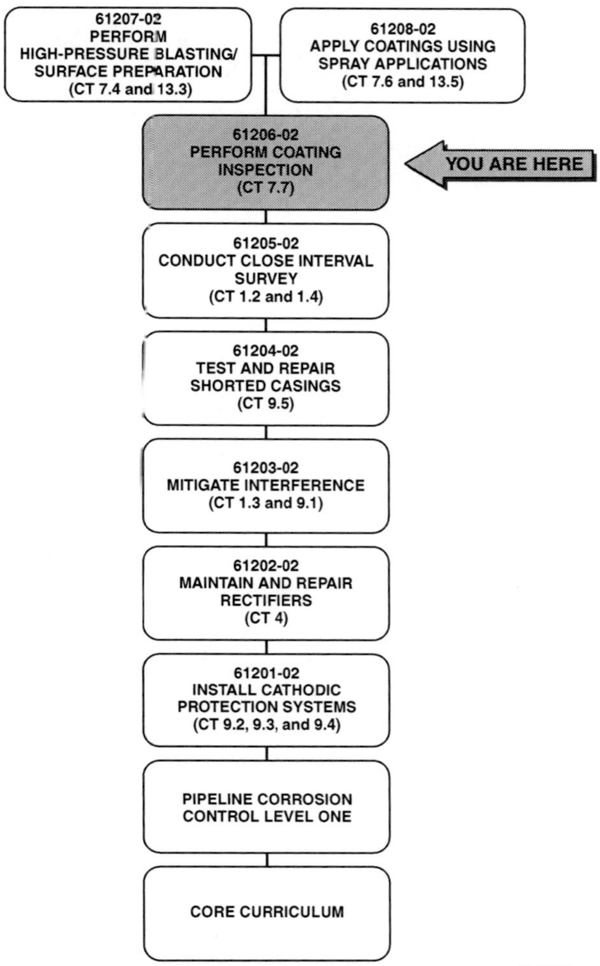

206CMAP.EPS

Homework

Assign reading of Module 61206-02.

Instructor's Notes:

MODULE 61206-02 CONTENTS

Figures

Table

Instructor's Notes:

MODULE 61206-02

Perform Coating Inspection (CT 7.7)

Objectives

When you have completed this module, you will be able to do the following:

1. Describe the required pre-inspection activities.
2. Perform quality inspections (CT 7.7).
3. Identify and explain causes of coating failures (CT 7.7).

Prerequisites

Before you begin this module, it is recommended that you successfully complete the following: Core Curriculum; Pipeline Core, Modules 66101-02 and 66102-02; Pipeline Corrosion Control Level One; Pipeline Corrosion Control Level Two, Modules 61201-02 through 61205-02.

Required Trainee Materials

1. Pencil and paper
2. Appropriate personal protective equipment

1.0.0 ◆ INTRODUCTION

All professional coating jobs incorporate procedures or checks to verify the quality of the work. These must be performed at the appropriate points throughout all phases of the job. Quality cannot be controlled unless all work is done to meet a standard. To meet quality standards in every component of work being done, have a clear idea of the standard of quality you wish to achieve. Always set high standards of quality. More importantly, you must have a clear idea of what quality looks like in every aspect of your work. Many of these quality standards, such as how to properly prepare a surface or apply spray coating to a surface, are learned through your formal apprenticeship, other training programs, and job experience.

Industrial coating jobs normally have formal quality standard requirements, which are spelled out in the job's contract and/or specifications. It is the coating contractor's responsibility to perform all applicable inspections and tests needed throughout the various stages of the job to ensure that the quality of the work and the environmental working conditions meet the standards specified for the job. This process is called **quality control (QC)**. Similarly, the owner, an engineering company, or other third-party group performs inspections and tests throughout the job to ensure that the coating contractor is indeed performing the job according to specification. This process is called **quality assurance (QA)**.

For large industrial jobs involving many people, it is common to have the QA tasks performed by a third-party inspector or group of inspectors. This can lead to the success of a project and reduce needless delays, costs, general disruptions, and tension between owners, contractors, and material suppliers.

2.0.0 ◆ QUALITY INSPECTIONS

The material presented in this module is intended to familiarize the technician with the diversity of tasks involved with the formal QA/QC process. It also introduces the procedures and associated instrumentation used to accomplish the more commonly performed QA/QC inspection tasks, with an emphasis placed on steel substrates. The tasks involved with QA/QC inspections are basically the same—only the persons who are responsible for doing them are different.

Materials

Ensure that you have everything required to teach the course. Check the Materials and Equipment List at the front of this module.

Classroom

Explain that terms shown in bold are defined in the Glossary at the back of this module.

Audiovisual

Show Transparency 1, Course Objectives.

Show Transparency 2, Performance Task.

Teaching Tip

See the general Teaching Tip at the end of this module.

Classroom

Ask a trainee to explain the quality control process. Have a trainee define quality assurance.

Classroom

Have trainees explain the advantages of performing inspections. Ask a trainee to explain when inspections are performed.

Ask a trainee to distinguish between formal and informal inspections. Discuss the importance of witnessing and documenting inspections.

Audiovisual

Show Transparency 3 (Figure 1). Ask trainees to identify the steps in the inspection process. Discuss the importance of each step.

Classroom

Have a trainee explain how to inspect a surface to ensure that it is ready to be prepared or coated.

2.1.0 Inspection

Inspections are performed to ensure quality control over prepared substrates and coatings and their applications. For contract work, inspections ensure that all specification requirements are met. Properly performed, inspections can prevent conditions that may result in a highly expensive premature coating failure. Without the use of appropriate instruments, it is difficult to determine the coating thickness, the number of coats that have been applied, and other characteristics. However, during the course of surface preparation and application of a coating, these items and many others that might affect the final quality of the coating can be readily witnessed and easily verified.

In its simplest form, an inspection occurs when, after a certain portion of the job is completed, work is stopped and examined for adequacy. Have any areas been missed? Are there any runs or **sags**? Is the blast cleaning pattern uniform and the cleanliness adequate? In the case of hand or power tool cleaning, are there any loose mill scale or rust deposits remaining? Such an inspection while the work is being performed is considered to be an informal one. The emphasis of this module is on formal inspections. This means that the inspection procedures are written, and the quality of work is witnessed and documented on a regular basis. It also means that the person or group of persons performing the inspections are specially trained and are equipped with the required instruments needed to verify the quality of the work. The scope of coating inspections should include:

- Inspection of materials
- Inspection of surface preparation
- Inspection of coating application
- Final inspection for acceptance

The inspection process normally begins with a pre-job conference during which the ground rules for the inspection are set. During this pre-job conference, questions should be addressed and any potential problems identified. As the job progresses, both the coating contractor's QC person and the designated QA inspector are responsible for witnessing, verifying, inspecting, and documenting their observations at various inspection points. As shown in *Figure 1*, the inspection process should include at least the following steps or points:

- Pre-surface preparation inspection
- Measurement of ambient conditions
- Evaluation of compressor, air cleanliness, and surface preparation equipment
- Determination of surface preparation, cleanliness, and profile
- Witness the mixing and thinning of coating materials
- Evaluation of application equipment
- Inspection of coating application and cleanliness between coats
- Determination of wet-film and dry-film thickness
- Pinhole and holiday testing
- Evaluation of adhesion/cure

2.2.0 Pre-Surface Preparation

Before starting work on a surface (substrate), the surface should be inspected to make sure that it is in a suitable condition to be prepared or coated. All contaminants such as grease, oil, dust, dirt, and cement splatter must be removed before starting blast cleaning. This prevents the contaminants from being redeposited onto the freshly blast-cleaned surfaces. Removal of contaminants

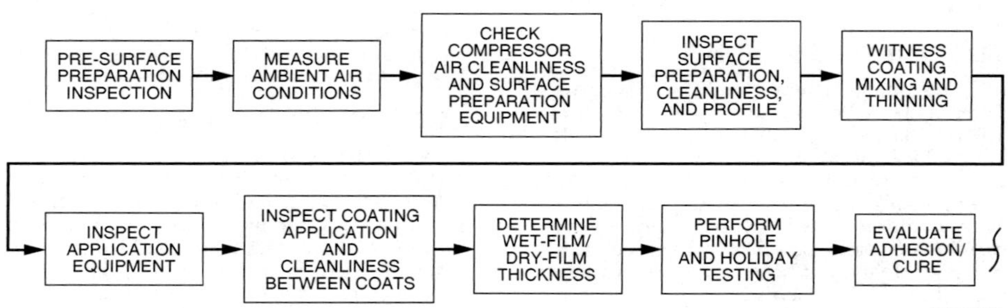

Figure 1 ◆ Sequence of inspection for a typical industrial coating job.

Instructor's Notes:

should be done as described in The Society for Protective Coatings (formerly the Steel Structure Coating Council) (SSPC) specification *SSPC-SP1, Solvent Cleaning*. The requirements of *SSPC-SP1* are included by reference in all SSPC cleaning specifications, including *SSPC-SP2, Hand Tool Cleaning*, and *SSPC-SP11, Power Tool Cleaning*.

Some specifications require that any weld splatter be ground or otherwise removed and the sharp edges rounded off. Pitting in a steel substrate should be examined and a determination made to either accept the pits as they are or to grind or fill them. Surfaces and areas adjacent to the work that are not required to be cleaned or coated should be masked or otherwise covered at this time to protect them from possible damage.

2.3.0 Ambient Conditions

Surface preparation and coating work should be done only under suitable ambient (surrounding) air conditions of temperature, **relative humidity (RH)**, and **dew point (DP)**. RH is the ratio of the amount of moisture present in a given sample of air to the amount it can hold at saturation. RH is expressed as a percentage. Dew point is the temperature at which the moisture vapor in the air will condense on a surface. For example, if the dew point is 50°F, moisture will condense on a surface if its temperature is 50°F or colder. A thin film of condensed moisture present on the surface of a substrate may not be visible, but a quick check of the dew point and the substrate surface temperature will determine if a moisture condition exists. For the application of most catalyzed coatings, specific minimum temperatures must also be met. In addition, many zinc-rich coatings have minimum humidity requirements for operation. Ambient conditions of air temperature, RH, dew point, and surface temperature are typically measured using psychrometers, digital thermometers, and/or hygrometers. These instruments and their use are described later in this section. Some general guidelines that apply when making temperature/ambient condition measurements follow:

- Ambient conditions should be checked before work begins each day and monitored periodically throughout the day. If weather conditions are worsening, the ambient conditions should be checked a minimum of every four hours or as needed.
- Take all readings at the actual work location. For general readings, however, the measurement should be made on the coldest point on the structure because a surface temperature/ dew point relationship problem will occur there first. Air and surface temperature considerations are important to make sure that coatings are not applied outside of their temperature limitations. Accordingly, readings for this purpose should be made at the coolest or warmest areas.
- Use direct reading humidity recorders to measure ambient conditions of temperatures below 32° Fahrenheit (F), or 0° Centigrade/Celsius (C).
- Final blast cleaning or the application of a coating should only be done if the substrate surface temperature is a minimum of 5°F higher than the dew point temperature. This precaution takes into account instrument inaccuracies, changing weather conditions across a structure, and human error.

2.3.1 Measuring Relative Humidity and Dew Point

Relative humidity and dew point are typically determined using psychrometers or digital thermometers/hygrometers. Sling and battery-operated psychrometers (*Figure 2*) use two identical thermometers: one to measure the dry-bulb (DB) temperature, and the other to measure the wet-bulb (WB) temperature. The sensing bulb of the wet-bulb thermometer is covered with a clean sock or wick saturated with clean distilled water before taking a reading. Dry-bulb temperatures are a measure of the sensible heat of the air being measured. Wet-bulb temperatures are a measure of the degree of moisture content of the air. It results from the latent heat loss of water evaporation from the wetted sock. Dry-bulb and wet-bulb temperatures, in conjunction with a psychrometric chart or table, are used to find the corresponding RH and dew point of the sampled air.

To use a sling-type psychrometer, the wet-bulb thermometer sock is first saturated with water, then the psychrometer is whirled rapidly for about 20 to 30 seconds, and a reading of the wet-bulb temperature is taken. This process of spinning and reading the wet-bulb temperature without additional wetting of the sock is repeated until the wet-bulb temperature is stabilized. This occurs when three consecutive readings of the wet-bulb temperature remain the same. After the wet-bulb temperature has stabilized, both the wet-bulb and dry-bulb temperature readings are recorded.

Battery-operated psychrometers have a motor-driven fan that pulls air across the thermometer bulbs. To use the battery-operated psychrometer, the wet-bulb thermometer sock is first saturated with water, then the instrument is turned on. About two minutes of operation are required for the wet-bulb temperature to stabilize. After it has stabilized, both the wet-bulb and dry-bulb temperature readings are recorded.

Demonstration

Distribute copies of the *SSPC-SP1* specifications for removal of contaminants. Have trainees discuss the requirements.

Classroom

Have a trainee define relative humidity. Ask a trainee to define dew point. Have trainees identify the specific requirements for catalyzed coatings and zinc-rich coatings.

Ask trainees to explain each of the guidelines for making temperature/ambient condition measurements.

Audiovisual

Show Transparency 4 (Figure 2). Ask a trainee to distinguish between wet-bulb and dry-bulb temperatures.

Audiovisual

Show Transparency 5 (Figure 3). Ask a trainee to explain how to use the psychrometric chart to determine the RH and DP.

Teaching Tip

See the Teaching Tip for Section 2.3.1 at the end of this module.

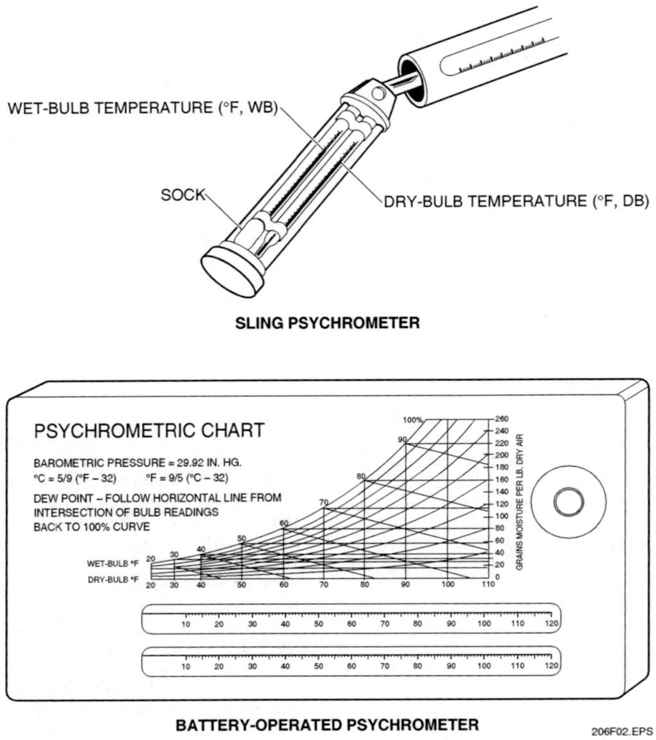

Figure 2 ◆ Psychrometers.

Once the wet-bulb and dry-bulb temperatures are known, a psychrometric chart or table is used in conjunction with the temperatures to find the relative humidity and dew point temperature of the air. A psychrometric chart (*Figure 2*) graphs the different air properties. To use it, the dry-bulb and wet-bulb temperatures are plotted on the appropriate lines, and the relative humidity and dew point are interpolated from their intersection. *Figure 3*, View A, shows a simplified representation of a psychrometric chart with lines plotted for dry-bulb temperature, wet-bulb temperature, dew point temperature, relative humidity, and the grains of moisture per pound of air. *Figure 3*, View B, shows an example of a plot that assumes that the dry-bulb temperature is 75°F and the wet-bulb temperature is 65°F for a given sample of air. Plotting these temperatures on the chart, you will find that the RH is about 58% and the DP is 59.5°F.

Psychrometric tables originally published by the U.S. Weather Bureau consist of individual tables for relative humidity and dew point. To use these tables, the WB temperature is subtracted from the DB temperature and the difference (WB depression) is found along the top row of the table. The DB temperature is found down the left column, and the intersection of the two yields either the humidity or DP, depending on which table is used. Unless the exact barometric pressure at the time of the reading is known, tables corresponding to 30" of barometric pressure should be used. Psychrometric tables published by other organizations may have both the RH and DP shown on the same table.

Instructor's Notes:

Obviously, it is important that the psychrometers used to measure ambient conditions be accurate. Sling and battery-operated psychrometers are factory-calibrated, but their thermometers should be compared periodically, preferably indoors, with ASTM International (formerly American Society for Testing and Materials) -certified thermometers to verify their calibration.

An electronic thermometer/hygrometer (*Figure 4*) can be used instead of a psychrometer to measure temperature and RH. Many models are available. The measured RH and/or temperature readings are provided as a direct reading on the display. Some thermometer/hygrometers are also capable of measuring and giving a direct reading of DP. Some, which use two or more temperature probes, can calculate the differential temperature measured between probes. Others are available which can display temperature and RH readings simultaneously. Calibration and operation of these instruments should be done in accordance with the manufacturer's instructions.

2.3.2 Measuring Ambient and Substrate Surface Temperatures

In addition to measuring ambient temperatures, electronic thermometers/hygrometers equipped with thermocouple or thermistor-type probes can be used to determine the surface temperature of substrates. Bimetallic-type thermometers (*Figure 5*) are also widely used. One type, called a surface temperature thermometer, is designed especially for that purpose. It is made with internal magnets for

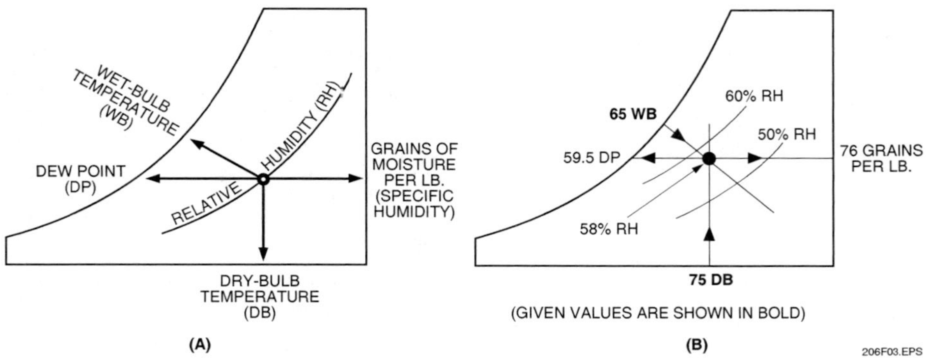

Figure 3 ◆ Simplified representation of air properties plotted on a psychrometric chart.

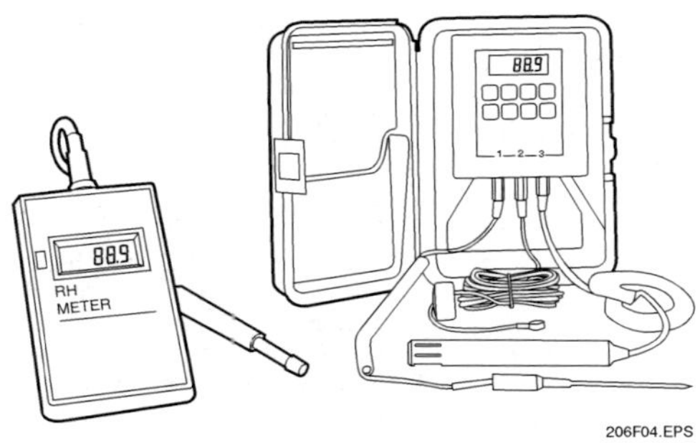

Figure 4 ◆ Electronic thermometer/hygrometer.

Audiovisual

Show Transparency 6 (Figure 4). Discuss the various types of electronic thermometer/hygrometers available.

Laboratory

Have trainees practice calibrating an electronic thermometer/hygrometer. Then have trainees practice measuring temperature and RH with an electronic thermometer/hygrometer.

Audiovisual

Show Transparency 7 (Figure 5). Ask a trainee to explain how the surface temperature thermometer is used.

Laboratory

Have trainees practice making surface temperature measurements using electronic and surface temperature thermometers.

Take the class outside and have trainees practice using a wind meter. Alternatively, have trainees measure wind created by an indoor fan.

Homework

Assign the reading of Sections 2.4.0–2.4.3 for the next session.

Materials

Ensure that you have everything required for the laboratory during this session.

Classroom

Ask a trainee to describe the compressed air supplied at the output of a compressor. Have trainees identify the coating problems that moisture contamination and oil contamination can cause.

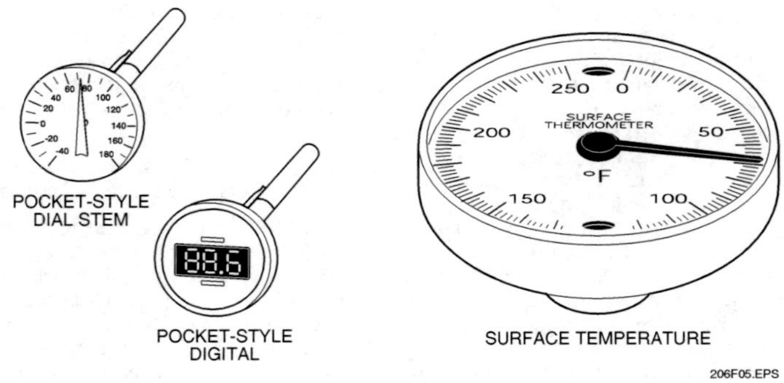

Figure 5 ◈ Bimetallic-type thermometers.

attaching the thermometer to steel and other ferrous surfaces. Thermometers should be operated and calibrated according to the manufacturer's instructions. Generally, calibration is done at the start of each job and at least once a week thereafter.

To make surface temperature measurements using an electronic thermometer:

Step 1 Connect the appropriate temperature range thermocouple/thermistor-sensor probe into the instrument receptacle.

Step 2 Turn the instrument on and select the correct temperature scale (F or C).

Step 3 Hold the sensor probe directly on the surface until the temperature reading stabilizes. Record the reading.

To make surface temperature measurements using a surface temperature thermometer:

Step 1 Make sure that the tab of the sensing element is in the slot of the encasing steel ring on the back side of the thermometer.

Step 2 Attach the thermometer to the surface such that the bimetallic sensing element is in good contact against the surface being measured. The internal magnets provide for its attachment to ferrous substrates. For nonferrous substrates, attach it with tape.

Step 3 Allow the thermometer to remain on the surface for about three minutes. Tap the glass gently to make sure that the needle is free, and view it directly face-on to avoid inaccurate readings because of parallax (an optical illusion resulting in an image being perceived differently when viewed from multiple angles).

2.3.3 Measuring Wind Velocity

Wind meters (anemometers) can be used to determine when wind speeds exceed specification requirements during the application of coatings. Excessive winds during coating applications may result in dry spray, **overspray**, excess wind load on the structure, or damage to nearby property. The wind meter being used should be operated according to the manufacturer's instructions. Position it vertically in the wind stream and read the wind speed in miles per hour (mph), or kilometers per hour (kph) from the meter scale. Always monitor the wind speed where the coating application is taking place because wind speeds at ground level can be greatly different than wind speeds several feet in the air.

2.4.0 Compressor Air Cleanliness and Surface Preparation Equipment

The compressed air supply, related air compressor, and all air-powered (pneumatic) tools and equipment used for surface preparation and coating application should be inspected to determine their suitability.

2.4.1 Compressed Air Cleanliness

The compressed air supplied at the output of a compressor contains concentrated water vapor, oil, dirt particles, and other contaminants. If not removed, these contaminants can cause severe coating problems. Moisture contamination can cause flash rusting of prepared surfaces, spray gun spitting, spatter, fisheyes, bubbles, pinholes, and other film defects. Oil contamination can cause adhesion failures, drying problems, soft tacky films, and various other

Instructor's Notes:

defects. Moisture and oil contamination in the air supply can also cause damage to air-powered spraying equipment and tools used for surface preparation. For jobs requiring that absolutely no moisture or oil be present in the compressed air, oil-less compressors and sophisticated air-drying equipment is normally used.

The cleanliness of a compressed air supply should be tested at frequent intervals during the day. It should be tested before use and at least once a shift during continuous compressor operation. When the RH is high, more frequent testing is necessary because there is more moisture vapor in the air. If compressor operation is stopped during breaks, lunchtime, or for some other reason, the air supply should be retested before use, because condensation can form in the air distribution lines.

A clean, white, absorbent material, such as a cloth, can be used to test the air supply for the presence of oil or moisture. To avoid false indications, the test should be done at an outlet that is downstream from the air line moisture/oil filter and as close as possible to the equipment using the air. Note that if the air is being discharged from several outlets fed from a common manifold source, it normally is not necessary to test the air cleanliness at each location. Testing of the air at each location would be necessary only if more than 100' of air hose is used, or if several distant separate outlet locations are used. Tests at a blast nozzle should be avoided for safety reasons.

Testing compressed air to detect the presence of oil or water is done in accordance with standard *ASTM D 4285*. The procedure is outlined here:

Step 1 Allow the compressor and air line pressure to reach operating conditions before testing. Do not open the air discharge outlet valve under test at this time.

Step 2 Mount the white cloth (or other absorbent material) used to collect the sample on a rigid backing.

Step 3 Secure the white cloth in place within 18" to 24" of the air-discharge outlet and centered in the airstream.

Step 4 While avoiding any personal contact with the airstream, open the air discharge outlet valve under test and allow the airstream to impact the white cloth for a minimum of one minute.

Step 5 Remove the white cloth from the airstream and immediately examine it for signs of moisture or oil. Visually inspect the cloth. Also, feel and smell it. If there is no discoloration on the cloth, the air quality is acceptable.

If the test indicates the presence of moisture or oil in the air supply, the condition of the air line moisture and oil filters (*Figure 6*) should be checked. Any clogged filter element should be replaced, and the filter bowl should be drained of accumulated moisture so that it does not fill up, allowing the moisture to overflow back into the air line. If these actions do not correct the problem, the compressor should be checked for a malfunction as directed in the manufacturer's service literature.

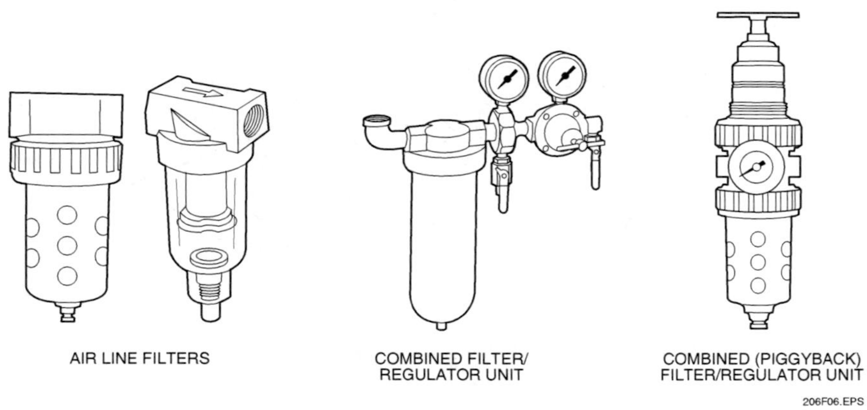

Figure 6 ◆ Service air line moisture/oil filters.

Classroom

Have a trainee explain how often the cleanliness of a compressed air supply should be tested. Ask a trainee to identify the location(s) of the outlet where the test should be done.

Laboratory

Have trainees practice checking a compressed air supply for oil or moisture. Ask trainees to practice checking the filters as well. Have trainees identify possible causes of moisture or oil in the air supply.

Audiovisual

Show Transparency 8 (Figure 6). Have trainees identify each type of filter unit.

Audiovisual

Show Transparency 9 (Figure 7). Have trainees identify each part of the blast machine. Ask a trainee to explain the dangers of moisture and oil or too much abrasive in the blasting system.

Classroom

Name some abrasives listed in Table 1 in the Trainee Guide and have trainees describe the benefits, usage properties, and physical properties for each.

Homework

Assign the reading of Sections 2.5.0–2.7.0 for the next session.

Materials

Ensure that you have everything required for the laboratory and demonstration during this session.

2.4.2 Evaluation of the Blast Cleaning Machine and System

The blast machine (*Figure 7*) mixes an abrasive with the airstream. Moisture and oil in the blasting system can create havoc with the operation. Moisture can cause abrasives to cake together and clog the lines, and lubricating oils from the compressor can contaminate the blasted surface. For this reason, the machine should be equipped with moisture and oil separators, or external separators should be provided. For safety, a blast cleaning machine must be equipped with a dead-man switch at the nozzle so that the machine shuts down in the event that the nozzle is dropped. Also, the blast machine must be constructed in compliance with the prevailing pressure vessel codes and American Society of Mechanical Engineers (ASME) standards.

The setting of the blast machine abrasive metering valve and size of the nozzle orifice should be checked to make sure that not too much abrasive is being used. Use of too much abrasive results in both decreased production and increased abrasive costs.

2.4.3 Checking Blast System Abrasives

There is a wide variety of metallic and non-metallic abrasives that can be used for blast cleaning. *Table 1* gives a summary of the properties for a number of different types of abrasive media, that with the exception of sand, can be recycled a varying number of times. Standard *SSPC-AB 1*, Paragraph 4, defines the requirements for selecting and evaluating mineral and slag abrasives used for blast cleaning. It also describes sampling methods and qualification and conformance tests to conduct on non-recyclable abrasives.

2.5.0 Determination of Surface Preparation, Cleanliness, and Profile

The two major criteria involved with the inspection of a surface are its degree of cleanliness and the height of the anchor profile. Preparation and evaluation of surfaces are addressed in the following sections.

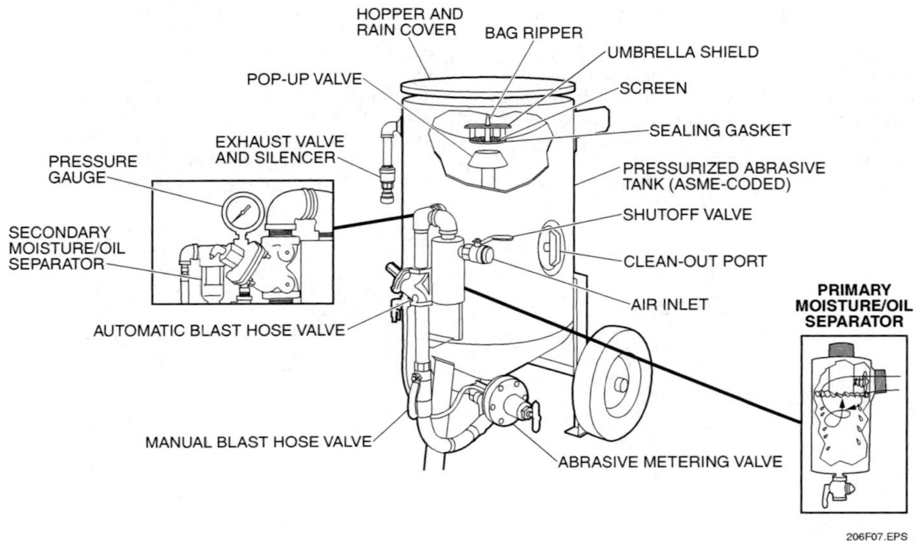

Figure 7 ◆ Typical components of a single-tank, high-production, pressurized blast machine.

Instructor's Notes:

Table 1 Properties of Various Abrasive Media

Abrasive	Benefits				Usage Properties					Physical Properties			
	Finishing	Cleaning and Removal	Peening	Etching	Average Cleaning Speed	Reusability	Dust Level(s)	Chance of Metal Removal	Typical Blast Pressures (psi)	Hardness MOH Scale (Rockwell C)	Bulk Density ($ft.^3$)	Mesh Sizes	Angular (◊) Spherical (○)
Glass Beads	X	X	X		MED	HI–LO	LO–HI	Very LO	20–55	5.5	100	30–440	○
Ceramic Shot	X	X	X		MED	HI	LOW	Very LO	20–90	7 (57–63)	150	8–46	○
S.S. Cut Wire	X	X	X	X	MED	HI	LOW	MED–HI	20–90	6–7.5 (35–55)	280	20–02	◊
Steel Shot	X	X	X		MED	Very HI	LOW	Very LO	20–90	6–7.5 (20–66)	280	8–200	○
Steel Grit	X	X		X	MED–HI	Very HI	LOW	MED	20–90	6–7.5 (40–66)	230	10–325	◊
Aluminum Oxide	X	X		X	HI	MED–HI	LO–MED	MED–HI	20–90	8–9	125	12–325	◊
Silicon Carbide	X	X		X	Very HI	LO–MED	MED–HI	MED–HI	20–90	9	95	36–220	◊
Crushed Glass	X	X		X	HIGH	LO–MED	MED–HI	LO–MED	20–50	5.5	100	30–400	◊
Plastic Media		X		X	MED–HI	MED	MED	Very LO	20–60	3–4	45–60	12–80	○
Wheat Starch		X			MED	MED	MED	Very LO	20–35	3	40	12–100	◊
Walnut Shell		X			LO–HI	LOW	MED–HI	Very LO	10–40	3	40–80	Many	◊
Corn Cob		X			LO–MED	LOW	MED–HI	Very LO	20–60	1	30	8–40	◊
Sodium Bicarb.[1]		X			MED–HI	NR[3]	HI	Very LO	20–90	3	62	54–150	◊
Silica Sand[2]	X	X		X	HI	NR[3]	HI	HI	30–90	7.5	100	12–100	◊

NOTES:
Above information is intended as a general reference guide.
1 Sodium bicarbonate must be treated with a flow agent.
2 Do not use silica sand in a hand cabinet.
3 Not recommended

206T01.EPS

Classroom

Have a trainee describe various types of surface preparation. Name various examples of poor surface preparation and ask trainees to explain the problems with each.

Distribute a copy of SSPC standards *SSPC-SP1* through *SSPC-SP13*. Have trainees read through some of the standards. Discuss the information contained in the standards.

Demonstration

Show trainees how to measure surface profile. Ask trainees to explain the problems with a too-high profile and a too-shallow profile.

Classroom

Have trainees explain important points regarding mixing and thinning of coating materials. Compile a master list on the board.

2.5.1 Surface Preparation

Depending on surface conditions, surface preparation can involve anything from a simple dry wiping and blowdown with air to a very thorough cleaning using power tools, abrasive blasting, or water jetting. Surface preparation is a major factor affecting the performance of a coating system. Some harmful conditions resulting from poor surface preparation, which affect the performance of a coating, include:

- Oil and grease prevent good coating or lining adhesion.
- Salts of various chemicals may increase moisture permeation through the coating and speed up the rate at which corrosion will occur.
- Dust and dirt particles on the surface prevent the application of smooth, uniform film and weaken the adhesion of the coating to the substrate.
- Rust cannot be penetrated by most coatings. The result is an uneven film with exposed areas of metal where further corrosion can take place. Rust scale or mill scale also cannot be penetrated, and if coated over, will eventually break loose from the substrate, taking the coating or lining with it.
- Old coating must be feathered into newer, sound adhesion. Otherwise, it may lift from the surface, causing peeling or delamination of the new material.
- Sharp ridges, burrs, or cuts left in the surface of the metal are difficult to coat over to obtain the full thickness and film continuity. They should be removed before coating.

2.5.2 Surface Preparation Standards

All surfaces should be inspected after preparation to ensure they comply with specified standards. The surface preparation standards most used in the United States are the SSPC standards. Surface preparation standards developed by the National Association of Corrosion Engineers (NACE) are also widely used.

Standards *SSPC-SP1* through *SSPC-SP13* each contain a narrative description for a different method and degree of surface preparation, percentage of the surface area to be cleaned, types of residues permitted to remain on the surface, and other criteria.

2.5.3 Evaluating Surface Profile

For metals, the **surface or anchor profile** is a measurement of the surface roughness that results from abrasive blasting. The profile is a measurement of the peak to valley depth (or height) caused by the impact of the abrasive onto the substrate. This rough area of peaks and valleys increases the area of the metal surface for bonding of the coating. As a general rule, thick coatings require a deeper surface profile than thin coatings. Ideally, the surface profile should be 20% to 25% of the total coating thickness. If the profile is too high for the specified coatings system, the peaks will project above the coating film, allowing pinpoint rusting to occur. If the profile is too shallow, poor adhesion can result, particularly for heavier, thicker coatings. Surface profile can be determined by using one of following methods:

- Surface profile comparators
- Replica tape with spring micrometer
- Profile depth gauge

2.6.0 Mixing and Thinning of Coating Materials

Improper mixing or thinning of coating materials can affect the coating's ability to resist the environment. The mixing process begins by making sure that the correct materials are available for the job. For two or more component materials, also make sure that the correct components are available and in the correct ratio and quantity. The manufacturer, product numbers, colors, and batch dates of all materials should be checked and recorded. Also, it is important to make sure that the shelf life of the material has not expired.

Coating products should be mixed in accordance with the instructions given in the coating manufacturer's application bulletin. The mixing of the material must be thorough so that it is smooth, homogeneous, and free of surface swirls or pigment lumps. Many coatings settle out during prolonged storage, so thorough mixing of these coatings is necessary to ensure that all pigment settled on the bottom of the container is incorporated in the mixed coating. When a thinner is used, the thinning should be done in accordance with the manufacturer's recommendations for both type and quantity. The amount and type of thinner used must be recorded to ensure documentation of the actual volume of solid content and volatile organic compound (VOC) level.

For multicomponent products, it is important to ensure that all components are mixed in the proper sequence and ratio as recommended by the manufacturer. Also, ensure that any required induction times (sweat-in times) stated in the product application bulletin are met. Pot life must also be monitored. Pot life varies from coating to coating and is usually temperature dependent. Generally, the warmer the temperature, the shorter the pot life.

Instructor's Notes:

2.6.1 Measuring the Temperature of Coating

The temperature of coatings during both mixing and subsequent application can be measured using a coating thermometer. Coating temperature at the time of mixing can provide valuable information related to pot life and thinning requirements. As shown in *Figure 8*, a coating thermometer is a typical bimetallic-type thermometer; however, it is equipped with a clip used to attach the thermometer to a coating container, allowing the user to constantly monitor the temperature of the coating. All temperatures should be taken at the mixing site.

If the mixed coating material is moved from the mixing site to the actual work location, the temperature should be taken again. After each use, the thermometer stem should be thoroughly cleaned.

2.6.2 Measuring Viscosity of Coating Materials

The viscosity (flow rate) of a coating is measured before application to ensure that proper thinning ratios are used and that thinning has not changed significantly from one container to another or from one mixed batch to another. A common method for measuring viscosity involves the use of a Zahn viscosity cup. As shown in *Figure 9*, it consists of a small cup of known volume with a precisely sized orifice in the center. Generally, five orifice sizes are available, numbered 1 through 5.

To find the correct viscosity for a specific material, the coating manufacturer should be consulted to determine what size viscosity cup to use, and the time in seconds for the volume of properly thinned material held by the cup to pass through the orifice. For example, the manufacturer might specify that the material be thinned such that it will pass through a No. 2 Zahn cup in 30 to 40 seconds at a given liquid coating temperature. The general procedure for using a Zahn cup follows:

Step 1 Select the cup with the proper size orifice as recommended by the coating manufacturer.

Step 2 Dip the cup into the coating and fill it to capacity. Make sure to immerse the cup in a location free from bubbles or foam.

Step 3 Quickly remove the filled cup from the coating and simultaneously start a stop watch. Be careful not to lift the cup more than about 2" above the surface of the coating in the container. This is important in order to keep the cup in the solvent atmosphere and away from all drafts.

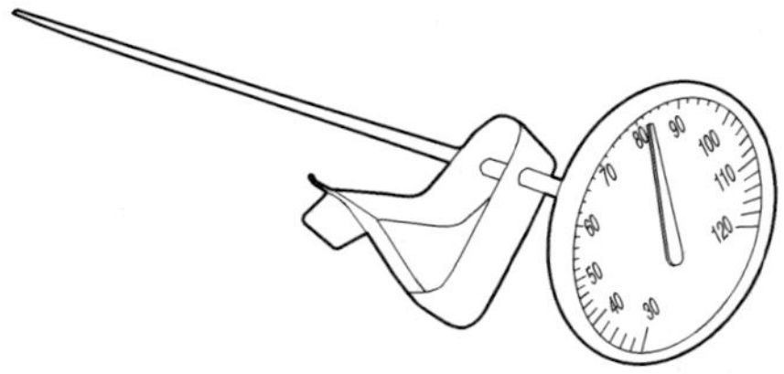

206F08.EPS

Figure 8 ◆ Coating thermometer.

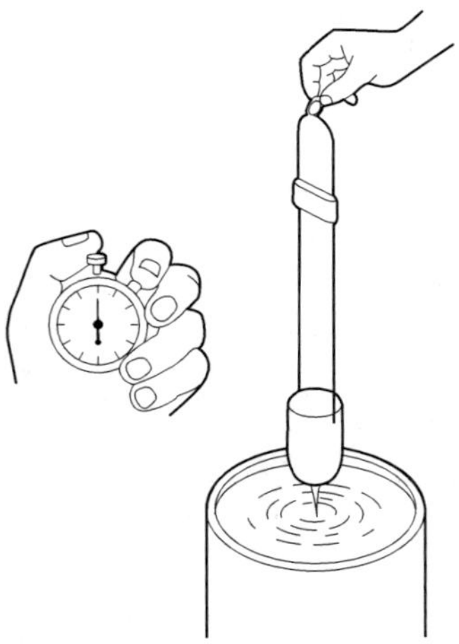

206F09.EPS

Figure 9 ◆ Measuring viscosity with a zahn viscosity cup.

Step 4 When the constant stream of coating material stops flowing, turn off the stop watch and note the seconds elapsed. This represents the viscosity in Zahn seconds.

If the viscosity of the coating being measured does not meet the manufacturer's recommendations, the amount of thinner should be adjusted so that the coating flows through the cup in the specified time. To ensure accuracy, promptly clean the Zahn cup after each use, making sure that no material residue exists in the cup or orifice. Note that some high-build thixotropic coatings cannot be measured using a Zahn cup, but they can be measured using other types of viscosimeters.

Audiovisual

Show Transparency 10 (Figure 8). Ask a trainee when coating temperatures should be taken.

Laboratory

Have trainees practice measuring viscosity of a coating using a Zahn cup. Ask a trainee how to ensure accuracy.

Classroom

Ask a trainee to explain when coating application equipment should be inspected. Have trainees identify the tasks involved in checking coating application equipment.

Homework

Assign the reading of Sections 2.8.0–2.12.0 for the next session.

Materials

Ensure that you have everything required for the demonstration and testing during this session.

Classroom

Have trainees discuss the inspection process between coating applications.

Audiovisual

Show Transparency 11 (Figure 10). Ask a trainee to explain how to read the thickness gauge in the figure.

2.7.0 Evaluation of Application Equipment

A wide variety of equipment and methods can be used to apply coatings. The specific equipment used is determined both by the coating materials and the type of substrate involved. As a consequence, QA/QC inspectors assigned to a particular job must be thoroughly familiar with the equipment and methods required for use on that job. The relevant coating manufacturer's product sheets and application bulletins should be consulted and the manufacturer's recommendations followed when selecting application equipment. In the case of spray equipment, the manufacturer's literature will recommend the type of equipment, such as conventional, airless, tip sizes, and pressure.

Cleanliness of spray equipment and its compressed air supply (if applicable) should be verified before the coating is mixed.

Clogged coating equipment resulting from dirt may cause the loss of the coating material because of elapsed pot life or the presence of contamination. The cleanliness and working order of spray tanks (pots) and other components should be checked before use. For conventional and some HVLP spray equipments, the compressed air supply should be clean, dry, and free of moisture and oil. Moisture contamination can cause such problems as spray gun spitting, spatter, fisheyes, bubbles, pinholes, and other film defects. Oil contamination can cause adhesion failures, drying problems, soft tacky films, and various other defects.

2.8.0 Inspecting Coating Applications and Cleanliness Between Coats

QA/QC of coating applications verifies that the proper application equipment and techniques are being used. This is especially important with spray applications, including maintaining the proper distance of the gun from the substrate, the spray angle, and overlaps.

Besides verifying that proper application techniques are being used, the existing environmental factors must also be monitored, especially when working at sites such as fossil fuel power stations, chemical plants, and other facilities where atmospheric contamination is likely to be present. When a multicoat system is being applied, it is important to verify and document that the coating to be top-coated is dry, clean, and free of foreign matter before the application of the next coat in the system. Dirt, dust, dry spray, or overspray on the previously coated surface can cause problems. All contaminants must be removed because they can cause reduced adhesion between coats. They can also reduce porosity, making the coating less resistant to the effects of the environment. Often, water washing between coats or application of the topcoat within a minimum time interval is necessary. Otherwise, contaminants often invisible to the eye may be coated over, leading to premature coating failures.

When applying multicoat systems, it is important to ensure that all subsequent applications of a coating are done within the recoat window as specified by the product manufacturer. Also, each coating should be checked to make sure that its thickness is neither too thin nor too thick, and that there is complete coverage. Methods for measuring wet-film and dry-film thickness of coatings are given in the following sections. One common method used to help ensure that complete coverage is achieved is to use coatings of a different color or tint for each coat. Each coat that is applied must be free of runs, sags, drips, streamers, and other film deficiencies. Ideally, any such deficiencies should be corrected during the application. If corrected after the coating has dried, grinding, sanding, or in extreme cases, blast cleaning may be required.

2.9.0 Determination of Wet-Film and Dry-Film Thickness

In contract coating work, achieving the coating thickness as specified by contract is normally a requirement. It is also important because a coating thickness that is either too thin or too thick can cause problems. A coating needs a minimum thickness to prevent corrosion, stand up to abrasion, and look good. A coating that is too thick may crack, flake, blister, or fail to cure properly. Two methods are used to determine the thickness of coating: wet-film and dry-film.

2.9.1 Measuring Wet-Film Thickness

To help achieve the specified dry-film thickness for each applied coat, the thickness of the coating should be periodically checked during its application, using a wet-film thickness gauge (*Figure 10*).

The gauge has four edges. The edge of each side has two end points of equal length with progressively shorter teeth (notches) in between. The typical gauge has two sides. One side has a 1 to 80 mil scale; the other side has a 25 to 2,032 micron scale. The number marked on each tooth represents the distance (in mils or microns) between the tooth and end point. To use the gauge, one edge is placed firmly and squarely onto the surface to be measured immediately after applying the coating. The gauge is then removed. The thickness of the wet film lies between the highest

Instructor's Notes:

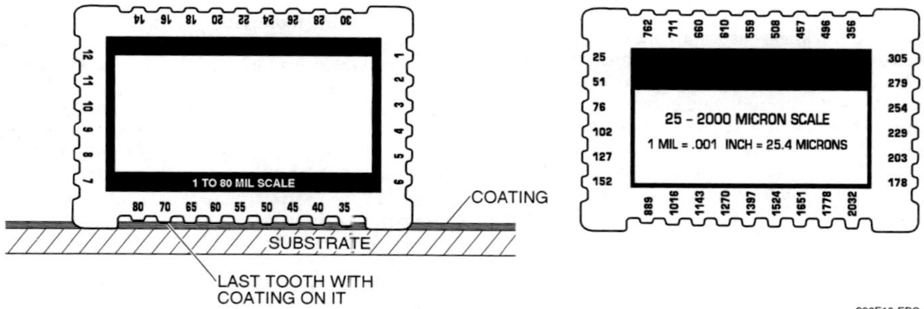

Figure 10 ◆ Wet-film thickness gauge.

tooth with coating on it and the next uncoated tooth. For the example shown in *Figure 10*, the wet-film thickness is between 70 mils and 80 mils. Note that 1 mil equals 0.001" or 25.4 microns (1 mil = 0.001" = 25.4 microns). Follow these guidelines for making accurate measurements:

- Do not drag, slide, or twist the gauge in the wet coating, because this will give incorrect readings.
- Avoid surface irregularities that might distort readings.
- If measuring curved surfaces, use the gauge on the length, not on the width, of the surface.

The wet-film thickness (WFT) required to yield any desired dry-film thickness can be calculated using the formula:

$$\text{WFT} = \frac{\text{desired dry-film thickness}}{\text{\% solids by volume*}}$$

(* The % solids by volume are given in the manufacturer's product data sheet or application bulletin.)

For example, assume that you are using a material that contains 78% solids by volume. Also assume that you want to apply one coat at a dry-film thickness of 8 mils. The required wet-film thickness is determined as follows:

$$\text{WFT} = \frac{8}{0.78} = 10.25 \text{ mils}$$

When thinner is added to a coating material, the result is that the thinned material will have a lower percentage of solids remaining on the surface. When thinner is added, the wet-film thickness required to yield the desired thickness can be calculated using the formula:

$$\text{WFT} = \frac{\text{desired dry-film thickness}}{\left(\frac{\text{\% solids by volume}}{100\% \times \text{\% thinner added}}\right)}$$

If the coating in this example is thinned 20%, the new required wet-film thickness is calculated as follows:

$$\text{WFT} = \frac{8}{\left(\frac{0.78}{1.2}\right)} \quad \frac{8}{= (0.65)} = 12.3 \text{ mils}$$

2.9.2 Dry-Film Thickness

The QA/QC inspector is responsible for verifying that the dry-film thickness of the coating meets specifications. If there are no specifications, use the coating manufacturer's recommended thickness on the product data sheet. Correct coating thickness is essential to the life and protective quality of the coating. Both too-thick and too-thin coatings cause premature coating failure, and also, coating that is too thick is expensive and wasteful.

Destructive testing can be used when necessary, but several dry-film thickness gauges and nondestructive methods of checking thickness can be used for certain tests, including magnetic pull-off, constant pressure probe (magnetic flux), and constant pressure probe (eddy current).

Classroom

Have a trainee identify the person responsible for verifying that the dry-film thickness of the coating meets specifications.

Classroom

Ask trainees to identify and describe various destructive thickness-testing methods.

Have trainees name and discuss three types of nondestructive dry-film testing. Ask a trainee to explain why the calibration of electronic magnetic flux constant pressure probes should be verified each time.

Teaching Tip

See the Teaching Tip for Sections 2.9.0–2.9.4 at the end of this module.

Classroom

Ask trainees to name the three methods for finding nicks, scrapes, and pinholes in coating film.

2.9.3 Destructive Dry-Film Testing

Destructive thickness-testing methods are used to measure specific characteristics, such as flexibility, adhesion, and hardness. One method is to place a piece of tape on the substrate before applying the coating. After the coating has dried, the strip of tape is carefully removed, lifting the coating off the surface. The thickness of the tape must be measured, and then the thickness of the coating can be measured. Flexibility is also tested by the tape method. The most important factor in testing for flexibility is consistency: the test must be performed exactly the same way each time. *ASTM D-3359* provides details and a rating scale for this test.

Film removal by a sharp blade or a mechanical cutting edge is often used to test adhesion, abrasion resistance, and hardness. Some cutting devices are equipped with gauges to measure the force necessary for removing coating. A micrometer is used to measure the thickness. Cross hatching involves two sets of parallel cuts made at 90-degree angles to each other. The technician estimates the amount of coating that remains on the substrate inside the checkerboard cuts. Scratch tests are performed by mechanically operated sharp or pointed objects. Hardness may be tested by pushing a pencil point into the coating; the hardest lead that fails to penetrate the surface is considered to indicate the coating hardness.

2.9.4 Nondestructive Dry-Film Testing

Nonmagnetic coatings, including galvanizing, epoxy, and powder coatings, that are used on magnetic substrates such as carbon steel, cast iron, and ferretic stainless steel are measured with magnetic gauges. One type of magnetic pull-off gauge works by holding it against the surface to be tested and pushing a thumbwheel forward until a magnet makes contact with the surface. Then the wheel is slowly turned until a spring or other device is able to pull the magnet off the surface. The thicker the coating, the easier it is to pull off. If the coating is thin, the magnet would be much more difficult to pull off. Calibration scales can be adjusted for the different magnetic strengths of various substrates, and a gauge on the magnetic pull off indicates the coating thickness.

Electronic magnetic flux constant pressure probes are available for measuring nonmagnetic coating thicknesses on ferrous metal surfaces. These gauges should be calibrated in accordance with *SSPC-PA2, "SSPC Method for Measurement of Dry Coating Thickness with Magnetic Gauges."* The calibration should be verified each time the gauge is used because there may be flux leakage problems from the gauge. In some units with a permanent magnet, such as the Elcometer Thickness Gauge, the magnetic flux produced across an air gap varies with the proximity a ferrous surface to two contact points in the device. The change of flux is indicated by a pointer assembly that pivots within the air gap. The pointer indicates the thickness on a graduated scale. The number of magnetic flux lines can be calculated by multiplying the magnetic field by the perpendicular area that the field penetrates. The unit of measure for magnetic flux is the weber, and the symbol for flux is the Greek letter phi, or π.

Where B is the magnetic field and A is the perpendicular area,

Magnetic flux = π = B A

Flux is shown as lines in a plane that makes contact with electric charge poles or magnetic poles. Electrostatic flux density is inversely proportional to the distance from the charge center. Flux lines cannot be seen, but because they produce certain effects, a graphic representation is possible. Iron filings on a sheet of paper above a magnet cause the filings to move and arrange themselves in a pattern. Computer representations use colors and fine black lines to indicate magnetic flux patterns.

Nonferrous (nonmagnetic) substrates with nonmagnetic coatings are thickness-tested with gauges that use the eddy current principle. An eddy, or circulating, current is an electric current induced in the body of a conductor when the conductor either moves through a nonuniform magnetic field or is in an area where there is a change in the magnetic flux. Eddy current is also know as Foucault current. Eddy currents can be induced in the magnetic field of a coil when an AC current flows in a coil that is close to a conducting surface. The eddy current phase and magnitude change the loading on the coil and, therefore, the impedance. Monitoring the voltage across the coil indicates changes in the subject material, which is the basis for eddy current testing.

2.10.0 Pinhole and Holiday Testing

After all coats have been applied, holiday, pinhole, or spark testing can be used to find the nicks, scrapes, and pinholes in the coating film on ferrous substrates. Pinhole testing is normally mandatory if the coating is intended for immersion service. A pinhole is a small, pore-like hole in a coating film, caused by solvent or moisture release or a porous substrate. Holiday testing may be required after application of either the next to last coat or the last coat. A holiday is an application defect in which small areas (skips or misses) are left uncoated. Usually, when pinhole and

Instructor's Notes:

holiday testing is specified, it is done before final cure of the coating has occurred so that any repair material applied will successfully bond to the underlying coats. Detectors used to locate pinholes and holidays are of three general types: low-voltage wet sponge, direct current (DC) high-voltage, and alternating current (AC) electrostatic.

A low-voltage holiday (pinhole) detector is used when working with nonconductive coatings, up to 20.0 mils in thickness, applied to conductive metal substrates. It consists of the detector, a ground cable, and a sponge electrode (*Figure 11*). To use a low-voltage detector, the ground cable is firmly attached to the bare substrate. For testing coatings up to 10 mils thick, the sponge electrode is saturated with tap water. For coatings 10 to 20 mils thick, the sponge should be saturated with water containing a non-sudsing wetting agent (such as photographic development fluid) to increase the wetting properties. Next, the sponge electrode is moved across the entire surface being tested, using a systematic overlapping method. If a pinhole or holiday is detected, a small current flows through the defect to the substrate, completing a circuit to the detector. This causes the detector to emit an audible signal. When this occurs, the sponge can be turned on its end to help pinpoint the exact location of the defect.

Low-voltage holiday detectors are factory calibrated. However, proper operation should be verified before use. This can be done by touching the dampened sponge to a section of bare metal to which the ground cable has been attached; this should activate the audible signal. If not activated, touch the sponge directly to the clamp or ground wire. If the audio signal is activated, the problem is the ground connection to the substrate. If not, the detector is not operational.

DC high-voltage detectors basically work on the same principle as the low-voltage detector, except that no sponge is used. These detectors are made in models capable of producing voltages of 20,000 volts (V) and more. They can be used to test coatings of any thickness, provided the proper voltage setting is used. However, the low-voltage detector is the preferred instrument of use when the coating thickness is 20 mils or less. During use, the DC high-voltage detector is grounded and the electrode passed over the coated surface. A spark will arc through the air gap at discontinuities to the substrate, simultaneously activating a signaling device in the unit. When using the high-voltage detector, it is important to use only the voltage level recommended by the coating manufacturer for the coating thickness. Otherwise, damage to a good coating can occur. It is a good rule of thumb to use 100 to 125V for each mil of coating thickness.

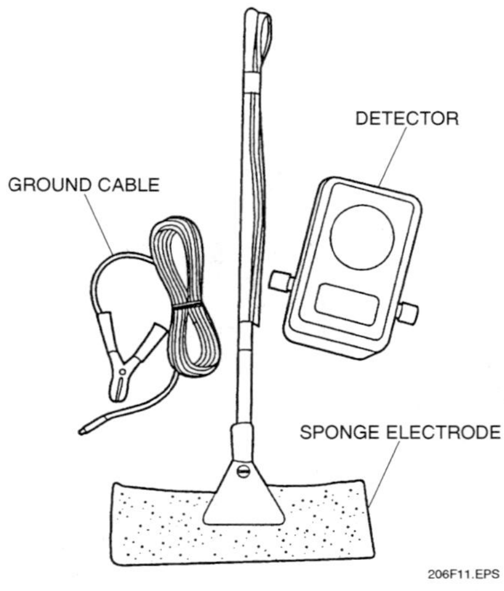

Figure 11 ◆ Low-voltage holiday detector.

When testing conductive coatings applied over steel substrates, AC electrostatic testers are generally used. They have a variable voltage output but do not use a ground wire. The detector constantly emits a corona, but when a break in the lining is passed over, a white spark will jump to the substrate.

2.11.0 Evaluating Coating Adhesion/Cure

Specifications may require that the adhesion characteristic(s) of applied coatings or coating systems be determined. Two test methods in accordance with ASTM standards are commonly used: *ASTM D 4541, Method of Pull-Off Strength of Coatings Using Portable Adhesion Testers*, and *ASTM D 3359, Test Method for Measuring Adhesion By Tape Test*.

2.11.1 Adhesion Testing Using Portable Adhesion Testers

When testing to standard *ASTM D 4541*, portable adhesion testers are used (*Figure 12*). They test the tensile adhesion strength of coatings and display the result in pounds per square inch (psi). There are three basic types of portable testers: mechanically operated, pneumatically operated, and hydraulically operated. All consist of the test unit itself and aluminum or stainless steel pull stubs

Audiovisual

Show Transparency 12 (Figure 11). Have trainees explain how to use a low-voltage holiday detector for testing coatings of 7-mil and 15-mil thickness.

Show Transparency 13 (Figure 12). Ask a trainee to explain how coating adhesion testers operate.

Classroom

Have trainees name and describe the three types of coating adhesion failures.

Ask trainees to explain the two methods of adhesion testing using a tape test.

Demonstration

Demonstrate how to use the pencil testing method to determine the cure of a coating.

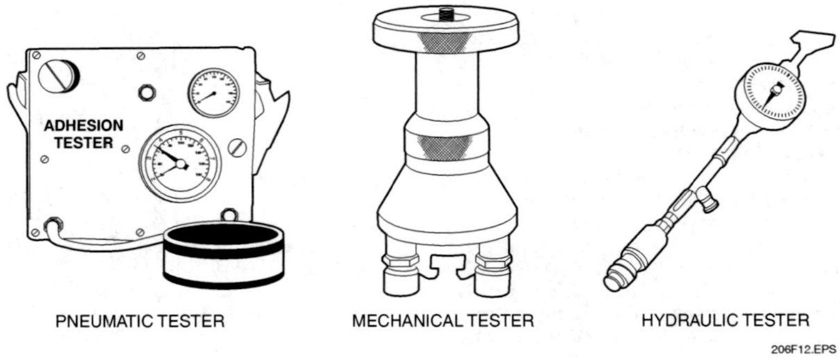

Figure 12 ◆ Coating adhesion testers.

(sometimes called dollies or loading fixtures). Regardless of the type of tester, the procedure for using them is basically the same. The pull stubs are cemented or bonded to a clean coating surface using an epoxy adhesive. After the adhesive has cured, the tester piston or claw is placed over the pull stub. The tester is then operated per the manufacturer's instructions to apply an increasing pulling force on the pull stub until the stub breaks from the surface (destructive testing), or until a predetermined specified value of force has been reached (nondestructive testing). At this point, the force indicated by the dial or indicator in psi should be recorded. Also, the pull stub and test surface should be examined to determine the type of failure. Failures are defined as follows:

- *Adhesive failure* – A failure between two distinct coating layers or between the substrate and the first layer of coating.
- *Cohesion failure* – A failure or break within a given coat or material (the coating breaks within itself).
- *Glue failure* – The applied psi exceeds the strength of the adhesive used to bond the pull stub.

2.11.2 Adhesion Testing Using a Tape Test

When testing to standard *ASTM D 3359*, ASTM-specified, pressure-sensitive tape is applied to a pattern cut into a coating and the results compared to written or visual standards. One method, designated Method A, is intended mainly for field use. It involves cutting an "X" with a utility knife or razor blade through the coating down to the substrate. The X is then covered with pressure-sensitive tape, rubbed vigorously, then rapidly and smoothly pulled off. Following this, the coating adhesion is evaluated by comparison to a scale of 0A to 5A as follows:

- *5A* – No peeling or removal
- *4A* – Trace peeling or removal along the edges
- *3A* – Up to 0.0625" jagged removal along most of either cut
- *2A* – Up to 0.125" jagged removal along most of either cut
- *1A* – Removal of most area of X under the tape
- *0A* – Removal beyond area of X

A similar method, designated Method B, is used mainly in a laboratory setting, but it can be used in the field. A lattice pattern of six or eleven cuts is made in the coating in each direction, then pressure-sensitive tape is applied over it and pulled off. The coating adhesion is then evaluated on a scale of 0B to 5B by comparison to a narrative and figures contained in *ASTM D 3359* showing these levels. To aid testing by Method B, a cross-cut guide kit is available. It contains a precision, chrome-plated steel template for cutting the lattice pattern, ASTM-specified tape, a utility knife, and testing instructions.

2.11.3 Evaluating Cure

Determining the cure of coatings is generally difficult. *ASTM D 1640* outlines one method, but there are no universally reliable field tests for such purposes. Just because a coating is hard does not mean it is cured. However, some job specifications may require that the hardness or solvent resistances of applied coatings be determined. Hardness can be tested by two methods: pencil testing and durometer testing. Durometer testing is done in accor-

Instructor's Notes:

dance with *ASTM D 2240*. It is more commonly used to determine the hardness of rubber, cellular materials, elastomeric materials, thermoplastic elastomers, and some plastics.

Standard *ASTM D 3363* governs the pencil testing method. To perform the test, a series of slightly sharpened pencils (actually leads) of different degrees of hardness are pressed downward into the coating at a 45° angle and forward about 0.25". The process is repeated until a pencil is found that will not cut through the coating and one is found that will not scratch the coating (they may be the same). The hardness of the pencil that leaves the film uncut is defined as the gouge hardness; the hardness of the pencil that does not scratch the film is defined as the scratch hardness.

To perform a solvent-resistance test, a solvent-resistant coating is rubbed with a medium pressure for 50 up-and-down strokes (called double rubs) with a solvent-soaked (not dripping) cloth. The solvent used is typically methyl ethyl ketone (MEK). After the 50 strokes are completed, the cloth is visually inspected to determine if any coating is removed. Fully cured thermosetting coatings are not affected by this test, but thermoplastic and uncured thermosetting coatings will be dissolved.

2.12.0 Quality Control/ Assurance Documents

Results of all QA/QC and inspection tasks should be thoroughly documented in writing to verify that the specified requirements have been met. Future maintenance or the removal and maintenance of a coating system may depend on the factual reporting of every phase of work. Proper documentation can help you make a case in the event of a dispute, but more importantly, the recording of the check or measurement helps ensure that no quality control task is overlooked and that no work goes unperformed. Records can be kept in a daily diary, or they can be kept on preprinted forms. However, preprinted forms are recommended because they are clearly organized with entries for specific kinds of information. An example of a basic quality control record is shown in *Figure 13*.

3.0.0 ◆ CAUSES OF COATING FAILURES

The causes of coating failures can be grouped into three broad categories. However, as would be expected, many coating failures can be a result of more than one of these categories:

- Application problems
- Specification-related problems
- Coating-related problems

3.1.0 Application Problems

Improper surface preparation and poor application of coatings are the major causes of most coating failures. Failure to properly prepare the surface either by blast cleaning, acid etching, or other applicable procedures is the most common factor, and it leads to loss of adhesion. Other problems occur because of poor workmanship or application techniques. Most coating materials are normally applied by brush, roller, or spray equipment. For example, during spray applications, the spray gun should be moved at the appropriate speed, held perpendicular to the surface being coated, and not held too close or too far away. In roller operations, the proper roller nap should be used, and care should be taken not to roll out the coating too much. Similarly, when brushing, the proper size and shape brush and suitable bristle should be used. It is also important to use the proper technique for dipping the brush and spreading the coating on the surface. Strict attention to the instructions and recommendations given on coating manufacturer's product data sheets (see *Figure 14*) and application bulletins help eliminate most application-related coating failures.

In addition to improper application techniques, application faults are also caused by:

- Coating when the weather or ambient conditions are too cold, too hot, too wet, too dry, or too windy
- Coating over wet or dirty surfaces
- Failure to observe proper recoat intervals
- Applying the coating too thin or too thick
- Over-thinning or using an unauthorized thinner
- Coating substitution
- Failure to properly disperse and/or mix multi-component coatings
- Improper tape wrap techniques

Coating failures that can result from poor workmanship or application techniques include (*Figure 15*):

- Runs or sags
- **Orange peel**
- Overspray
- Cratering
- Fisheyes
- Blushing
- **Wrinkling** and lifting
- Blistering
- Cobwebbing
- Rust, pinholing, and pinpoint rusting
- Holidays and spatter coating
- Flaking, peeling, and delamination
- Structural difficulties

Classroom

Ask a trainee to explain how to perform a solvent-resistance test. Ask trainees if they have any questions.

Audiovisual

Show Transparency 14 (Figure 13). Discuss the types of information requested in the quality control record.

Teaching Tip

See the Teaching Tip for Sections 2.0.0–2.12.0 at the end of this module.

Performance Testing

Have each trainee complete Performance Task 2 to your satisfaction. Fill out a Performance Profile Sheet for each trainee.

Homework

Assign the reading of Sections 3.0.0–3.4.0 for the next session.

Materials

Ensure that you have everything required for teaching this session.

Classroom

Ask trainees to name various causes of coating failures.

Quality Control Record

Job # ________________
Client ________________
Date ________________
Activity ________________
Items and Location of Work ________________
Time Start ________________
Time Complete ________________
Client Representative ________________

Foreman ________________
Crew ________________

3rd Party Inspector ________________

	Method of Verification or Measurement	Results or Readings
Weather		
Temperature		
Humidity		
Dew point		
Surface temperature		
Surface Preparation		
Degree of cleaning specified		
Method		
Equipment check		
Abrasive check		
Degree of cleaning achieved		
Profile achieved		
Coating Material		
Product and amount		
Batch #		
Color		
Conformance with shelf life		
Thinner type & amount		
Mixing according to specification		
Induction according to specification		
Induction according to data sheet and label		
Application		
Type equipment		
Equipment check		
Compressed air		
Pot agitation		
Wet film thickness required		
Wet film thickness achieved		
Dry film thickness		

206F13.EPS

Figure 13 ◆ Example of a quality control record.

Instructor's Notes:

Figure 14 ◆ Typical manufacturer's product data sheet and bulletin.

- RUNS OR SAGS
- ORANGE PEEL
- OVERSPRAY
- CRATERING
- FISHEYES
- BLUSHING
- WRINKLING AND LIFTING
- BLISTERING
- COBWEBBING
- RUST, PINHOLING, AND PINPOINT RUSTING
- HOLIDAYS AND SPATTER COATING
- FLAKING, PEELING, AND DELAMINATION

Figure 15 ◆ Coating failures that can result from poor workmanship or application techniques.

Audiovisual

Show Transparency 15 (Figure 15). Ask trainees to explain each of the coating failures listed in Figure 15.

Classroom

Discuss the importance of properly specifying coatings.

3.2.0 Specification-Related Problems

The second most frequent cause of coating failures is an improperly specified coating for the specific application requirements (for example, the wrong coating for a given environment or substrate). Poor coating specifications also include those that improperly specify surface preparation, coating thickness, application methods, or other specifications. Each kind of substrate has its own characteristics, including hardness, porosity, flexibility, coefficient of thermal expansion, roughness, pH level, and moisture sensitivity. No single coating system is designed to work over all substrates. An example of an improperly specified coating is one that would call for the use of an alkyd over concrete. Alkyds are essentially oil-modified polyesters and, as such, are sensitive to moisture at high pH. Under such circumstances, the coating will chemically degrade by a process known as saponification.

A common cause of coating failure is the use of coatings made by different manufacturers within a coating system. The coatings may or may not be chemically compatible, resulting in a failure. The best source of information about a particular coating, or about compatibility of different coatings, is the coating manufacturer's product sheets and/or application bulletins.

Poor specifications include those that fail to point out the exposure conditions such as temperature extremes, chemical contamination, unanticipated abrasion, or mechanical damage to which a coating will be subjected. Coatings are developed for use in various exposures and for application on various substrates. Some types of exposure include immersion service, severe outdoor exposure, heavy fume areas, moderate outdoor exposure, and indoor exposure. Each coating's chemical and physical properties and application requirements should be considered when coating systems are specified. Those individuals responsible for specifying coatings must have a thorough understanding of the job requirements and the coating product's performance characteristics and limitations (*Figure 16*).

3.3.0 Coating-Related Problems

Coating failures can result from characteristics inherent in the coating itself (*Figure 17*). Some of these include:

- Chalking
- Color change
- Checking, cracking, alligatoring, and mudcracking
- Manufacturer's defects

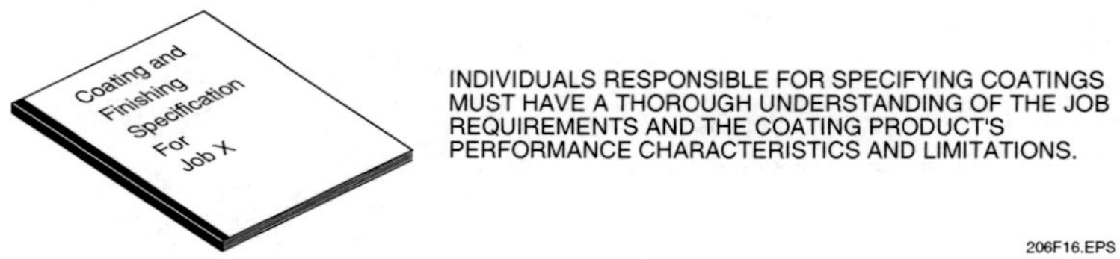

Figure 16 ◆ Improperly specified coatings cause premature coating failures.

Figure 17 ◆ Coating failures can result from properties inherent to the coating material itself.

Instructor's Notes:

3.4.0 Coating Failures on Steel

Steel surfaces are relatively uniform, impervious, hard, durable, and comparatively easy to coat. All contaminants that will interfere with the adhesion of coating must be removed. These include oil, grease, dust, dirt, debris, rust, mill scale, and deteriorated old coating. A bright, shiny, slightly roughened steel surface is ideal for coating. Such a surface can be readily created by blast cleaning or most methods of power tool cleaning. Virtually all coating systems are compatible with such a properly cleaned surface, and if applied properly and in sufficient thickness, will provide long-term corrosion protection and good appearance.

If, on the other hand, contaminants are not sufficiently removed or the coating is not applied thickly enough, rusting, peeling, disbonding, or other forms of coating failure can occur.

Adequate coating coverage requires surface irregularities to be minimized or eliminated by grinding, smoothing, welding, or other methods. Sharp edges should be ground smooth to a radius of at least $\frac{1}{16}$" (or greater, if possible), particularly in immersion or other environmentally severe service. Rough welds should be ground smooth to the touch, if not flush, prior to coating. Protruding weld spatter should be removed. Crevices, such as those at lap seams where coating cannot penetrate and seal, should be caulked or welded to prevent corrosion and rust bleed-out.

Rolling slivers or laminations on steel plates should be removed or spot seal-welded prior to coating application.

The coating should be applied in a smooth, uniform manner with sufficient thickness to ensure that it does not draw thin at edges and protrusions. Some specifications call for a stripe coat to all edges, bolts, welds, and irregularities immediately prior to the full cover coat over the entire surface. This stripe coat builds additional thickness at these areas so that when the coating dries, it is sufficiently thick.

The most common causes of coating failure on steel surfaces are deficient surface preparation and insufficient (or sometimes excessive) coating thickness.

Deficient surface preparation can lead to blistering, rusting, or loss of adhesion when oil, moisture, or soluble salts are not removed before coating. It can also cause delamination when the steel substrate or a prior coating surface is not sufficiently roughened to assure adhesion.

Insufficient coating thickness can lead to pinpoint rusting or to general rusting across a substrate if the coating is uniformly thin. Excessive thickness can lead to solvent entrapment or checking, cracking, or flaking of the coating film.

Another cause of coating failure on steel substrates (or any surface) is improper thinning or mixing, especially when proportioning two or more components of a catalyzed coating is required. When coating is not properly thinned, inducted, mixed, or proportioned, failure can manifest itself as poor adhesion, checking, cracking, pinholing, or other similar defects.

Coatings on steel can fail for other reasons as well, including poor spray application techniques or poor atmospheric conditions during application.

3.5.0 Diagnosing Coating Failures

Analyzing coating problems, like any diagnosis, requires a systematic and organized approach. The process begins by determining the kind of problem, how widespread it is, and what background information is available. It is important to point out that if it is suspected that the problem is a product failure, the coating manufacturer's representative should be notified immediately in order to participate in the evaluation of the failure.

Background information is critical when trying to determine the cause of a failure. This is true both throughout the performance of a job and after the fact. QA/QC inspection documents and records are a good source of background information. Typically, they contain the following types of information:

- The applicable specifications
- The materials used, batch numbers, other applicable data
- The history of the structure
- Coating history (when, by whom, with what)
- Recent coating sequence (cleaning, surface prep, first coat, or second coat)
- Application equipment (brush, roller, spray)
- Weather conditions during application (temperature, relative humidity, or wind)

3.5.1 Guidelines for Analysis Data Gathering

The background information, coupled with an on-site investigation of the problem area, can help categorize the problem as to type. Following this, the problem is investigated in detail to pinpoint the specific cause. Guidelines for obtaining the information needed for fault analysis based on category of fault or conditions are given here. Remember to keep an open mind and be objective.

Classroom

Ask a trainee to describe a steel surface that is in ideal condition for coating. Have trainees identify causes of coating failure on steel surfaces.

Homework

Assign the reading of Sections 3.5.0–3.5.2 and the completion of the Review Questions for the next session.

Materials

Ensure that you have everything required for teaching and testing during this session.

Classroom

Have trainees identify the types of background information that are useful when diagnosing coating failures.

Classroom

Review the guidelines for analysis data gathering. Ask trainees if they have any questions.

Ask a trainee to explain how to take coating samples that properly represent the problem. Have a trainee explain why several samples should be taken.

Teaching Tip

See the Teaching Tip for Sections 3.5.0–3.5.2 at the end of this module.

- *Application problems:*
  - Low film build (make dry-film measurements)
  - Poor hiding (look for holidays)
  - Viscosity differences (look for differences between batches)
  - Determine concentration of bubbles (sparse or widespread)
  - Hardness (hard, soft, spongy)
  - Flow (even, uneven)
  - Color (in line with desired color; consistent throughout)
- *Occurrence of problem:*
  - North, south, east, west; by area, or in all areas
  - Problem in direct sun, constant shade, other
  - Problem in high areas, ground level, other
  - Substrate type (steel, galvanized, concrete, other)
  - Time of day when problem was first noticed
- *Take samples:*
  - Coating on failing areas
  - Coating on nonfailing areas
  - Wet sample of material used
  - Thinner used
  - Photos (including closeups) of both failing and nonfailing areas
- *After application problems (rust):*
  - Determine surface preparation methods used
  - Determine dry-film thickness (adequate coverage)
  - Any identifiable spray patterns (fingering, other)
  - Any inclusions in dry film, such as dirt, oil, dust, other
- *After application problems (blisters):*
  - Break blisters to determine content (liquid, dry, gas)
  - White salts enclosed
  - Any visual pattern to surface
  - Determine vessel contents
  - pH reading of blister liquid
  - Sample of blistered structure
  - Photos of blistered areas
- *After application problems (peeling/delaminating):*
  - Coating beneath (dry, hard, other)
  - Attack by substrate (anything seeping through such as oil, oxides, other)
  - Note any patterns
  - Samples of failing and nonfailing coatings and substrate
- *Other problems:*
  - Color constant or faded and/or drifting
  - Describe the nature of the structure
  - Failing areas versus nonfailing areas
  - Physical characteristics (dry-film thickness, number of coats, condition of substrate, pH, evidence of contamination)
  - Both wet and dry samples of coating and thinner
  - Background information (noticeable history of previous failures or problems)
  - Photos of problem areas

Use of the checkpoints just outlined, in conjunction with the applicable coating manufacturer's product data sheets and application bulletins as a reference, will aid in solving most coating failures at the site level. If it is determined that the problem cannot be solved at the site, laboratory tests and analysis of various samples will be necessary.

3.5.2 Guidelines for Sample Taking

Fault analysis often involves laboratory testing of samples taken at the job site. To obtain a correct analysis, the samples submitted must properly represent the problem. Taking samples relies mainly on common sense. For example, if the problem is one of delamination or poor adhesion, do not just pick up a coating chip that has fallen on the ground and send it to the laboratory. If the laboratory finds contamination on the back of the chip, it is difficult to tell whether the contamination happened before or after the chip touched the ground.

Instead, you should try to find an area where the coating still appears to be intact on the structure but where the adhesion is actually poor. Cutting with a knife is usually enough to disbond such coatings, allowing the backside of the sample chip to represent the actual condition. To prevent contamination from another source, the chip should be placed in a sealed plastic container and labeled.

It is recommended that several samples be taken, not only of failing coatings, but also of nonfailing coatings. In this way, the laboratory can look for differences between the good and bad areas. They may also be able to observe failure-related patterns, such as coating thickness.

Because many coating failures involve delamination or blistering, the evidence of the problem is not always on the disbonded topcoat, but it can also be on the underlying intermediate coat or prime coat. Even if intact chips of primer or intermediate coat cannot be removed, scrapings can almost always be obtained, and they should be placed in a separate plastic container from the disbonding topcoat.

Instructor's Notes:

An intact piece of substrate with the failing coating in place is the best sample. On steel structures, such samples might be available by removing small plates or bolts and sending them to the laboratory.

The location where each sample is taken should be recorded and a photo of the location taken. This not only helps the laboratory but, if a dispute should ever reach the stages of litigation, such documentation can be important in court.

If blistering is the problem, the blisters often contain liquid. If large enough, an attempt should be made to obtain a sample of some of the liquid from inside them. This is because the composition of the liquid often provides the evidence to solve the problem. To take a sample, first clean the coating surface with a clean rag and water. This is necessary so that any grime or other contamination on the surface of the blisters is not transferred to the liquid during sampling. Following this, a small hole should be punctured in the blister with a needle.

A small syringe is then inserted into the blister, and the liquid is removed. Note that the syringe needle itself should not be used to puncture the hole in the blister. If the blisters are very small, several of them may need to be sampled to get enough liquid for analysis (the more liquid the better).

Summary

Pipeline coating is a major component of the pipeline protection system. The quality and integrity of the coating determines its effectiveness in this protective function. Throughout the coating application and curing process, and after completion, inspection and measurements are made to ensure that the final product will meet the necessary specifications.

Many coating variables can affect the quality. These include thickness, viscosity, hardness, and consistency. Problems with coatings include lack of adhesion or cohesion, holes and other breaks, permeability, weakness, and cracking or flaking.

Most coating problems begin with the application process. Other causes are using the incorrect specifications for the job and inherent coating defects. The causes of coating failures must be diagnosed to prevent recurrent problems after repairs. Diagnosis techniques include study of background information, sampling, analyzing, and testing.

Performance Testing

Have each trainee complete Performance Task 1 to your satisfaction. Fill out a Performance Profile Sheet for each trainee.

Classroom

Go over the Review Questions for this module, and answer any questions trainees may have.

Review Questions

1. For large industrial jobs involving many people, it is common to have the quality assurance tasks performed by _____.
   a. company inspectors
   b. contractor inspectors
   c. third-party inspectors
   d. government inspectors

2. The reason for inspections is to ensure _____.
   a. productivity
   b. quality control
   c. contractor reliability
   d. safety

3. The inspection process normally begins with a(n) _____.
   a. pre-job conference
   b. test of materials
   c. instrument survey
   d. measurement of ambient conditions

4. All surface _____ must be removed before starting blast cleaning.
   a. moisture
   b. paint
   c. finish
   d. contaminants

5. Some specifications require that any _____ be ground or otherwise removed.
   a. scratches
   b. weld splatter
   c. cracks
   d. coating

6. Surface preparation and coating work should only be done under _____.
   a. suitable ambient conditions
   b. protective cover
   c. company supervision
   d. temperatures of 50°F or less

7. Temperature readings should be made on the _____ of the surface to be coated.
   a. highest point
   b. lowest point
   c. midsection
   d. coldest point

8. Final blast cleaning should only be done if the surface temperature is a minimum of _____ higher than the temperature of the dew point.
   a. 5°
   b. 8°
   c. 10°
   d. 15°

9. Sling and battery-operated psychrometers use two identical _____.
   a. wet bulbs
   b. wicks
   c. dry bulbs
   d. thermometers

10. Wet-bulb temperatures are a measure of the _____.
    a. sensible heat of the air
    b. cooling properties
    c. evaporation index
    d. air moisture content

11. Sling and battery-operated psychrometers are _____.
    a. factory-calibrated
    b. ASTM certified
    c. uncalibrated
    d. field calibrated

12. _____ can be used to determine when wind speeds exceed specification limits.
    a. Hygrometers
    b. Psychrometers
    c. Anemometers
    d. Thermocouples

13. Moisture contamination can cause _____ in coatings.
    a. adhesive failure
    b. flash rusting
    c. drying problems
    d. tacky films

14. Testing compressed air to detect the presence of oil or water is done in accordance with _____ standards.
    a. ASME
    b. API
    c. ASTM
    d. SSPC

Instructor's Notes:

15. The blast machine mixes a(n) _____ with the air stream.

a. abrasive
b. fluid
c. acid
d. cleaner

16. One of the two major criteria involved with the inspection of a surface is its _____.

a. smoothness
b. temperature
c. thickness
d. cleanliness

17. As a general rule, thick coatings require a _____ surface profile than thin coatings.

a. broader
b. smoother
c. shallower
d. deeper

18. Generally, the warmer the temperature, the _____ the pot life of wet coating.

a. longer
b. shorter
c. better
d. more unstable

19. A common method for measuring coating viscosity involves the use of a _____.

a. comparator
b. densitometer
c. hygrometer
d. Zahn cup

20. A coating that is too _____ may crack, flake, blister, or not cure properly.

a. thin
b. hard
c. thick
d. soft

21. The thickness of the coating being applied should be checked periodically, using a _____.

a. wet-film gauge
b. dry-film gauge
c. micrometer
d. pressure probe

22. After all coats have been applied, _____ testing can be used to find the nicks, scrapes, and pinholes in the coating film.

a. ultrasonic
b. holiday
c. adhesion
d. magnetic

23. When testing conductive coatings applied over steel substrates, _____ testers are generally used.

a. low-voltage
b. wet-sponge
c. AC electrostatic
d. DC high-voltage

24. _____ failure is a failure or break within a given coating or material.

a. Cohesion
b. Adhesion
c. Glue
d. Tensile

25. Hardness can be tested by two methods: pencil testing or by using a _____.

a. tape test
b. pressure probe
c. densitometer
d. durometer

Examination

Administer the Module Examination. Be sure to record the results of the Exam on Craft Training Report Form 200 and submit the results to the Training Program Sponsor.

Performance Testing

Ensure that all Performance Tasks have been completed and Performance Profile Sheets for each trainee are filled out. Be sure to record the results of the testing on Craft Training Report Form 200 and submit the results to the Training Program Sponsor.

GLOSSARY

Glossary

This glossary will help you learn some terms specific to performing coating inspections. Trade Terms introduced in this module appear in italic text.

Dew point: The temperature at which the moisture vapor in the air will condense on a surface.

Dry film thickness: The thickness of a coating after it has completely dried or cured, usually measured in mils (1 mil = 0.001 inch).

Orange peel: Hills and valleys in a coating that resemble the skin of an orange.

Overspray: Dry, flat, pebbly surface resulting from paint particles falling outside the spray pattern.

Quality assurance (QA): The testing and inspection process in which the owner or a third party is responsible for ensuring that the contractor is preforming to specifications.

Quality control (QC): The inspection and testing process in which the contractor is responsible for ensuring throughout the job that specifications are met.

Relative humidity (RH): The ratio (in percent) of the amount of moisture present in a given sample of air to the amount it can hold at saturation.

Sags: Also called runs or curtains, defects caused by excess coating flow.

Surface or anchor profile: A measure of the surface roughness that results from surface blasting, measured from peak to valley depth.

Wet Film Thickness: The thickness of a fresh coating before it dries, usually measured in mils.

Wrinkling: A rough, crinkled surface.

Instructor's Notes:

MODULE 61206-02 — TEACHING TIPS

The following are suggested activities or instructional methods to help you teach the material in this AIG.

General

When you call on someone to answer a question, the rest of the class relaxes or even tunes out because they expect that the question and answer will take place only between you and the trainee you called on. Instead, use this technique to involve more trainees in answering questions and to keep them on their toes.

1. Ask trainees to define a term or explain a concept.
2. After one trainee has answered, ask a trainee seated nearby if the answer is right. Then ask whether a trainee in the back of the room agrees.
3. Ask trainees to explain why they think an answer is right or wrong.
4. Use the session to clear up incorrect ideas and encourage trainees to learn from their mistakes.

Section 2.3.1 *Measuring Relative Humidity and Dew Point*

In this exercise, trainees will learn how to calibrate and use a sling-type and a battery-operated psychrometer. Trainees will need appropriate personal protective equipment, pencils, paper, and their Trainee Guides. You will need to provide sling-type and battery-operated psychrometers, water, a psychrometric chart, ASTM International-certified thermometers, and any other necessary materials and equipment. Alternatively, you may have trainees simulate the use of the psychrometers. Allow 30 to 60 minutes for this exercise.

1. Have trainees verify the accuracy of each type of psychrometer by comparing their indoor-temperature readings with that of an ASTM International-certified thermometer.
2. Ask trainees to take the wet-bulb and dry-bulb measurements using both types of psychrometers.
3. Have trainees use one set of measurements to find the relative humidity and dew point temperature of the air.
4. Discuss the results. If trainees did not arrive at the correct RH and DP, work with them to get the right figures. Ask trainees if they have any questions.

Sections 2.9.0–2.9.4 *Determination of Wet-Film and Dry-Film Thickness*

In this exercise, trainees will practice measuring wet-film and dry-film thickness using various methods. Trainees will need appropriate personal protective equipment, pencils, paper, and their Trainee Guides. You will need to provide necessary testing equipment for each type of test (wet-film and destructive and nondestructive dry-film testing) as well as coated materials. Allow 2 to 3 hours for this exercise.

1. Set up a station for each type of testing. Divide the class into small groups.
2. Explain how to conduct each type of test. Stress any safety requirements.
3. Supervise trainees as each group takes a turn at each station. Have them record their findings.
4. Verify the accuracy of trainees' results. Answer any questions trainees may have.

Sections 2.0.0–2.12.0

Quality Inspections

This exercise will familiarize trainees with the entire coating inspection process. Trainees will need pencils, paper, and their Trainee Guides. You will need to describe a scenario in which a coating is applied and provide manufacturer's data sheets and application bulletins as well as blank quality control/assurance documents. Allow 1 to 2 hours for this exercise.

1. Describe a scenario in which a coating is applied; provide as much information as is needed, including manufacturer's data sheets and application bulletins.
2. Have trainees take turns explaining in detail each step in the coating inspection process. Encourage trainees to ask each presenter questions.
3. Have trainees practice filling out the required documentation using the data sheets and application bulletins. Alternatively, have trainees review and discuss completed sample quality control records.
4. Answer any questions trainees may have.

Sections 3.5.0–3.5.2

Diagnosing Coating Failures

This exercise will introduce trainees to the systematic method of diagnosing coating failure. Trainees will need appropriate personal protective equipment, pencils, paper, and their Trainee Guides. You will need to provide some examples of a coating failure and any necessary equipment, including manufacturer's data sheet/application bulletins. Allow 1 to 2 hours for this exercise.

1. Review the guidelines for sample taking. Discuss the importance of being systematic and organized when diagnosing coating failure.
2. Set up stations at which you have placed at least one example of coating failure. Provide the following:
   - coating manufacturer's product data sheets
   - coating manufacturer's application bulletins
   - a written explanation of any special circumstances or information needed to determine the coating failure
3. Divide the class into small groups. Assign each group to a station.
4. Supervise each group as trainees follow the guidelines in their Trainee Guides to determine the source of the failure. Have them take notes on their observations and write up a summary of their final determination and the evidence for it.
5. Have groups make presentations to the class in which they describe the coating failure and explain how they determined its probable cause.

MODULE 61206-02 — ANSWERS TO REVIEW QUESTIONS

Answer	Section
1. c	1.0.0
2. b	2.1.0
3. a	2.1.0
4. d	2.2.0
5. b	2.2.0
6. a	2.3.0
7. d	2.3.0
8. a	2.3.0
9. d	2.3.1
10. d	2.3.1
11. a	2.3.1
12. c	2.3.3
13. b	2.4.1
14. c	2.4.1
15. a	2.4.2
16. d	2.5.0
17. d	2.5.3
18. b	2.6.0
19. d	2.6.2
20. c	2.9.0
21. a	2.9.1
22. b	2.10.0
23. c	2.10.0
24. a	2.11.1
25. d	2.11.3

CONTREN™ LEARNING SERIES — USER UPDATES

The NCCER makes every effort to keep these textbooks up-to-date and free of technical errors. We appreciate your help in this process. If you have an idea for improving this textbook, or if you find an error, a typographical mistake, or an inaccuracy in NCCER's Contren™ textbooks, please write us, using this form or a photocopy. Be sure to include the exact module number, page number, a detailed description, and the correction, if applicable. Your input will be brought to the attention of the Technical Review Committee. Thank you for your assistance.

Instructors – If you found that additional materials were necessary in order to teach this module effectively, please let us know so that we may include them in the Equipment/Materials list in the Instructor's Guide.

Write: Curriculum Revision and Development Department
National Center for Construction Education and Research
P.O. Box 141104, Gainesville, FL 32614-1104

Fax: 352-334-0932

E-mail: curriculum@nccer.org

Craft ______ Module Name ______

Copyright Date ______ Module Number ______ Page Number(s) ______

Description

______

______

______

(Optional) Correction

______

______

(Optional) Your Name and Address

______

______

Module 61207-02

Perform High-Pressure Blasting/Surface Preparation

(CT 7.4 and 13.3)

Perform High-Pressure Blasting/Surface Preparation Annotated Instructor's Guide

Module 61207-02

MODULE OVERVIEW

This module introduces trainees to basic abrasive blast system equipment. Trainees will learn how to prepare the surface, choose the appropriate abrasive system, and operate the blast system.

PREREQUISITES

Please refer to the Course Map in the Trainee Module. Prior to training with this module, it is recommended that the trainee shall have successfully completed the following:

Core Curriculum; Pipeline Core, Modules 66101-02 and 66102-02; Pipeline Corrosion Control Level One; Pipeline Corrosion Control Level Two, Modules 61201-02 through 61206-02

OBJECTIVES

Upon completion of this module, the trainee will be able to:

1. Identify and explain basic abrasive blast system equipment.
2. Effectively and safely operate basic abrasive blast system equipment (CT 7.4 and CT 13.3).
3. Describe the characteristics of blast cleaning media (CT 7.4 and CT 13.3).
4. Explain preparation standards, profiling, and inspection techniques (CT 7.4 and CT 13.3).
5. Prepare the surface according to manufacturer's specifications and in accordance to company safety and environmental guidelines (CT 7.4 and CT 13.3).

PERFORMANCE TASKS

Under the supervision of the instructor, the trainee should be able to:

1. Prepare the surface in accordance with appropriate specifications and guidelines (CT 7.4 and 13.3).
2. Operate power-cleaning tools in a safe and effective manner (CT 7.4).
3. Visually inspect or repair defects (CT 13.3).

NCCER STANDARDIZED CRAFT TRAINING PROGRAM

The National Center for Construction Education and Research (NCCER) provides a standardized national program of accredited craft training. Key features of the program include instructor certification, competency-based training, and performance testing. The program provides trainees, instructors, and companies with a standard form of recognition through a National Craft Training Registry. The program is described in full in the *Guidelines for Accreditation*, published by the NCCER. For more information on standardized craft training, contact the NCCER by writing us at P.O. Box 141104, Gainesville, FL 32614-1104; calling 352-334-0911; or e-mailing info@nccer.org. More information may be found at our Web site, www.nccer.org.

HOW TO USE THIS ANNOTATED INSTRUCTOR'S GUIDE

Each page presents two sections of information. The larger section displays each page exactly as it appears in the Trainee Module. The narrow column ties suggested trainee and instructor actions to each page and provides icons to call your attention to material, safety, audiovisual, or testing requirements. The bottom of each page includes space for your notes.

Teaching Tip

If you see the Teaching Tip icon, that means there is a teaching tip associated with this section. Also refer to any suggested teaching tips at the end of the module.

SAFETY CONSIDERATIONS

Ensure that the trainees are equipped with appropriate personal protective equipment. Emphasize the importance of following all safety precautions and procedures when working with power tools.

PREPARATION

Before teaching this module, you should review the Module Outline, Objectives, Performance Tasks, and the Materials and Equipment List. Be sure to allow ample time to prepare your own training or lesson plan and gather all required equipment and materials.

MATERIALS AND EQUIPMENT LIST

Overhead projector and screen

Whiteboard/chalkboard

Transparencies

Markers/chalk

Blank acetate sheets

Transparency pens

Pencils and scratch paper

Module Examinations*

Performance Profile Sheets*

Copies of the Performance Verifications for Covered Tasks 7 and 13**

Appropriate personal protective equipment

Copies of your company's policies and procedures manual

Whip checks, hoses, couplings, and coupling gaskets

Photo or drawing of a National Board label on a pressurized tank blast machine

Samples of these types of abrasives: metallic, mineral, by-product, and special-purpose

Copies of *SSPC-VIS 1* photographs

Profile comparator, profile gauge, and replica tape

Materials and equipment necessary for blast system start-up, operation, and shutdown, including manufacturer's operation and instruction manual

* Located in the Test Booklet.

**Available for download from www.nccer.org.

TEACHING TIME FOR THIS MODULE

An outline for use in developing your lesson plan is presented below. Note that each Roman numeral in the outline equates to one session of instruction. Each session has a suggested time period of 3 hours. This includes 10 minutes at the beginning of each session for administrative tasks and one 10-minute break during the session. Approximately 18 hours are suggested to cover *Perform High-Pressure Blasting/Surface Preparation (CT 7.4 and 13.3).* You will need to adjust the time required for hands-on activity and testing based on your class size and resources.

Topic	Planned Time
Session I. Basic Abrasive Blast System Equipment, Part One	
A. Air Compressors	______
B. Compressor Air Pipes/Hoses and Couplings	______
C. Blast Machines	______
Session II. Basic Abrasive Blast System Equipment, Part Two	
A. Blast Pipes/Hoses and Couplings	______
B. Air/Abrasive Consumption and System Sizing	______
1. Laboratory	______
C. Operator Protective Equipment	______
D. Portable Centrifugal Blast Cleaning Equipment	______
Session III. Blast System Accessories	
A. Wet Blast Cleaning Equipment	______
B. Spin Blast Nozzles	______
C. Abrasive Loading/Recovery Equipment	______
D. Closed-Circuit Abrasive Recovery Tools	______
E. Abrasive Cleaning/Recycling Systems	______
F. Blast Rooms and Blast Cabinets	______
Session IV. Blast Cleaning Media	
A. Media Types	______
1. Laboratory	______
B. Productivity and Characteristics of Media	______
C. Media Specifications	______
Session V. Preparation Standards, Profiling, and Inspection Techniques	
A. Blast Surface Cleanliness Standards and Production Rates	______
B. Blast Surface Profiling and Measurement	______
1. Laboratory	______
2. Performance Testing	______
C. Containment and Temperature/Humidity Control	______
D. Safety Guidelines	______
E. Blast System Start-Up, Operation, and Shutdown Guidelines	______
1. Performance Testing	______
Session VI. Chemical Strippers	
A. Pickling	______
B. High- and Ultra-High-Pressure Water Jetting	______
C. Environmental Humidity/Temperature Control	______
D. Summary	
1. Summarize module	______
2. Answer questions	______

E. Module Examination
   1. Trainees must score 70% or higher to receive recognition from the NCCER.
   2. Record the testing results on Craft Training Report Form 200 and submit the results to the Training Program Sponsor.

F. Performance Testing
   1. Trainees must perform each task to the satisfaction of the instructor to receive recognition from the NCCER.
   2. Record the testing results on Craft Training Report Form 200 and submit the results to the Training Program Sponsor.

Module 61207-02

Perform High-Pressure Blasting/Surface Preparation

(CT 7.4 and 13.3)

Instructor's Notes:

Instructor's Notes:

Introduction to the Pipeline Operator Qualification Rule, API, and NCCER

It could be said that petroleum products keep America running. Petroleum products like oil and natural gas are the principal sources of energy in the United States and are used in many manufacturing and chemical processes. Petroleum products are primarily transported by pipeline. There are more than a million miles of liquid and natural gas pipeline in this country, ranging in diameter from two inches to four feet. These pipelines bring crude oil and natural gas from production fields to processing plants and refineries—and ultimately to the consumer.

Pipeline Operator Qualification

Petroleum products are considered hazardous, so pipeline personnel must take extra precautions and follow specific procedures when working with them. To help ensure the safe handling and transportation of petroleum products, the United States Department of Transportation (DOT) mandates safe practices as federal regulations.

One such regulation, the Operator Qualification (OQ) rule, requires pipeline operators to develop a written qualification plan to evaluate their personnel and their contract personnel (see *49 CFR Part 192 Subpart N—Gas*, and *49 CFR Part 195 Subpart G—Liquid*). The evaluation hinges on the operator's ability to perform certain covered tasks and to be able to recognize and respond to abnormal operating conditions that may be encountered while performing these covered tasks.

Covered Tasks

A pipeline performance task is deemed *covered* if the activity passes the following four-part test:

1. Is it performed on a pipeline facility?
2. Is it an operations or maintenance task?
3. Is it performed as a requirement of *49 CFR Part 195 (Liquid)* or *49 CFR Part 192 (Gas)*?
4. Does it affect the operation or integrity of the pipeline facility?

If the answer to all four questions is *yes*, then the activity is considered *covered* and must figure into the evaluation process of pipeline personnel.

American Petroleum Institute

The OQ rule and its related covered tasks provided the catalyst to develop a unified pipeline training and assessment program—so that pipeline operators and owners have ready access to the assessment and training tools they need. This effort was the direct result of the vision and leadership provided by The American Petroleum Institute (API) and API members.

Efforts to unify the industry are nothing new to API. Indeed, the impetus for forming the Institute in 1919 was the need to standardize engineering specifications for drilling and production equipment. Today, the most pressing issues revolve around public perceptions and government policies toward the industry. Speaking with one voice on these issues has become as essential as having interchangeable parts in the field. API's members determine the petroleum industry's positions on public policy as well as the standards that govern its day-to-day operations worldwide.

Consortium on Operator Qualification

API assembled a focus group whose primary purpose was to identify covered tasks and provide compliance recommendations to pipeline operators.

The Consortium on Operator Qualification (COOQ) became the guiding force for the identification and development of covered tasks for the liquid pipeline industry. Their work product, API publication 1161, *Guidance Document for the Qualification of Liquid Pipeline Personnel*, formed the basis for the resulting Pipeliner Training and Assessment Program.

The covered task list created in API publication 1161 has now evolved to include gas covered tasks and is maintained by NCCER. The most current covered task list can be downloaded at www.nccer.org.

National Center for Construction Education and Research

With the guidance and support of API, its members, and the Consortium on Operator Qualification, the National Center for Construction Education and Research (NCCER) was able to design and implement the Pipeliner Training and Assessment Program.

NCCER is a not-for-profit 501(c)(3) education foundation established in 1995 by the world's largest and most progressive construction and maintenance contractors and national contractor associations. NCCER was founded to address the

severe workforce shortage facing the construction and maintenance industries and to develop a standardized training process and curricula. Today, NCCER is supported by hundreds of leading construction and maintenance companies, manufacturers, and national associations and developed more than 35 construction and maintenance titles.

NCCER also maintains a National Registry, which contains the records of trainees and instructors who have received NCCER training. All completion information for every trainee is sent to the NCCER and filed within the Registry. The Registry can then confirm training and skills for workers as they move from company to company, state to state, or even within their own company.

Standardized Curriculum (NCCER's Contren™ Learning Series)

NCCER curricula are developed using the expertise of industry subject matter experts (SMEs). The use of SMEs as an integral part of every new development or revision ensures that the curriculum is current, industry-driven, and indicative of industry practices. The development and/or revision process is facilitated from start to finish by NCCER project managers, who work closely with the SMEs and a select team of technical writer(s) on the development and presentation of the curriculum.

Pipeline: Five Training Paths

The pipeline curriculum focuses on essential skills and competencies within five skill areas:

- Corrosion Control
- Electrical and Instrumentation
- Maintenance
- Mechanical
- Operations (Field and Control Center)

SMEs from the pipeline industry have determined the skills and competencies needed based on API's covered tasks and the consensus of the group/industry. Specific company information is generally not incorporated into the curriculum unless the SMEs determine that the information is the industry standard.

Training Modules and Covered Tasks

The smallest unit of instruction in the curriculum is a *module*. Each curriculum is composed of modules (the equivalent of stand-alone chapters) that are logically sequenced into complete *levels* of instruction. Every module is task-oriented and requires the successful completion of a written (knowledge-based) examination and a performance (hands-on) test. Modules may be individually selected and/or combined for efficient task training. Modules in the pipeline curricula, to the extent possible, will align with API-developed covered tasks. While not all modules will relate to covered tasks, all covered tasks will relate to at least one module.

Benefits of Training to the Pipeline Marketplace

Applying NCCER's standardized training process allows pipeline owners and contractors in the United States to establish the benchmark for acceptable skill and competency levels within the pipeline industry. This process will assist training organizations in developing and/or improving standardized pipeline training programs that effectively match their resources and operating conditions while responding to regulatory demands and to the needs of the industry. The process will also recognize training organizations that meet the minimum standards of competence, warranting regulatory, public, and professional confidence. This in turn will foster national unity within the pipeline industry for the mutual benefit of individual technicians, owners, contractors, and ultimately pipeline-users and society.

Additional Resources

To review the Pipeliner Training and Assessment Program Implementation and Resource Guide and the latest version of pipeline covered tasks, visit NCCER's Web site at www.nccer.org, or contact:

National Center for Construction Education and Research
PO Box 141104
Gainesville, FL 32614-1104
(888) 622-3720
Web site: www.nccer.org

For more information on API, please contact:

American Petroleum Institute
1220 L Street, Northwest
Washington, D.C. 20005
(202) 682-8000
Web site: www.api.org

To order *API Pub 1161*, contact Global Engineering Documents by calling 1-800-854-7149 or visiting their Web site at www.global.his.com.

Additional sources of information on the pipeline operator qualification regulation include:

DOT – Office of Pipeline Safety
www.ops.dot.gov

Transportation Safety Institute
www.tsi.dot.gov

Instructor's Notes:

COURSE MAP

This course map shows all of the modules in the second level of the Pipeline Corrosion Control curriculum. The suggested training order begins at the bottom and proceeds up. Skill levels increase as you advance on the course map. The local Training Program Sponsor may adjust the training order.

PIPELINE CORROSION CONTROL LEVEL TWO

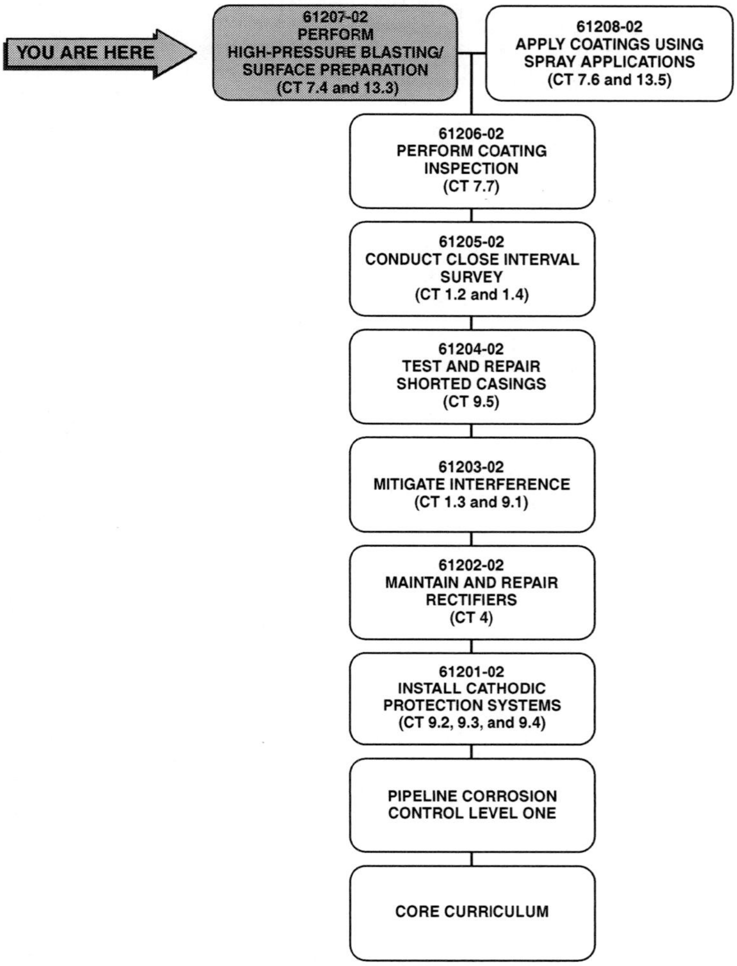

207CMAP.EPS

Homework

Assign reading of Module 61207-02.

Instructor's Notes:

MODULE 61207-02 CONTENTS

MODULE 61207-02 CONTENTS (CONTINUED)

Figures

Tables

Instructor's Notes:

MODULE 61207-02

Perform High-Pressure Blasting/Surface Preparation (CT 7.4 and 13.3)

Objectives

When you have completed this module, you will be able to do the following:

1. Identify and explain basic abrasive blast system equipment.
2. Effectively and safely operate basic abrasive blast system equipment (CT 7.4 and CT 13.3).
3. Describe the characteristics of blast cleaning media (CT 7.4 and CT 13.3).
4. Explain preparation standards, profiling, and inspection techniques (CT 7.4 and CT 13.3).
5. Prepare the surface according to manufacturer's specifications and in accordance to company safety and environmental guidelines (CT 7.4 and CT 13.3).

Prerequisites

Before you begin this module, it is recommended that you successfully complete the following: Core Curriculum; Pipeline Core, Modules 66101-02 and 66102-02; Pipeline Corrosion Control Level One; Pipeline Corrosion Control Level Two, Modules 61201-02 through 61206-02.

Required Trainee Materials

1. Pencil and paper
2. Appropriate personal protective equipment

1.0.0 ◆ INTRODUCTION

Applying the finish coat is the least time-consuming part of a pipeline coating job. The most important part, and perhaps the most time consuming, is the surface/substrate preparation. A coating cannot, and will not, make a bad surface or substrate into a good surface. A finish applied over a poor surface or substrate will not provide a lasting coating. It may temporarily make the surface or substrate look good, but in some cases, it will not even do that.

Because protection is a primary purpose of coating, you should not invest a large amount of money in coating a surface/substrate that is not prepared properly when new, or has deteriorated to the point where it is not worth protecting. A coating is only as good as the surface or substrate to which it is attached.

This module covers three general actions necessary to prepare a surface or substrate for coating:

- Cleaning the surface/substrate
- Repairing the surface/substrate (if necessary)
- Pretreating the surface/substrate (if necessary)

All three procedures may be performed, or only one or two may need to be done. They may or may not need to be repeated. The surface and its condition determine which of the three actions should be used, the sequence in which they should be done, and whether they should be repeated. The desired end result is to obtain a surface or substrate that is clean, dry, and dull.

Cleaning and preparing of surfaces/substrates are defined in a set of standards developed by The Society for Protective Coatings (**SSPC**) and **NACE** International. These standards are referenced by coating manufacturers or in job specifications developed by engineering or architectural firms as universally defined preparation methods and requirements for various coating applications on steel.

SSPC-SP2, Hand Tool Cleaning; SSPC-SP3 and *SSPC-11, Power Tool Cleaning;* and *SSPC-SP13/NACE 6, Preparation of Concrete Surfaces* are covered in this module.

Materials

Ensure that you have everything required to teach the course. Check the Materials and Equipment List at the front of this module.

Classroom

Explain that terms shown in bold are defined in the Glossary at the back of this module.

Audiovisual

Show Transparency 1, Course Objectives.

Show Transparency 2, Performance Tasks.

Teaching Tip

See the general Teaching Tip at the end of this module.

Classroom

Ask a trainee to identify the most important part of a pipeline coating job and explain why it is important.

Have trainees name the two groups that have defined standards for the cleaning and preparation of surfaces/substrates.

Audiovisual

Show Transparency 3 (Figure 1). Ask trainees to point out the six major components of the basic blast system.

2.0.0 ◆ BASIC ABRASIVE BLAST SYSTEM EQUIPMENT

Essential basic blast system equipment (*Figure 1*) used in **abrasive blast cleaning** consists of the following components:

- A large-capacity air compressor to furnish clean, dry, high-volume compressed air
- A large air pipe/hose to convey the compressed air to the equipment
- A blast machine to mix the compressed air and abrasive
- A large blast pipe/hose to convey the air-abrasive mixture to the work area
- A blast nozzle to direct the air-abrasive mixture
- Operator personal protective equipment (PPE)

The blast machine stores an adequate quantity of abrasive and mixes it with the compressed air in the correct proportions. The blast nozzle delivers the mixture at speeds up to 500 miles per hour (mph). These nozzles are equipped with on/off remote safety controls in case of a problem. The operator must be fully protected from the flying particles and the dust-laden air. Either a breathing air compressor or a breathing air pump with a filter and carbon monoxide detector is required.

To operate blast system equipment, the operator aims the nozzle at the surface to be cleaned or etched, depending on the abrasive. The abrasive cuts into the substrate surface, loosening and blowing away contaminants such as old coating, dirt, oil, and oxidation. Depending on the hardness and the impact of the abrasive, a portion of the substrate surface may be removed as well.

To the average person, abrasive blast cleaning appears to be a simple operation with reasonably uncomplicated equipment. However, every component of abrasive blast system equipment must be specifically designed and selected to carry out its function. Each equipment component is important to the overall operation and safety of the system. Neglecting to learn about the strengths and limitations of the equipment can severely affect the selection and operation of the equipment. This can result in hampered blast cleaning operations or hazards to the operator(s). The following paragraphs discuss each of these typical equipment components.

2.1.0 Air Compressors

For compressed air blast systems, the air compressor is the most critical component of the system. Productivity is directly linked to the pressure and volume of air that is available to the blast nozzle of the system.

2.2.0 Compressor Air Pipes/Hoses and Couplings

Compressor air outlets and air hoses (sometimes called *bull hoses*) used for blast cleaning should be equipped with large, externally coupled fittings (*Figure 2*) to reduce friction, turbulence, and wear (*Figure 3*). To prevent pressure drop from friction, the inside diameter (ID) of the hoses and couplings should be three times, for less than 50' or 15 meters (m), and four times, for over 50' or 15m, the sum total of all the blast nozzle orifice diameters that will be used at one time. If that is not possible, then

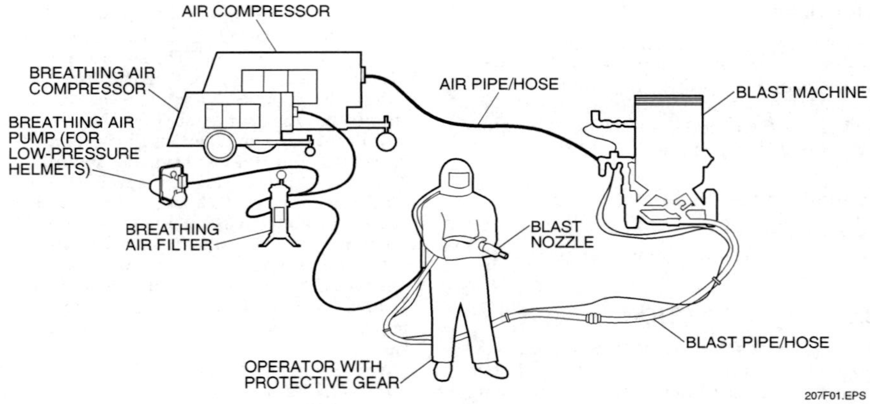

Figure 1 ◆ Basic blast system equipment.

Instructor's Notes:

the output of the compressor should be from a manifold equipped with multiple smaller outlets and hoses with inside diameters whose sum total equals this calculation. The smaller, parallel hoses are then run to the blast machine, where they can be connected to a similar manifold on the input of the blast machine. The compressor outlet or the outputs of multiple outlet manifolds should be equipped with check valves to prevent accidental backflow of air/abrasive from the blast machine. Most standard compressors are equipped with internally coupled (quick-disconnect) fittings, which should be replaced with the external type unless their inside diameters meet the requirements mentioned. Remember, you cannot use too large an air line. Large lines reduce friction loss. For every 1-pound loss in air pressure, a 1½% loss in productivity occurs. In all cases, the pressure at the blast machine should be checked with all nozzles operating. If the pressure drop to the blast machines is below the range necessary to obtain proper pressure at the nozzles, then the diameter of the air line will have to be increased.

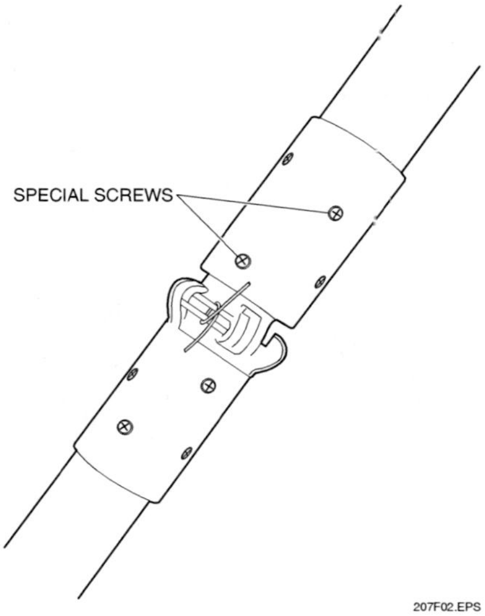

Figure 2 ◆ Typical external-type coupling for hoses.

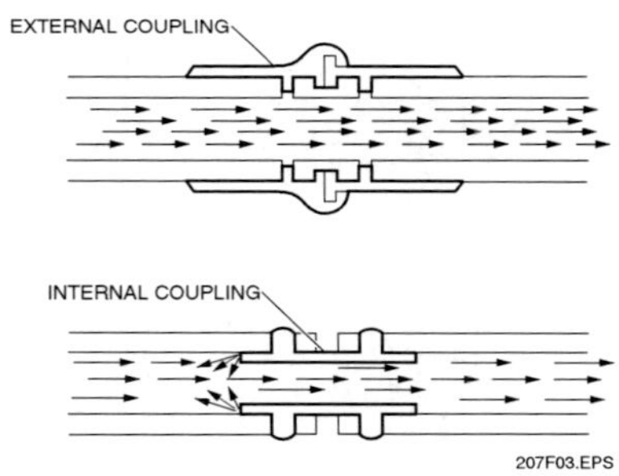

Figure 3 ◆ Turbulence, friction, and wear caused by internal couplings.

Compressors should be placed as close to the blast machine as possible, and the air hoses should be run with as few bends as possible. However, if a choice must be made between a long blast hose or a long air hose, use a long air hose. Try to place the air hoses so they will not be damaged or run over. If they must cross a high-traffic area, try to suspend them overhead using cables and slings along the length of the hose and at the couplings for support, or bury them in a shallow trench with an appropriate hard cover, or cover them with earth. For vertical runs, hoses should be supported by rope with four or five wraps around the hose below each coupling connection.

Air hoses and couplings should have a working pressure rating that meets or exceeds the maximum output pressure rating of the compressor. The bursting strength rating should be about four times the working pressure rating. Couplings should have safety pins or wire to prevent them from separating accidentally. In addition, a **whip check** (safety cable) should be used to connect hoses near the coupling to keep the hoses from flying around if the joint breaks. The whip checks, hoses, couplings, and coupling gaskets should be checked daily for signs of damage or leaks as follows:

Step 1 Inspect the hose for damage or thin/soft spots caused by wear. Thin/soft spots can be detected by using your fingers to squeeze the unpressurized hose along its entire length.

Step 2 Ensure that the hose ends are cut square and inserted (screwed) completely into the couplings and are flush against the internal coupling collar.

Step 3 Ensure that when securing the hose, only the correct size of special screws is used in the couplings and that the screws are not penetrating through the hose (see *Figure 2*).

Classroom

Ask a trainee to identify factors that can affect productivity of compressed air blast systems.

Audiovisual

Show Transparencies 4 and 5 (Figures 2 and 3). Ask a trainee to explain how externally coupled fittings reduce friction, turbulence, and wear.

Classroom

Describe various compressor installation scenarios, and have trainees discuss any special considerations for each.

Teaching Tip

See the Teaching Tip for Section 2.2.0 at the end of this module.

Audiovisual

Show Transparencies 6 through 8 (Figures 4 through 6). Have trainees identify each blast machine and explain how it works.

Step 4 Ensure that whip checks are securely installed around the hose, not the coupling.

Step 5 After the hoses are connected and pressurized, ensure that the couplings are safety-wired or pinned and that no leaks can be detected.

In fixed-site or long-term field locations, metal or plastic pipe can and should be used to provide the largest part of the air path to the blast machine. Pipe is subject to fewer failures than hose and has much less friction resistance, which results in less pressure drop. However, more extensive support and protection or covering must be provided for pipe than for hose to prevent damage caused by traffic or other operations in the area.

2.3.0 Blast Machines

There are two types of blast machines (sometimes called **blast pots**) in use today. One, which is basically a gravity-flow machine (*Figures 4, 5, and 6*), uses equal air pressure on top of and underneath the abrasive inside a pressurized tank. In this type, the abrasive literally falls through a metering valve at the bottom of the tank into the blast hose where the air stream carries the abrasive to the nozzle. The other type of machine employs one of two forms of pressure feed. One form is a jet feed blowing abrasive into a standpipe and out through an elbow into the blast hose.

The other form, used on very small machines, is a siphon feed in which a vacuum (suction pressure) carries the abrasive from a nonpressurized tank up a separate feed hose to an injector at the nozzle. Nearly all the machines used for commercial work today are pressurized-tank, gravity-feed-type blast machines.

Like air compressors, pressurized-tank blast machines must also have a National Board label that certifies the machine meets the American Society of Mechanical Engineers (ASME) code. If it does not, do not use the machine.

Commercial pressurized blast machines are available in a variety of capacities with blast times (before refill) of ½ to 1 hour and with media delivery rates of 150 pounds (lbs.) (68 kilograms [kgs]) to well over 3,000 lbs. (1,361 kgs) per hour. Since the variety of media used today varies greatly in

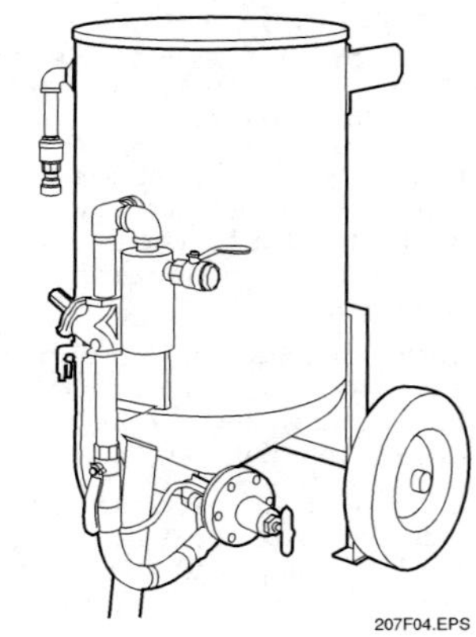

Figure 4 ◆ Typical portable, high-production, pressurized blast machine.

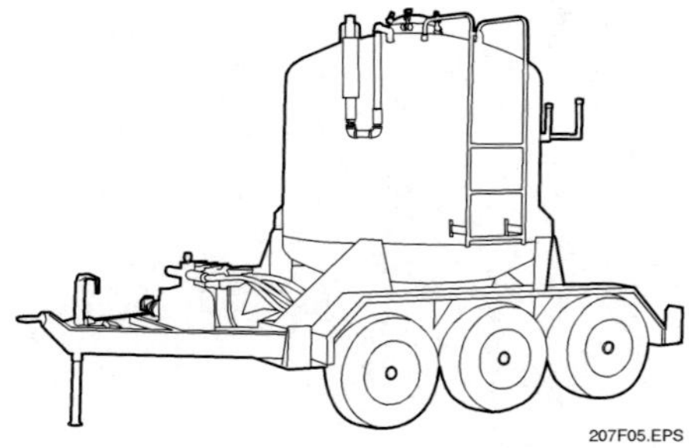

Figure 5 ◆ Typical bulk-abrasive, towable, single- or dual-tank, pressurized blast machine.

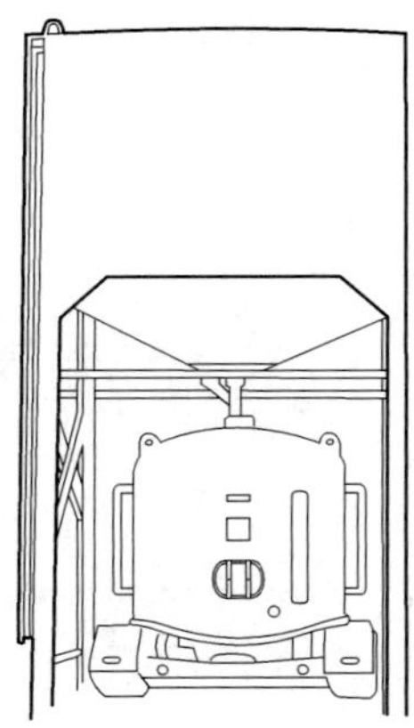

Figure 6 ◆ Typical bulk-abrasive, portable/stationary, single- or dual-tank, pressurized blast machine with separate overhead hopper.

Instructor's Notes:

weight for a given volume, most blast machines are now rated for storage and/or flow of media by volume rather than by the weight of the media. They are generally classified as light-duty, medium-duty, high-production, or bulk-abrasive machines.

Light-duty machines have ½ to 1 cubic foot (0.01 to 0.02 m^3) of abrasive storage and the same cleaning power as larger units. Equipped with small piping, short blast hoses, and a straight-bore nozzle, they are ideal for spot cleaning and small jobs.

Medium-duty machines have 1½ to 4 cubic feet (0.04 to 0.11 m^3) of abrasive storage. These machines offer pickup-truck portability. They usually have 1", 25-millimeter (mm), piping and blast hoses suitable for use with $\frac{3}{16}$" through $\frac{5}{16}$" (5 through 8 mm) venturi nozzles. They are useful for jobs that take an hour or two to complete.

High-production machines have 6 to 20 cubic feet (0.17 to 0.57 m^3) of abrasive storage. These machines are available as stationary units and as two- to four-wheel portable units. They are usually equipped with 1¼" or 1½" (32 or 38 mm) piping and blast hoses suitable for use with ⅜" to ½" (10 to 12.5 mm) venturi nozzles. These machines are the most popular and are suitable for use in extended blast cleaning operations.

Bulk-abrasive machines have 60 to 800 cubic feet (1.7 to 23 m^3) of abrasive storage and operate the same as high-production machines but they have more storage capacity. These machines are large and are available as skid-mounted stationary or two-, four-, six-, or eight-wheel towable units. Most are rated to be towed when filled with abrasive that weighs less than 100 pounds per cubic foot (1.5 kg/liter). They can be equipped to use multiple nozzles up to ¾" (19 mm) inside diameter. Some are dual-tank units that allow continuous blast cleaning. They all demand enormous quantities of air along with large, efficient air drying and air storage equipment, especially the dual-tank units. They must be centrally located at the work site and, as a result, the blast hoses must be longer and larger to eliminate friction losses. The advantage of these units is the lower cost of bulk abrasive versus bagged abrasive and the elimination of hand loading. These units are all loaded by external overhead hoppers or trailer level hoppers using blowers or by vacuum-loading or mechanical conveyor systems. The hoppers are filled directly from trucks using blowers or some other filling method. In some cases, the versatility of a number of smaller machines equipped with hoppers and transfer equipment may be preferable. Each smaller machine can be positioned closer to its work area, thus reducing the blast hose length. Each machine can also be shut down for filling or maintenance without affecting the operation of other machines.

The size of the machine selected is determined by how many nozzles will be used, the size of the nozzles, the individual work-site conditions and requirements, and by the size and capacity of the air supply.

Figure 7 illustrates the various components of a typical high-production, pressurized, single-tank blast machine with one nozzle. They are described in detail in the following paragraphs. Larger machines use essentially the same types of components.

2.4.0 Blast Pipes/Hoses and Couplings

Like compressor air hoses and air hose couplings, blast hoses and blast hose couplings must be sized with an internal diameter that is at least three or four times larger than the orifice size of the nozzle to be used.

Metallic abrasive, such as steel or iron grit, can provide the exception to the three or four times size rule for blast hose. Steel grit is about 2½ times heavier than common nonmetallic abrasive. When blast hoses exceed 50' (15 m), some contractors use slightly smaller hoses to keep the heavy abrasive moving at a steady speed. To maintain pressure at the nozzle with the same size orifice and a smaller hose, the compressor and blast machine may have to be operated at higher pressures to allow for friction and restriction losses in the blast hose. This, in turn, will usually require a larger capacity compressor.

The couplings should be of the external type, equipped with safety pins or wire to prevent the couplings from coming apart during blast cleaning operations. Blast hoses should be run in as straight a line as possible, avoiding all radical bends. If they must be curved around an object, use long-radius curves. Sharp curves cause rapid wear because of the abrasive flow and will result in a blowout at the bend. Try to place the air hoses so they will not be damaged or run over. If they must cross a high-traffic area, try to suspend them overhead, using safety cables and slings at each coupling for support, or bury them in a shallow trench with an appropriate hard cover, or cover them with earth. For vertical runs, hoses should be supported by rope with four or five wraps around the hose below each coupling.

Of all the components in a blast system, the blast hose is subject to most of the abuse and is the item that requires replacement most often. To obtain the maximum life from a blast hose, purchase only the

Demonstration

Show trainees a photo or drawing of a National Board label on a pressurized-tank blast machine. Stress the importance of using a machine that has a National Board label.

Classroom

Have trainees list the four classifications of blast machines on a blank sheet of paper. Ask trainees to discuss each type of blast machine and list important points about each on their papers.

Audiovisual

Show Transparency 9 (Figure 7). Point to each component and ask trainees to identify each.

Homework

Assign the reading of Sections 2.4.0–2.7.0 for the next session.

Materials

Ensure that you have everything required for teaching this session.

Classroom

Ask a trainee to explain the sizing requirements for blast hoses and blast hose couplings. Have a trainee explain the exception to this rule.

Classroom

Have trainees compare and contrast the three types of high-quality blast hose.

Ask a trainee to explain the internal tube requirements. Have a trainee describe the outer covering of a blast hose.

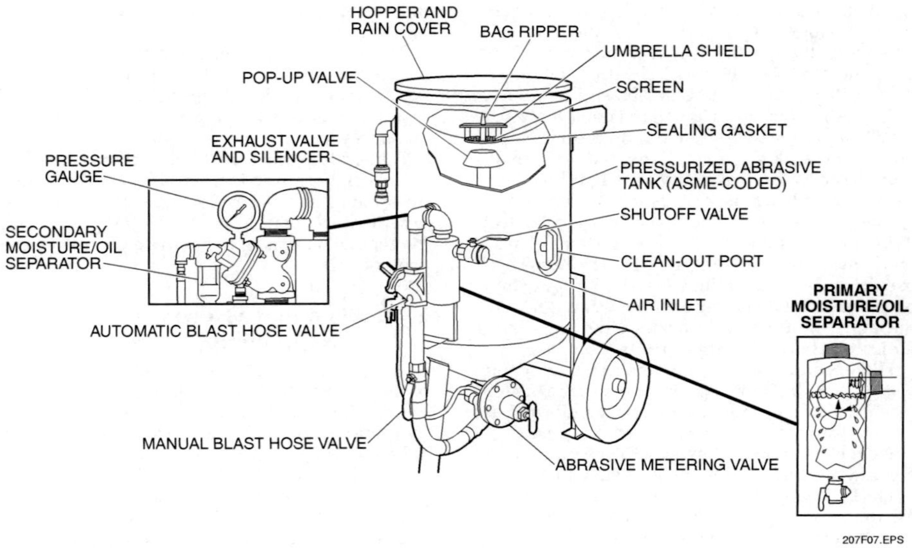

Figure 7 ◆ Typical components of a single-tank, high-production, pressurized blast machine.

highest quality available and take care of it. High-quality blast hose is available in two-ply, two-braid, and four-ply versions and is manufactured to tight tolerances for durability and for close fit in couplings.

- *Two-ply hose* – Consists of two layers of semi cross-woven fabric and is usually rated at 170 pounds per square inch (psi), or 1,172 kiloPascals (kPa) working pressure. This is a light, extremely flexible hose sometimes used as a short whip between a heavier duty hose and the nozzle. The inside diameter of a whip hose should be the same as the heavier duty hose to which it is attached. This type of hose will not withstand abuse as well as two-braid or four-ply hoses.
- *Two-braid hose* – Consists of two layers of cross-woven fabric and is usually rated at 160 to 170 psi (1,103 to 1,172 kPa) working pressure. This hose is moderately flexible with enough outer support to keep the hose round. It is almost as rugged as four-ply hose.
- *Four-ply hose* – Consists of four layers of straight-woven fabric and is usually rated at 100 to 125 psi (689 to 862 kPa) working pressure. This type of hose is stiff with great exterior durability. It rebounds to a fully round shape and is used in high-traffic areas to withstand the weight of motor vehicles.

The internal tube thickness, where internal wear occurs, should be the same no matter what outer ply construction is used. The internal tube should be long-wearing, natural gum rubber or equivalent that is treated with a static-dissipating compound to prevent static buildup and electrical shock to the operator or electrical sparks from the nozzle to the work surface. Older-style hose was equipped with wire running the length of the hose to prevent static buildup, but the wire had to be grounded. The electric shock by itself does not harm the operator, but it may cause the operator to recoil and fall. Electric sparks, however, could cause an explosion or a flash fire if flammable vapor or dust is present when the sparks occur.

The outer covering of blast hose is covered with thousands of small pin holes to allow entrapped air to escape without causing a bubble between the plies and/or the internal tube. The ends of replacement blast hose should be square and sealed or capped to prevent air, abrasive, or moisture from being forced in between the outer plies and/or the internal tube when the hose is installed in the couplings.

Instructor's Notes:

External-type hose couplings are secured to the hose by special screws that only penetrate into the outer wall of the hose. Pressure inside the hose forces the hose against ribs inside the coupling and provides the seal between the hose and the coupling. External-type hose couplings are quick-connect type couplings with universal jaws that allow the coupling of various sizes of hose. They all have replaceable gaskets so that a perfect air seal can be obtained and maintained. The couplings are available in brass, aluminum, or nylon. The brass couplings will give longer life, but they are easily distorted by weight or misuse. The aluminum couplings are lighter, but they are easily broken. Nylon couplings are much lighter to handle, are not easily deformed or broken, and are the least expensive. The hoses, couplings, and coupling gaskets should be checked daily for signs of damage or leaks. As with air hoses, safety pins or wire should be used to secure the couplings, and whip checks (safety cable) should be used to prevent the hoses from flying around if the joint breaks.

In fixed-site or long-term field locations, metal pipe can and should be used to provide the largest part of the air/abrasive path to the blast nozzle. Metal pipe is subject to less wear and failure than hose and has much less friction resistance resulting in less pressure drop. However, more extensive support and protection or covering must be provided for pipe than for hose to prevent damage due caused by traffic or other operations in the work area.

2.5.0 Air/Abrasive Consumption and System Sizing

Air/abrasive flow through a nozzle is determined by the orifice size and the air pressure at the nozzle along with proper adjustment of the abrasive metering valve. It is vitally important in production blast cleaning that the desired air pressure is available and maintained at the nozzle. This is necessary in order to obtain the desired air/abrasive flow for the orifice size used and to achieve the desired cleaning/profiling results at the desired production rates. Previously, emphasis was placed on selecting adequately sized compressors, blast machines, hoses, and couplings, all to reduce friction losses and to provide adequate air/abrasive volume and pressure at the nozzle. Many users of blast equipment believe they are blast cleaning with the correct pressure because the gauge on the compressor and/or the blast machine is showing the correct pressure. Unfortunately, these gauges do not show the pressure at the nozzle where the work is being done. Without this knowledge, the user does not know whether the blast cleaning is being performed efficiently.

Nozzle pressure should be checked with a hypodermic needle pressure gauge (*Figure 8*) while blast cleaning is being performed. The needle should be carefully inserted through the blast hose into the air/abrasive stream behind the nozzle holder at a 45-degree angle pointing toward the nozzle. With an accurate measure of nozzle pressure, the compressor and/or the blast machine operating pressures can be adjusted up or down as necessary to obtain the desired nozzle pressure (typically 90 to 100 psi). If the pressure drop between the blast machine and nozzle is excessive, check the following immediately:

- Is the compressor functioning properly?
- Is the pressure more than adequate at the blast machine?
- Are the pipes and fittings on the blast machine large enough?
- Is the blast hose large enough and unrestricted by sharp turns?
- Is there a blockage in the system?
- Is the compressor capacity large enough for the project?

When operating with a known nozzle pressure and with media that weighs approximately 100 lbs./ft^3 (1.4 kg/liter), the average air and media consumption for the various commonly used nozzle orifice sizes given in *Tables 1* and *2* can be assumed if the abrasive metering valve is adjusted properly. The abrasive metering valve can be adjusted at the blast machine either manually or, in some cases, by remote control. Most

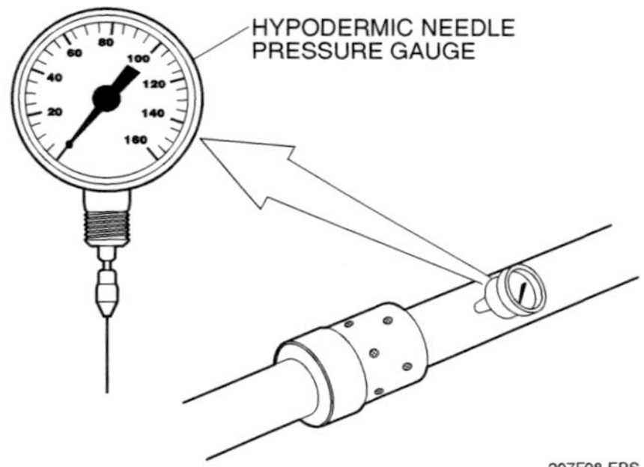

Figure 8 ◆ Hypodermic needle pressure gauge.

Classroom

Ask a trainee to explain why a user of blast equipment cannot rely on the gauge on the compressor or blast machine to determine whether the pressure is correct.

Audiovisual

Show Transparency 10 (Figure 8). Ask a trainee to explain how to check nozzle pressure. Have a trainee explain what to check if the pressure drop between the blast machine and nozzle is excessive.

Classroom

Have a trainee explain how an operator can tell whether the air/abrasive mixture is correct.

Laboratory

Explain how to use Tables 1 and 2 to approximate the required blast system abrasive tank size. Describe example blast systems scenarios and have trainees use the tables to determine required tank size.

Table 1 Typical Nozzle Consumption Rates

Nozzle Orifice	Pressure at the Nozzle							
	50 psi	60 psi	70 psi	80 psi	90 psi	100 psi	125 psi	
No. 2 1/8"	11	13	15	17	18.5	20	25	Air – CFM
	0.7	0.8	0.9	1	1.1	1.2	1.5	Media – CFH
	2.5	3	3.5	4	4.5	5	5.5	Compressor – HP
No. 3 3/16"	26	30	33	38	41	45	55	Air – CFM
	1.5	1.8	2	2.2	2.4	2.6	3.2	Media – CFH
	6	7	8	9	10	10	12	Compressor – HP
No. 4 1/4"	47	54	61	68	74	81	98	Air – CFM
	2.7	3.1	3.5	4.1	4.5	5	6.1	Media – CFH
	11	12	14	16	17	18	22	Compressor – HP
No. 5 5/16"	77	89	101	113	126	137	168	Air – CFM
	4.6	5.3	6	6.7	7.4	8.1	9.8	Media – CFH
	18	20	23	26	28	31	37	Compressor – HP
No. 6 3/8"	108	126	143	161	173	196	237	Air – CFM
	6.7	7.6	8.6	9.6	10.5	11.5	13.9	Media – CFH
	24	28	32	36	39	44	52	Compressor – HP
No. 7 7/16"	147	170	194	217	240	254	314	Air – CFM
	8.9	10.3	11.8	13.1	14.5	15.8	19.3	Media – CFH
	33	38	44	49	54	57	69	Compressor – HP
No. 8 1/2"	195	224	252	280	309	338	408	Air – CFM
	11.6	13.5	15.1	16.8	18.6	20.2	24.6	Media – CFH
	44	50	56	63	69	75	90	Compressor – HP

Based on abrasives with a density of 100 pounds per cubic foot.

This chart provides calculated consumption rates for new nozzles. When selecting a compressor, add 50% to the compressor rating shown to allow for normal nozzle wear, friction loss, etc.

207T01.QXD

operators tend to use too much abrasive, which actually cuts down the speed of blast cleaning, creates excessive dust, and increases the cost for abrasive use, cleanup, and disposal. The air stream from the nozzle should have only a slight discoloration from the abrasive being carried. Usually, this indicates the proper air/abrasive mixture. Experienced operators can hear a steady abrasive flow when the adjustment is correct. Too little abrasive causes a high-pitched sound; too much abrasive causes an erratic pulsating sound.

Given that a blast system will be operated with the desired air pressure present at the nozzle, the nozzle will have the desired orifice size, and that the abrasive flow will be adjusted correctly, the data in *Table 1* or *Table 2* may be used to roughly approximate the required blast system abrasive tank size. This is accomplished by adding up the average volumetric media consumption per hour, usually in cubic feet per hour (CFH) or kg/hr., shown in the appropriate table under the desired nozzle operating pressure for each nozzle orifice size to be used simultaneously and dividing the sum by 60 (minutes). This yields the total volumetric media consumption per minute.

Multiplying this consumption per minute by a desired average blast cleaning time in minutes (between refills) gives a rough approximation of the volumetric size of the abrasive pressure tank required.

Instructor's Notes:

Table 2 Typical Metric Nozzle Consumption Rates

Nozzle Orifice	Pressure at the Nozzle								
	350 kPa	420 kPa	490 kPa	560 kPa	630 kPa	700 kPa	860 kPa	1035 kPa	
5 mm ³⁄₁₆"	0.73	0.84	0.92	1.06	1.15	1.26	1.54	1.82	Air – cubic meters/min.
	85	97	111	123	136	150	180	216	Media – kilograms/hr.
	4.5	5.3	5.6	6.4	7.1	7.5	9	10.8	Compressor – kilowatts
6.5 mm ¼"	1.31	1.51	1.71	1.9	2.08	2.27	2.75	3.22	Air – cubic meters/min.
	152	178	200	231	254	280	336	403	Media – kilograms/hr.
	7.9	9	10.1	11.6	12.4	13.5	16.2	19.4	Compressor – kilowatts
8 mm ⁵⁄₁₆"	2.16	2.5	2.83	3.16	3.53	3.84	4.71	5.57	Air – cubic meters/min.
	265	302	342	380	420	460	552	662	Media – kilograms/hr.
	13.1	15	19.1	20.2	21	22.9	27.5	33	Compressor – kilowatts
9.5 mm ⅜"	3.02	3.53	4.00	4.50	4.85	5.50	6.64	7.79	Air – cubic meters/min.
	378	433	490	544	596	653	784	940	Media – kilograms/hr.
	18	21	24	27	28.9	33	39.6	47.5	Compressor – kilowatts
11 mm ⁷⁄₁₆"	4.12	4.76	5.44	6.09	6.73	7.11	8.80	10.48	Air – cubic meters/min.
	507	585	655	744	820	896	1075	1290	Media – kilograms/hr.
	24.8	28.5	32.6	36.4	40.1	42.4	50.9	61.1	Compressor – kilowatts
12.5 mm ½"	5.46	6.28	7.06	7.85	8.65	9.46	11.46	13.45	Air – cubic meters/min.
	657	756	856	951	1050	1148	1378	1653	Media – kilograms/hr.
	32.6	37.5	42	46.9	51.8	56.3	67.6	81.1	Compressor – kilowatts

Abrasive consumption based on abrasives with a density of 1.5 kg/liter.

This chart provides calculated consumption rates for new nozzles.

When selecting a compressor, add 50% to the compressor rating shown to allow for normal nozzle wear, friction loss, etc.

1035 kPa (150 psi) working pressure in last column relates to blast machine and components specifically built for higher pressures. Never exceed approved working pressure.

207T02.QXD

The compressor air delivery requirements can be approximately determined from the appropriate table by adding up the volumetric air consumption per minute, usually in cubic feet per minute (CFM) or m^3/min., under the desired nozzle pressure for each nozzle orifice size to be used simultaneously, then adding 50% for reserve capacity plus any amounts required for ancillary use. The same approach can also be used to determine the approximate compressor power requirements, horsepower (hp) or kiloWatts (kW).

2.6.0 Operator Protective Equipment

Blast cleaning propels abrasive media to speeds in excess of 400 mph or 644 kilometers per hour (km/hr.), turning paint, rust, contaminants, and part of the substrate and/or the media into microscopic dust particles. To an unprotected, untrained operator, blast cleaning can be a dangerous, even deadly occupation. First, the air/abrasive jet from a blast nozzle can cause serious injury if it strikes unprotected flesh at close range. Second, rebounding abrasive can cause eye injury. Third, inhalation of the dust from blast cleaning operations can cause permanent lung injury and may also be toxic, especially if silica, asbestos, and lead or other

Classroom

Ask trainees to name the potential dangers to an unprotected, untrained operator doing blast cleaning.

Safety

Stress the importance of using replacement parts that are approved by the manufacturer.

Audiovisual

Show Transparency 11 (Figure 9). Ask trainees to identify the various pieces of protective equipment.

Classroom

Ask trainees to compare and contrast centrifugal blast cleaning with other blast cleaning methods. Have trainees name advantages of centrifugal blast cleaning.

Audiovisual

Show Transparency 12 (Figure 10). Ask a trainee to name typical uses for portable blast cleaning machines.

toxic metal particles are present in the dust. For these reasons, an operator working in an average blast cleaning environment must be equipped with safety equipment which has been approved by the National Institute of Occupational Safety and Health (NIOSH), such as that shown in *Figure 9*. This equipment is explained in the following paragraphs.

Operators should read the system equipment owner and operator manuals. They should also read and memorize the warning labels affixed to each component of the blast system from the compressor to the blast nozzle. Each label instructs the operator on precautions required to avoid a specific hazard.

Safety equipment used for abrasive blast cleaning must conform to rigid standards and be certified to meet NIOSH approval.

> **WARNING!**
> Using any replacement parts other than those approved by the manufacturer of the safety equipment voids NIOSH approval and exposes the blast operator to increased risk of injury or death.

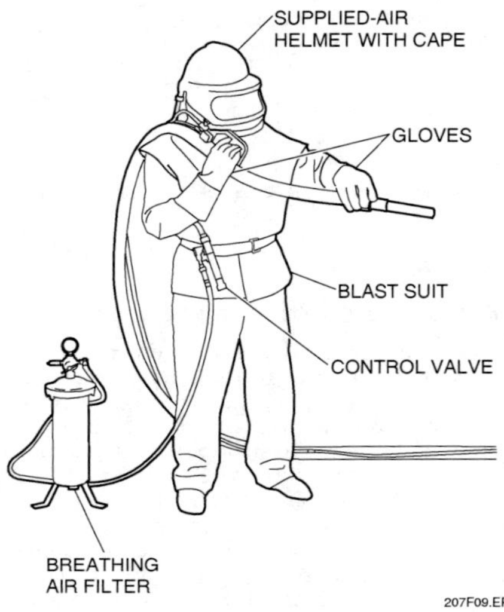

Figure 9 ◆ Typical operator protective equipment.

2.7.0 Portable Centrifugal Blast Cleaning Equipment

Centrifugal blast cleaning is related to the other blast cleaning methods discussed in this module in that the goal of producing a clean surface by pounding it with abrasive material is the same. However, the method of achieving this goal is much different.

Instead of pressurized air or water, centrifugal blast cleaning uses a motorized wheel to propel abrasives at very high velocity against a surface. The abrasive material is fed to the center of the wheel, and from there onto blades that radiate from its hub. As the wheel turns, centrifugal force accelerates the abrasive to the ends of the blades, which hurl it against the surface. The rate and degree of cleaning depend on the number of wheels in each machine, as well as the size, type, and quantity of abrasive. The higher the horsepower turning the wheel, the more efficient the operation becomes. The abrasive media commonly used is recyclable steel shot, which is available in various sizes and can be tailored to a specific project.

The self-contained abrasive recycling system of centrifugal blast cleaning machinery also distinguishes it from other blast cleaning methods. The rebound energy of the abrasive, along with the vacuum of a dust and debris collection system, force the dust, debris, and abrasive up a recovery chute and into a recovery storage box where the dust and debris are separated and the recovered abrasive is fed back into the blast wheel.

Centrifugal blast cleaning will clean, profile, and remove burrs, flashing, or defects. It is used on different forms of steel and on other surfaces, such as stone and concrete.

In addition to large, stationary blast cleaning units found in factories and fabricating ships, portable machines (*Figure 10*) are used for on-site centrifugal blast cleaning during construction or maintenance work, such as cleaning the decks of ships, parking garages, and other flat steel or concrete surfaces. These units are available in sizes ranging from a small, electric hand-held unit that can clean approximately 40 $ft.^2$/hr. of horizontal or vertical surface up to a large, self-contained, gasoline-powered rider unit that can clean up to 5,000 $ft.^2$/hr.of horizontal surface. The unit shown in *Figure 10* is capable of cleaning up to 2,500 $ft.^2$/hr. and uses a separate, portable dust collector for dust and debris removal.

Centrifugal blast cleaning is cleaner than air abrasive blast cleaning. It can also be quicker and easier, making it more cost efficient. Since the abrasive action is automated, it can clean with more consistency and

Instructor's Notes:

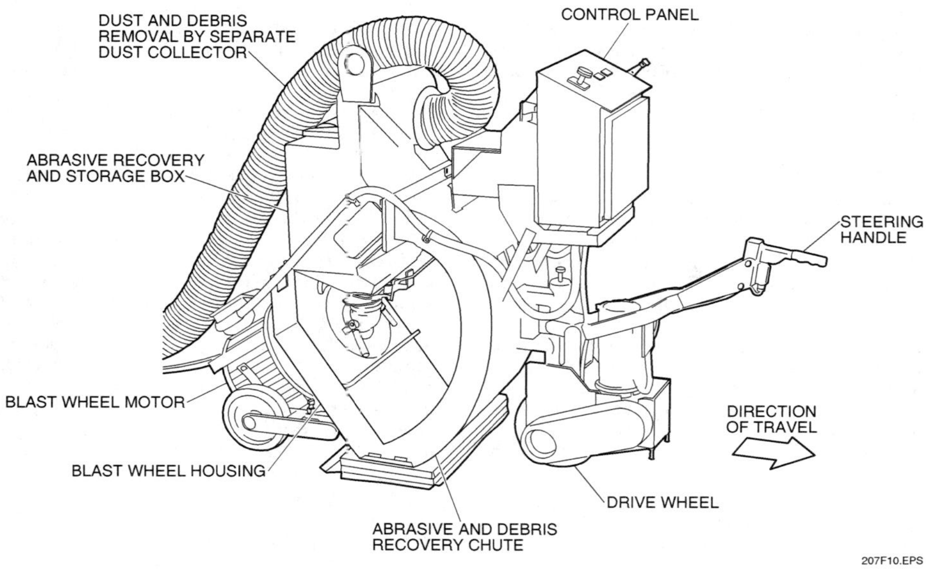

Figure 10 ◈ Portable centrifugal blasting machine.

uniformity than hand-held blast cleaning systems in which air pressure and volume, speed, nozzle distance, and other variables determine the quality of the operation. On the other hand, the equipment is cumbersome and expensive and has difficulty cleaning irregular surfaces.

Usually, portable centrifugal blast cleaning is limited to large, accessible, and relatively smooth, flat areas. These systems are generally the only acceptable preparation method for concrete floors prior to the application of high-performance floor coatings.

2.8.0 Blast System Accessories

A number of accessories are available for use with blast system equipment. Some of the more common accessories are described in the following paragraphs.

2.8.1 Wet Blast Cleaning Equipment

Wet blast cleaning is a method used with a conventional blast machine to suppress dust. It makes use of precisely controlled amounts of water, with or without a rust inhibitor, added externally or internally to the air/abrasive stream at the blast nozzle in just enough quantity to suppress dust. The blast system produces the same results as with dry abrasive blast cleaning but with much less dust. If the wet blast cleaning equipment is equipped with a shutoff/regulator valve and the blast system has an abrasive cutoff control, the blast operator can perform dry blast cleaning, wet blast cleaning, water washoff, and dry blowoff all with one nozzle.

> **WARNING!**
> Wear an approved supplied-air respirator and some form of waterproof blast suit along with adequate warm clothing when wet blast cleaning operations are being conducted. Wet or dry, inhaled contaminants will cause lung damage. In cool or cold temperatures, a risk of hypothermia may also exist when wet blast cleaning.

When blast cleaning steel or concrete, an added benefit of using wet blast cleaning is that water-soluble salts, chlorides, and other chemicals that might not be removed by dry blast cleaning alone will be washed away by the water. This can reduce subsequent coating failures that might be

Homework

Assign the reading of Sections 2.8.0–2.8.7 for the next session.

Materials

Ensure that you have everything required for teaching this session.

Safety

Stress the importance of wearing appropriate personal protective equipment when conducting wet blast cleaning. Ask trainees to explain why a respirator, waterproof blast suit, and warm clothing are required during blast cleaning.

Classroom

Ask a trainee to explain the advantage of using wet blast cleaning on steel or concrete.

Classroom

Have trainees name examples of mechanical loading/recovery equipment. Ask a trainee to explain why vacuum-loading/recovery equipment is more efficient than manual loading and recovery equipment.

Audiovisual

Show Transparency 13 (Figure 12). Ask a trainee to explain how vacuum-blasting tools work.

caused by these contaminants. However, rust inhibitors may have to be used to prevent rust-back of steel substrates. Any inhibitor used must be compatible with any subsequent coating applied to the steel, as recommended by the coating manufacturer.

2.8.2 Spin Blast Nozzles

Special nozzles in a variety of sizes are available for cleaning and profiling the interiors of pipes or other cylindrical objects or vessels. *Figure 11* shows one such device designed for use inside cylinders ranging from 7", or 17.8 centimeters (cm), to 42", or 106.8 cm in diameter. The device connects to standard blast hose. The pair of angled nozzles at the head of the device can rotate to clean the interior of the cylinder. Different-size standard nozzles can be installed in the head. The device is centered within the cylinder by the three adjustable rollers at the front and rear of the device and is manually moved back and forth within the cylinder.

2.8.3 Abrasive Loading/ Recovery Equipment

Abrasive loading/recovery equipment is used to load new, recovered, or recycled abrasives into blast machines, cleaning and recycling equipment, or waste storage hoppers. Some of the loading/recovery equipment in use today is of the mechanical variety. It consists of feed hoppers and storage hoppers employing screw conveyers, belt conveyers, or bucket elevators. They are usually manually loaded with the abrasive and may require manual cleanup of residual abrasive and dust. For recovery operations, spent abrasive and dust must be manually collected with brooms and shovels, then manually loaded into the feed hoppers. For high-production operations, most contractors consider these methods too costly in labor and time.

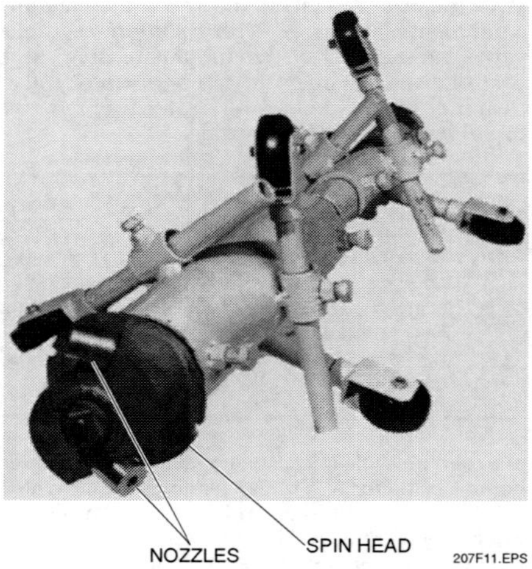

Figure 11 ◆ Typical spin blast nozzle.

Vacuum-loading/recovery equipment is more efficient and cost effective than manual loading and recovery in high-production blast cleaning operations. Vacuum equipment also captures and separately contains most of the dust, debris, and spent abrasive fines when used to load or recover abrasives. Vacuum equipment is classified as blower-powered or air-powered equipment. Of the two, the blower-powered equipment is the largest, the most efficient, and the most expensive. Both types of equipment are usually available with a variety of pickup equipment such as hoppers, collectors, crevice tools, and closed-circuit recovery tools.

2.8.4 Closed-Circuit Abrasive Recovery (Vacuum-Blasting) Tools

In areas where open air blast cleaning must be restricted, closed-circuit recovery tools (commonly called vacuum-blasting tools) are used to confine the abrasive, fines, dust, and other contaminants removed during blast cleaning. With these tools, abrasive is propelled against a surface and immediately captured along with the dust and other materials. Reasonable production rates can be achieved using this type of equipment. A typical high-production, closed-circuit blast tool is shown in *Figure 12*. These tools are coupled to standard-pressure blast systems and to blower-powered abrasive recovery systems. The abrasive recovered can be returned to the blast system or stored in a waste hopper depending on the type of abrasive.

These tools typically use a ¼" (6.5 mm) nozzle with 1¼" (32 mm) blast hose along with a larger diameter recovery hose. Some tools are equipped with rollers to help the operator maintain the correct distance. They are also equipped with a thick nylon capture brush that surrounds the nozzle and rests on the work surface. Air is drawn through the brush to keep the dust and other materials from escaping. On some tools, the rollers and the brush are mounted on a flexible rubber boot that helps them stay flat against the work surface as the operator moves. Other tools may be equipped with a separate vacuum on/off control so that the vacuum can be controlled separately from the abrasive

Instructor's Notes:

flow. The blast pattern from one of these tools is typically about 3" (76 mm) in diameter.

Used with recyclable abrasive, closed-circuit tools can provide environmentally safe blast cleaning with good cost savings in labor and abrasive. However, production rates are lower than those of other types of abrasive blast cleaning. The highest production tools are usually used only on flat surfaces. Smaller, low-production blast tools, usable with either blower-powered or air-powered recovery equipment, are available with interchangeable heads for different-shaped surfaces such as flat surfaces, inside corners, outside corners, and curves.

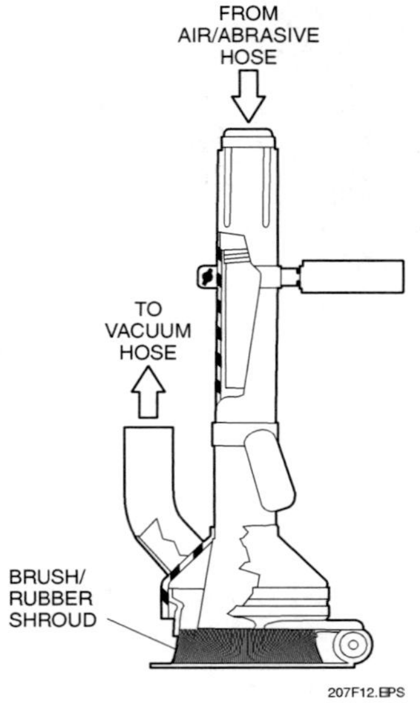

Figure 12 ◆ Typical high-production, closed-circuit blast tool.

2.8.5 Abrasive Cleaning/Recycling Systems

While vacuum abrasive loading/recovery equipment will clean certain fines, dust, debris, and contaminants from most recyclable abrasives, other systems are available that will perform additional cleaning and sorting of abrasives. Some of these systems are manually loaded and emptied. Others use vacuum or mechanical transport equipment for loading and emptying. These systems make use of various combinations of multistage vibratory and/or rotating sieve separation, magnetic separation, and air-wash cleaning. They are available in a variety of sizes, capabilities, and capacities and are usually classified as small, medium, and large systems.

2.8.6 Blast Rooms

The use of a blast room can offer improved safety and greatly reduced operating costs over free-air blast cleaning when cleaning, finishing, or surface-profiling large workpieces in an industrial or field environment. With the process contained in a specially designed enclosure, blast media and toxic debris are segregated from the external environment. Random media loss is nearly eliminated, noise is reduced, and automatic collection of expended media during blast cleaning can be implemented.

Figure 13 shows a typical blast room with an air-wash abrasive cleaning/recycling system, blast machine, and a blower-powered ventilation system with a reverse-pulse, multiple-cartridge filter dust collector. Many blast rooms are available both as welded, packaged units or as bolt-together, field-erectable units. The packaged units are available in a number of sizes or can be custom built. The field-erected units can be expanded to almost any length by adding additional panels. They are equipped with high-illumination, shielded fluorescent lighting and wide access doors to allow loading of workpieces. They also have an OSHA-required emergency access and exit door for personnel. Rubber curtains are also available to shield the interior, reduce noise, and reduce media rebound. Various custom turntables, overhead monorail beams, and work carts are also generally available to ease workpiece handling. Some rooms also have a built-in exterior workstation equipped with a viewing window and gloves for occasional blast cleaning of small parts from outside the room.

These rooms are available with a number of different floor designs for expended media and debris collection and recovery. The designs vary from the simple to the sophisticated, based on media volume and frequency of use. They may be shovel or sweep chutes, pneumatic and/or screw-feed recovery troughs, or full-floor automatic vibratory and/or pneumatic and/or screw-feed type systems, depending on the type of cleaning/recycling/recovery equipment used.

Sweep/shovel chutes require the operator to sweep the entire floor manually and load the media and debris at one spot. Grated trough systems can be designed to extend across one end, or one end and both sides, or both sides and across

Classroom

Have trainees identify the advantages of using a blast room.

Classroom

Ask a trainee to describe a blast cabinet. Have a trainee explain what blast cabinets are used for.

Homework

Assign the reading of Sections 3.0.0–3.3.0 for the next session.

Materials

Ensure that you have everything required for the laboratory during this session.

Classroom

Have trainees name the factors to consider when choosing an abrasive for a job.

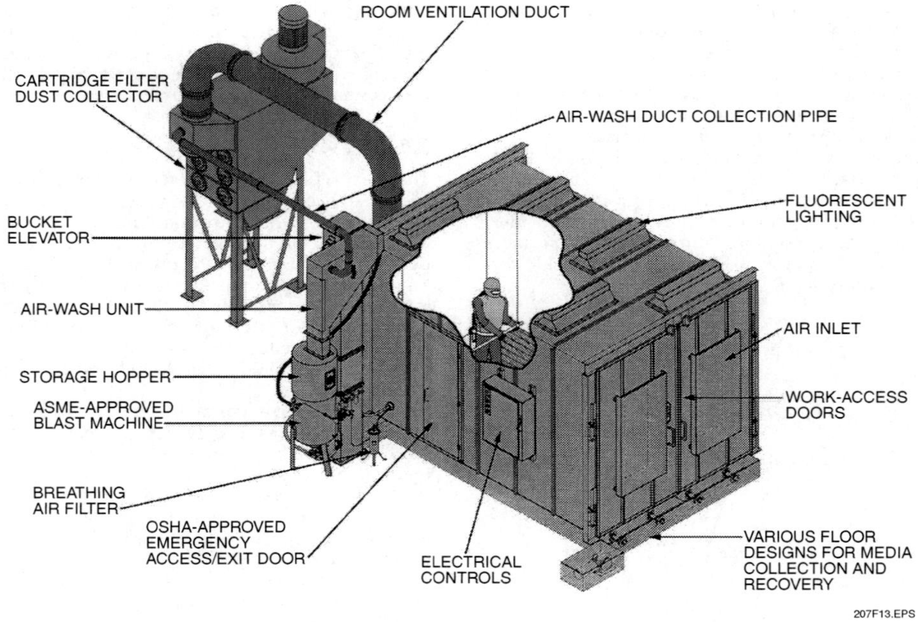

Figure 13 ◆ Typical blast room with air-wash cleaning/recycling system.

the middle, etc. Each larger configuration of the trough system requires less sweeping effort for the operator. The full-floor systems require no operator cleanup. The entire floor is grated, and the expended media and debris fall directly into the collection and recovery system.

2.8.7 Blast Cabinets

Blast cabinets, either bench mounted or free standing (*Figure 14*), are basically just small versions of blast rooms with external access. They are used for convenient cleaning and profiling of small parts. They are available in a variety of sizes and configurations and offer all of the advantages of confined blast cleaning versus free-air blast cleaning. They are almost always used inside industrial or commercial buildings in a fixed location. They are usually equipped with side access doors or a top-opening access cover. They should also be equipped with a large, safety glass viewing window, bright internal lighting, and some form of an air-curtain device that directs a sheet of high-velocity air across the interior side of the viewing window to prevent dust buildup and reduce frosting caused by abrasive.

3.0.0 ◆ BLAST CLEANING MEDIA

Abrasive, the actual cleaning material of the blast cleaning process, comes in a range of types, shapes, sizes, and degrees of hardness.

3.1.0 Media Types

Given the multitude of options, selecting the right abrasive for the job becomes a matter of determining which material will give a surface the desired finish, including the proper degree of cleanliness and roughness as well as the type of texture needed for coating. Other factors to consider in choosing an abrasive are the type of surface to be blasted, its present condition, the consumption rate of the abrasive, its cost, and any environmental regulations that may affect the blast cleaning operation.

Instructor's Notes:

There are four types of abrasives: metallic, mineral, by-product, and special-purpose.

The most common metallic abrasive is cast steel. Others include malleable iron and chilled cast iron. All come in the form of shot and grit (*Figure 15*) and in different degrees of hardness. Abrasive made from wire cut into short pieces is also available. Shot is rounded, which makes it appropriate for cleaning heavy, brittle deposits and producing a rounded profile or anchor pattern. Grit, which is made by crushing hardened shot pellets, and cut wire both have an irregular, angular shape that effectively removes rust, mill scale, and old coating and leaves a rough profile that feels like abrasive paper. Shot, grit, and cut wire are screened according to size.

Metallic abrasives can be recycled many times before breaking down into a size too small to be reused. The repeated blast cleaning impact causes the abrasive to fracture or flake. Generally, the harder the abrasive, the faster it breaks down. The particles become smaller and smaller until they are eventually filtered out in the recycling process. New abrasive must be added regularly to replace what is lost. The result is a mixture of sizes called a *work mix* or *operating mix* of large particles that do the heavy work of **cracking** and breaking up the contaminants, and small particles that do the cleanup work of scouring and removing them. Maintaining a properly balanced work mix is important for efficient blast cleaning. This can be done by hand or with a continuous, automated system of replenishing spent abrasives.

Laboratory

Set up several stations at which you have placed samples of each of the four types of abrasives. Have trainees identify and list three characteristics of each.

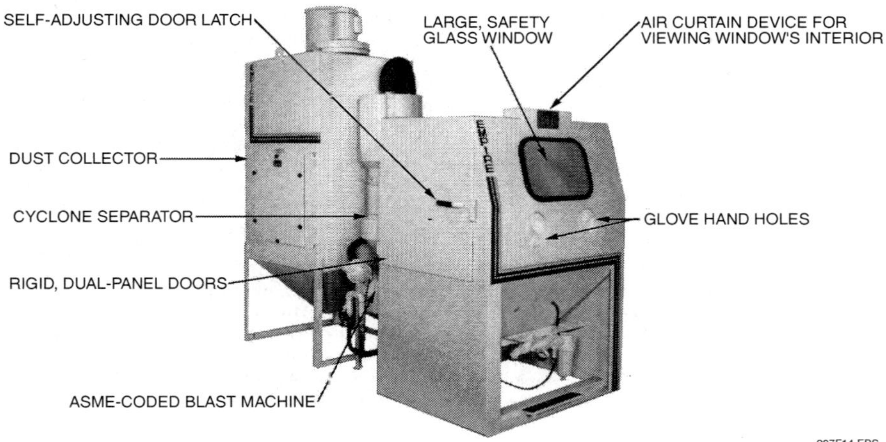

Figure 14 ◆ Typical blast cabinet with blower-powered abrasive loading/recovery system.

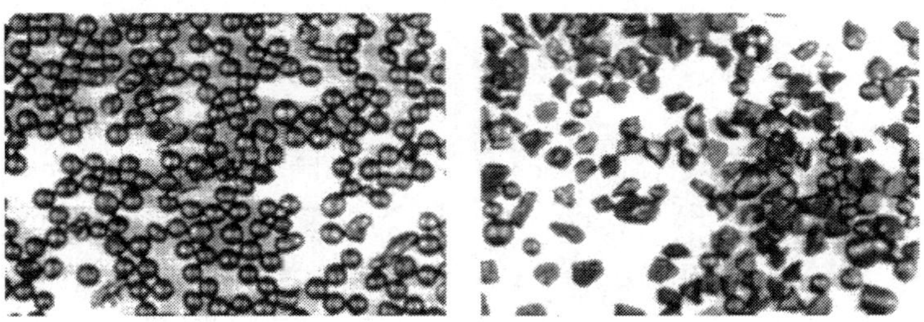

Figure 15 ◆ Steel shot (left) and grit.

Classroom

Have a trainee name the characteristics of abrasive media that influence the productivity of the operator.

Homework

Have trainees review Sections 4.0.0–4.5.5 for the next session.

Mineral abrasives include sand, silica, and nonsilica. From the beginning, sand has been used for abrasive blast cleaning. Silica sand in particular is readily available, which makes it economical. Its range of angular and spherical shapes gives it good cleaning action. However, it breaks down quickly and produces a large amount of silica dust, which causes a dangerous lung disease known as *silicosis* when the dust is small enough to be respirable (10 microns or less). Nonsilica or heavy mineral sands used in blast cleaning include magnetite, staurolite, and olivine rutile, which are tough, dense, and smaller than silica sand but contain no dangerous free silica. These abrasives are generally not recycled because they quickly break down in blast cleaning. On the other hand, garnet, a tough abrasive, can be recycled several times. Other mineral abrasives, such as zircon and novaculite, are generally used for specialty blast cleaning because of their cost and the finish they leave. Zircon, for example, gives a smooth, matte finish, and novaculite leaves a satin luster.

Some abrasives are made from by-products. Slag, the refuse from smelting or other processes, makes a good abrasive because of its availability, relatively low cost, and low silica content. Its sharp, angular shape gives it good cutting power, and it comes in a full range of sizes. Blast cleaning slags include coal slag (*Figure 16*), copper slag, and nickel slag. By-product abrasives are not recycled because they break down too rapidly.

Special-purpose abrasives are made from a variety of materials. There are specialty abrasives made with walnut shells or peach pits that will remove light deposits without damaging the surface. Corncob and sodium bicarbonate abrasives can also be used for this purpose, or for removing surface grease. Other special abrasives include sponge, silicon carbide, aluminum oxide, cut glass, and plastic or glass and ceramic beads. Plastic and glass abrasives will clean small, delicate surfaces. Plastic media blast cleaning is also used to remove old paint from aircraft, military vehicles, and radar dishes because it does not damage the substrate. Aluminum oxide and silicon carbide are tough, hard, aggressive abrasives that can be recycled. Aluminum oxide works on stainless steel, and silicon carbide does specialty etching. A polyurethane sponge, sometimes incorporating various abrasives, can be used to remove oil and other contaminants as it cleans and roughens a surface for coating. Most of these abrasives can be recycled a limited number of times.

Shape, size, density, and hardness are the other factors to consider in selecting the proper abrasive material. As noted, abrasives come in spherical and angular shapes, as well as irregular shapes that have features of both. Rounded or spherical abrasive particles clean by hitting the surface; sharp or angular abrasives clean by cutting or gouging it. Large, abrasive particles remove thick, heavy contaminants, while small ones work on the fine, residual debris. A mix of sizes is often best. Generally, the denser the abrasive, the more effective it is, and the harder it is, the deeper it will clean the surface. All abrasives should be kept clean and dry before blast cleaning.

Table 3 presents a condensed summary of the properties for a number of different types of abrasive media that, with the exception of sand and sodium bicarbonate, can be recycled a varying number of times. Their characteristics (physical properties) and benefits related to productivity and selection are discussed in the following paragraphs in more detail.

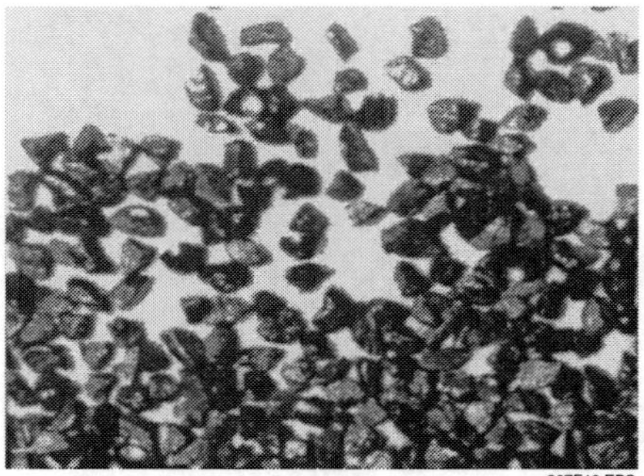

207F16.EPS

Figure 16 ◆ Coal slag abrasive.

3.2.0 Productivity and Characteristics of Media

Certain characteristics of an abrasive media influence the productivity of the operator. Work performed by the media is controlled by its size and density (weight), and more importantly, by its impact velocity.

3.3.0 Media Specifications

Some projects may use a specification to identify the abrasive required. Federal government agencies, the Society of Automotive Engineers (SAE), and others have abrasive specifications applicable to their operations. SSPC also has abrasive specifications that will be discussed briefly in the following paragraphs.

Instructor's Notes:

Table 3 Properties of Various Abrasive Media

	BENEFITS				USAGE PROPERTIES					PHYSICAL PROPERTIES			
	Finishing	Cleaning and Removal	Peening	Etching	Average Cleaning Speed	Reusability	Dust Level(s)	Chance of Metal Removal	Typical Blast Pressures (psi)	Hardness MOH Scale (Rockwell C)	Bulk Density (ft.^{3})	Mesh Sizes	Angular (◊) Spherical (○)
Glass Beads	*	*	*		MED	HI–LO	LO–HI	Very LO	20–55	5.5	100	30–440	○
Ceramic Shot	*	*	*		MED	HI	LOW	Very LO	20–90	7 (57–63)	150	8–46	○
S.S. Cut Wire	*	*	*	*	MED	HI	LOW	MED–HI	20–90	6–7.5 (35–55)	280	20–62	◊
Steel Shot	*	*	*		MED	Very HI	LOW	Very LO	20–90	6–7.5 (20–66)	280	8–200	○
Steel Grit	*	*		*	MED–HI	Very HI	LOW	MED	20–90	6–7.5 (40–66)	230	10–325	◊
Aluminum Oxide	*	*		*	HI	MED–HI	LO–MED	MED–HI	20–90	8–9	125	12–325	◊
Silicon Carbide	*	*		*	Very HI	LO–MED	MED–HI	MED–HI	20–90	9	95	36–220	◊
Crushed Glass	*	*		*	HIGH	LO–MED	MED–HI	LO–MED	20–50	5.5	100	30–400	◊
Plastic Media		*		*	MED–HI	MED	MED	Very LO	20–60	3–4	45–60	12–80	○
Wheat Starch		*			MED	MED	MED	Very LO	20–35	3	40	12–100	◊
Walnut Shell		*			LO–HI	LOW	MED–HI	Very LO	10–40	3	40–80	Many	◊
Corn Cob		*			LO–MED	LOW	MED–HI	Very LO	20–60	1	30	8–40	◊
Sodium Bicarb.[1]		*			MED–HI	NR[3]	HI	Very LO	20–90	3	62	54–150	◊
Silica Sand[2]	*	*		*	HI	NR[3]	HI	HI	30–90	7.5	100	12–100	◊

NOTES:
- Above information intended as a general reference guide.

[1] Sodium bicarbonate must be treated with a flow agent.
[2] Do not use silica sand in a hand cabinet.
[3] Not recommended.

207T03.EPS

4.0.0 ◈ PREPARATION STANDARDS, PROFILING, AND INSPECTION TECHNIQUES

The quality of the surface preparation has a direct bearing on how well a coating performs and how long it lasts. It is believed that most premature coating failures are due, all or in part, to faulty surface preparation and/or coating application. That being the case, it is important to determine how well a surface has been prepared and to catch any problems before they are covered with coating.

Standards and inspection techniques for surface preparation have been established to help protect against premature coating failures. Generally, the greatest number of standards and the most formal inspection processes are applied to industrial applications.

The following is a brief overview of some generalized standards for surface preparation of various substrates and how inspectors would evaluate them.

- Steel surfaces:
  - Slag, flux, weld spatter, and rough edges should be removed.
  - The surface should be washed with solvents to remove grease, dirt, oil, debris, and other loose contaminants.
  - The surface should be cleaned of all rust, dirt, mill scale, old coating, or other tightly adhered contaminants to the required degree according to the specification requirements, using one or more of the following: hand tools, power tools, wet or dry abrasive blast machines.
  - The surface should be profiled with abrasives to produce the specified anchor pattern.

The two primary areas of concern to inspectors in the surface preparation of steel are cleanliness and profile. SSPC and NACE have developed joint standards for evaluating four degrees of abrasive blast cleaning.

4.1.0 Blast Surface Cleanliness Standards and Production Rates

As a guide for comparing surface cleanliness to the following four standards, SSPC has developed reference photographs (*SSPC-VIS 1*) to illustrate each grade. These visual standards are for steel that has not previously been painted.

Materials

Ensure that you have everything required for the laboratory and testing during this session.

Classroom

Ask a trainee to identify the cause of most premature coating failures. Have a trainee explain how to prepare steel surfaces.

Distribute copies of *SSPC-VIS 1* photographs, and have trainees compare them to the standards for surface preparation of steel surfaces listed in Section 4.0.0 of the Trainee Guide.

Ask a trainee to explain the process of flash-rusting. Have a trainee explain a way of preventing flash-rusting.

Ask a trainee where the profiling requirement can be located. Have a trainee explain how Table 4 is used.

Abrasive blast cleaning can be used to achieve various degrees of surface cleanliness to meet job specifications. These specifications usually define the minimum level of cleaning required by referencing and supplementing the standards (SSPC/NACE) used for steel finishing. These standards, in decreasing order of severity, are summarized in the following paragraphs. They are based on dry abrasive blasting. If wet abrasive blasting is being used by mutual agreement between the parties specifying the work and the parties performing the work, then inspectors must be made aware of the agreement. In any case, inspectors should be made aware of the abrasive used and the appearance of the final blasted surface by a test demonstration before the job commences. The test demonstration area can be coated with clear urethane and used as a reference standard for cleaning and profiling by both blast operators and inspectors.

Operators should be aware that once steel surfaces are cleaned and bare metal exposed, flash-rusting of the surface (rust-back) may begin almost immediately. If the **primer** coating cannot be applied within a short period of time, a wash of the bare surfaces using a rust inhibitor may be required. One such inhibitor is water containing 0.32% sodium nitrate and 1.28% (by weight) secondary ammonium phosphate (dibasic). In any event, the coating system must be compatible with whatever inhibitor is used.

4.2.0 Blast Surface Profiling and Measurement

Profile is required to provide an anchor pattern for subsequent application of many high-performance coatings on new surfaces or previously coated surfaces that were not profiled.

In most cases, the project specifications will prescribe the amount of profiling required. If not, the coating manufacturer's data sheets will provide this information. *Table 4* provides a guide for obtaining various depths of profiling using assorted dry mineral and ferrous abrasives.

The data in the table should be used only for approximating the abrasive size required to obtain a specified anchor pattern. The standard metal used to obtain these results was hot rolled steel with tightly adhering mill scale. The resulting depth of anchor pattern will vary with the method used for measuring depths as well as with a number of other variables: for example, type and hardness of steel, thickness of mill scale, and degree of

Table 4 Profile Versus Abrasive Required

Profile	Abrasive Required	
1 mil	30/50 mesh silica sand G-80 steel grit S-110 steel shot*	80-mesh garnet 100-grit aluminum oxide Coal slag 3060
1.5 mil	16/35 mesh silica sand G-50 steel grit S-170 steel shot*	36-mesh garnet 60-grit aluminum oxide Coal slag 3060
2 mil	16/35 mesh silica sand G-40 steel grit S-230 steel shot*	36-mesh garnet 36-grit aluminum oxide Coal slag 2040
2.5 mil	8/35 mesh silica sand G-40 steel grit S-280 steel shot*	16-mesh garnet 24-grit aluminum oxide Coal slag 2040
3–4 mil	8/20 mesh silica sand G-25 steel grit S-330 or 390 steel shot*	16-mesh garnet 16-grit aluminum oxide Coal slag 1240

*Steel shot alone will not give a good angular anchor pattern and should be used in combination with steel grit for best results.

207T04.EPS

Instructor's Notes:

cleaning specified. This information can be used for centrifugal wheel blasting as well as pressure blasting. Pressure blasting should be done using 90- to 100-psi-nozzle pressure. The depth of anchor pattern shown in *Table 4* is an average and not a minimum or maximum depth obtainable.

To check the **surface profile**, use tools such as a profile comparator (*Figure 17*) to compare the surface to established standards; a profile gauge (*Figure 18*) to measure the depth of the profile; or a replica tape (*Figure 19*) to make an impression of the profile. How clean and profiled a surface needs to be will depend on the type of structure, its location, and the job specifications. Again, inspectors and blast operators should be made aware of the appearance of the final blasted and profiled surface by a test demonstration before the job commences.

4.3.0 Containment and Temperature/Humidity Control

Of the various surface preparation methods, abrasive blasting produces the greatest quantity of environmental contaminants. In most cases, local, state, or federal agencies require the use of containment to control the emissions of dust, debris, paint chips, and spent abrasives generated by abrasive blasting. The type, concentration, and toxicity of the contamination will determine the extent of the containment required.

Environmental conditions, such as humidity and temperature, have a very substantial impact on cleaning and coating operations. In some cases, the cleaning or coating of a substrate, especially ferrous materials (steel or iron) or concrete, must be delayed because of excess moisture, humidity, or temperature. In other cases (especially outdoors), weather, humidity, and temperature are controlling factors in the cycle of ferrous or concrete surface preparation and coating. In nearly all instances, the manufacturer's predictions of coating service life are based on optimum, ideal surface preparation and application conditions that are difficult, if not impossible, to attain in the field without some form of containment and environmental control.

4.4.0 Safety Guidelines

Review the following safety precautions before attempting to use any blast machine or accessories.

Nozzle blast is a serious hazard. The abrasive/air jet from a blast nozzle can cause serious physical injury including injection of abrasive particles, rust inhibitors if used, and other contaminants into the body. If this occurs, get emergency surgical treatment immediately.

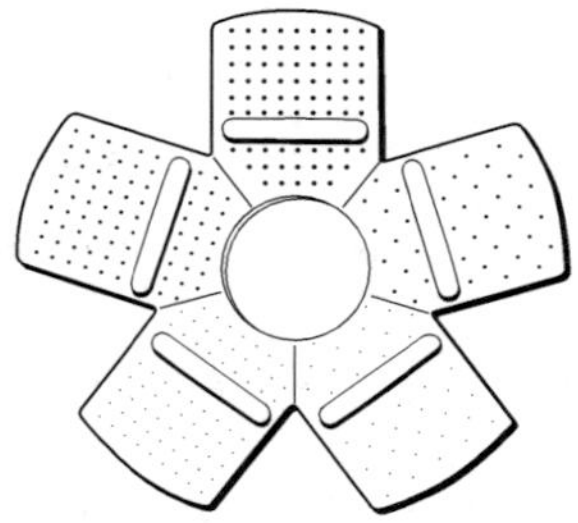

207F17.EPS

Figure 17 ◆ Profile comparator.

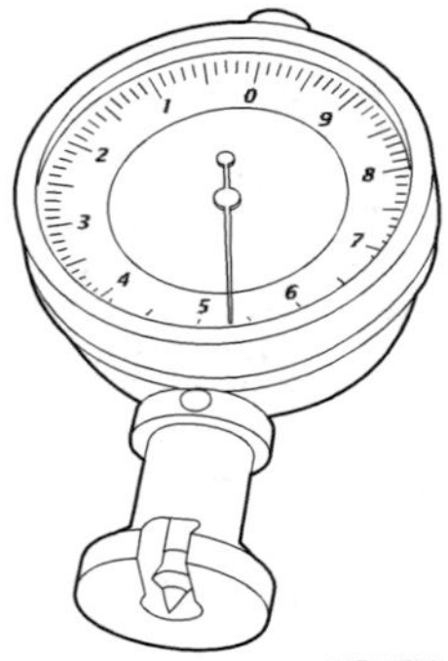

207F18.EPS

Figure 18 ◆ Profile gauge.

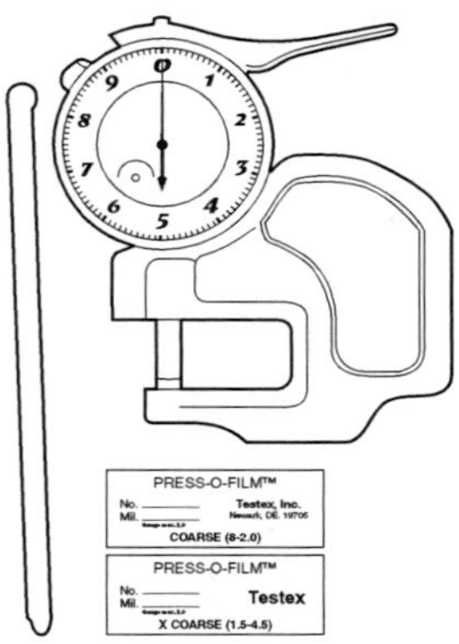

207F19.EPS

Figure 19 ◆ Replica tape and micrometer.

Audiovisual

Show Transparencies 14 through 16 (Figures 17 through 19). Ask trainees to explain how each of the tools depicted is used.

Laboratory

Have trainees practice checking surface profile using a profile comparator, a profile gauge, and a replica tape. Discuss the results.

Performance Testing

Have each trainee perform Performance Task 1 to your satisfaction. Fill out a Performance Profile Sheet for each trainee.

Classroom

Have trainees discuss the effects of temperature and humidity on cleaning and coating operations. Ask trainees if they have any questions.

Teaching Tip

See the Teaching Tip for Section 4.4.0 at the end of this module.

Safety

Emphasize the importance of keeping all body parts clear of the abrasive/air jet coming from the blast nozzle.

Other guidelines include:

- Do not point a blast nozzle at any part of your body or at any other person.
- Do not put your hand or fingers in front of the blast nozzle.
- Make sure the nozzle and hose connections are secure and safety-pinned or wired before applying pressure to the hoses and nozzle. In addition, make sure the hose connections have whip checks installed.
- Make sure the dead-man control handle is functioning correctly and that the safety latch engages properly before commencing blast cleaning operations. If you stop blast cleaning, even for a moment, ensure that the safety latch is fully engaged or it will not prevent the dead-man control from being accidentally triggered if the hose and nozzle are dropped or bumped.
- Make sure that you wear the proper blast suit, air-fed helmet, hearing, and respiratory protection when entering or working in a blast cleaning area. Make sure the inner safety lens of the helmet is in place and is not damaged.
- Check the hoses, couplings, nozzles, and gaskets daily for signs of damage, leaks, and wear. Do not attempt to repair damaged, leaking, or worn components—always replace them.

Hypothermia is a hazard to consider when conducting outdoor blast cleaning operations. In temperatures below 60°F (16°C), with any amount of wind, a risk of hypothermia exists, especially with wet abrasive blast cleaning. Always wear adequate protective clothing, head covering, and gloves.

Footing and balance may be impaired when conducting dry blast cleaning operations; vision may become temporarily impaired by dust, which could obscure hazardous footing conditions. When wet abrasive blast cleaning, footing surfaces could become slick. Either of these conditions can cause difficult footing and/or balance problems for the operator. Other considerations include:

- Any scaffolding or staging used must be secure and free of debris and equipment to prevent slipping and tripping. Metal scaffolding or staging must be properly grounded.
- A proper fall protection system should be used with any scaffolding or staging.
- Hoses and other materials must remain clear of your feet at all times.
- With weight on both feet and feet spaced apart in an open stance, grip both the control handle and the nozzle before triggering the control handle to commence blast cleaning. Be prepared for the reactive force from the nozzle which will occur shortly after the control handle is triggered.

Hazardous materials may be encountered during blast cleaning. Dust from disintegrating coatings and/or abrasives during blast cleaning operations must never be inhaled. In fact, no dust (including dust from abrasive handling, loading, and site cleanup operations) should ever be inhaled. Severe incurable health problems and/or eventual death may occur. To avoid respiratory injury:

- Always wear a NIOSH-approved air-fed helmet or hood before, during, and after blast cleaning, until cleanup is completed and the air has been tested and found clear of toxic dust.
- Always clean, maintain, and inspect your helmet, breathing hose, air control device, and breathing air filter daily. Replace any damaged components with approved parts; otherwise, the NIOSH approval of the equipment will be voided.
- Always use dry blast media that is prepared for that purpose and is properly sized and free of harmful substances such as free silica, cyanide, arsenic, or lead. Check material safety data sheets for presence of toxic or harmful substances.
- Surfaces to be blasted must be tested for toxic substances before blast cleaning commences. If toxic, additional, appropriate protective measures for operators and bystanders must be taken as required by OSHA and/or NIOSH.
- If blast cleaning operations are conducted in a confined area, proper sealing of the area and separate, adequate, negative pressure ventilation (in addition to any abrasive recovery equipment) with appropriate filtering must be in place to allow removal of airborne toxic dust from the blast cleaning area and to prevent leakage of the dust to other areas.

Breathing air compressors can, and do, produce deadly, colorless, odorless carbon monoxide gas that can be fatal to blast operators. In addition, carbon monoxide (CO) gas from the exhaust of other engine-driven equipment can be sucked in by breathing air compressors or breathing air pumps and transmitted to operators. To avoid death caused by carbon monoxide poisoning:

- Make sure the breathing air is not supplied from an oil-lubricated, piston-type compressor.
- Make sure the breathing air compressors are checked and maintained strictly according to the manufacturer's instructions and schedule; document the actions in an equipment log as required by OSHA.
- Calibrate and perform maintenance on any CO monitors or converters used for breathing air strictly according to the manufacturer's instruc-

Instructor's Notes:

tions and schedule; document the actions in an equipment log as required by OSHA. These devices, when used in the field, usually require more frequent calibration/maintenance. If monitors or converters are not used, frequent, periodic testing of breathing air must be accomplished and documented as required by OSHA.

- When working outdoors, if at all possible, locate all compressors and/or air pumps used to supply breathing air away from, and upwind of, any other engine-operated equipment and the dust created by the blast cleaning operation.
- Do not operate engine-powered equipment in a confined area without adequate ventilation to remove exhaust fumes and carbon monoxide from the area.

Fire/explosion is a common hazard around engines. For gasoline- or diesel-engine powered equipment, the fuel used is combustible. When spilled on a hot surface, it can ignite and cause a fire or explosion. Unless the fuel tanks are located remotely from the equipment, do not fill the fuel tanks while the engine is running or hot. Make sure that the blast machine and any conductive work surfaces are grounded and that the blast hose is of the static-suppressing or grounded type. Electrical static discharges from the blast nozzle to the work surface may cause a flash fire or explosion if flammable vapor or dust is present and if a discharge occurs. When working in confined areas such as tanks, make sure all volatile fumes and dust are removed by ventilation. Sparks created by abrasive striking a surface may cause a flash fire or explosion.

Electrical hazards should always be considered. If wet blast attachments or injectors are used, all electricity to motors, electrical boxes, or electric devices must be turned off and locked out; or, the motors, boxes, or devices must be completely water sealed by materials resistant to the abrasive blast jet to prevent accidental death from electrocution or fires caused by shorts in the electrical equipment.

Misuse, unauthorized repair, alteration, or modification of the equipment may cause violent ruptures or explosions resulting in serious physical injury or death. In addition, unauthorized repairs, alterations, or modifications may violate ASME, OSHA, or NIOSH certifications or regulations.

- Do not alter or modify any part of the equipment.
- Do not change any factory-set adjustments.
- Do not exceed the maximum rated pressure of any equipment or component.
- Do not repair ASME-coded pressure vessels or equipment without recertifying them.
- Do not alter the throttle settings of any engine-operated equipment unless recommended or allowed by manufacturer's instructions.
- Do not perform unauthorized repairs of equipment. Replace worn or damaged parts immediately with new, recommended, and approved parts.
- Do not allow the blast nozzles to be dropped, roughly handled, or used as hammers or signaling devices. The inner lining may crack causing premature failure of the nozzle.
- Do not allow the blast hoses or air hoses to be curved with sharp bends or to be driven over repeatedly by vehicular traffic. Premature wear or damage can occur that may result in an unexpected rupture of the hose.

4.5.0 Blast System Start-Up, Operation, and Shutdown Guidelines

The following guidelines are provided as a general method for the start-up, operation, and shutdown of a typical medium-size, high-production blast system with pneumatic blast machine controls, separate breathing air compressor, and loading/recovery/recycling equipment. These guidelines may not include all required actions for a specific blast system. For a specific blast system, any manufacturer's operation and instruction manuals, including any warnings, cautions, and notes, take precedence and must be read, understood, and used for start-up, operation, and maintenance of the system equipment.

4.5.1 Pre-Start and Daily Checklist Guidelines

Before starting and operating a blast system and on a daily basis, the operator must perform the following actions as a minimum:

Step 1 If a blast cleaning operation is outdoors, make sure the breathing air compressor is located away from, and upwind of, other engine-driven equipment and any dust that will be generated by the blast operation.

Safety

Warn trainees not to fill fuel tanks while the engine is running or hot. Ask a trainee to explain what can happen if fuel is spilled on a hot surface.

> **WARNING!**
> When spilled on a hot surface, fuel can ignite, causing a fire or explosion. Unless fuel tanks are located away from the equipment, do not fill fuel tanks while the engine is running or hot.

Step 2 Check for proper oil, fuel, and coolant levels in engine-driven equipment and fill as necessary.

Step 3 Check for proper oil level in oil-lubricated compressors and fill as necessary.

Step 4 Check that all oil and moisture traps and separators on blast or breathing air compressors, blast machines, and breathing air filters have been drained.

Step 5 If not automatically monitored, check the intake air filters on the engines and blast and breathing air compressors for excessive accumulation of dust and dirt.

Step 6 If used, check that CO monitors or converters are within their calibration dates. If not, they must be re-calibrated before blast cleaning operations can commence.

Step 7 Check that all air and blast pipes/hoses are the correct size and are laid out with no sharp bends.

Step 8 Check that the air and blast hose couplings and nozzle holders are securely fastened together and to the hoses, and that whip checks are in place. Also, check that the hoses, couplings, and nozzle holders are not worn, damaged, or cracked, and that the hoses are fully inserted and secured by the special screws in the couplings. Replace any worn, damaged, or cracked parts immediately with new, recommended, and approved parts.

> **NOTE**
> Air and blast hoses may be checked for internal abrasive wear by squeezing the hose along its length with your fingers. Soft spots indicate internal wear or a ply failure that can result in a hose rupture.

Step 9 Check the nozzle gasket for damage or wear and replace if necessary.

Step 10 Check the nozzle lining for cracks and the nozzle orifice for wear with gauge. Replace the nozzle if damaged or if the orifice is $\frac{1}{16}$ larger than its original size.

Step 11 Clean and inspect your NIOSH-approved supplied-air respirator, cape, breathing air hose, breathing air supply hose, and air control valve or climate control valve. Make sure that all parts are in place including the inner lens of the helmet. Replace any damaged or worn parts with approved parts only.

Step 12 If communication equipment will be used, check that the batteries are fully charged.

Step 13 Check that an adequate supply of the desired type and size of abrasive, specified for blast cleaning use, is on hand and available. If required, take samples (including samples from any additional incoming supplies) and have them tested for compliance with specifications.

Step 14 Check that adjacent surfaces are protected, if required. Containment enclosures may have to be erected to protect against dust infiltration or contamination of adjacent areas.

Step 15 Check that all required OSHA-approved rigging and scaffolding is in place, secure, and properly grounded.

Step 16 Check that the blast machine is properly grounded.

Step 17 Test any surface to be blasted for toxic materials. If the coating contains toxic materials, special containment as well as additional respiratory protection will be required as specified by OSHA and NIOSH.

Step 18 Clean any soluble salts or visible oil or grease from the surfaces to be blasted in accordance with *SSPC-SP1, Solvent Cleaning*.

4.5.2 Blast System Start-Up Guidelines

If not already accomplished, perform the pre-start and daily checklist, and follow these steps to start up the blast system:

Step 1 Close the air-inlet valve on the blast machine and the breathing air filter, and close the air-outlet valves on the compressors.

Instructor's Notes:

Step 2 Start the breathing air and blast air compressors and any ventilation equipment as directed by the manufacturer's operation manual. Observe the gauges and indicators on the engines and/or compressors for correct operation.

Step 3 Open the vent valves on the compressor receivers and allow the compressed air to escape. Operate the compressors for 15 to 30 minutes in this mode to allow them to warm up. Air being vented from the receivers should begin to feel warmer after the warm-up period.

Step 4 Close the compressor receiver vent valves after the warm-up period and allow the pressure to increase.

Step 5 Check that the compressors unload at the proper pressure and that the regulated pressure at the output of the compressors is correct.

Step 6 Open the compressor outlet valves and check for air leaks in the compressor air lines and couplings to the blast machine and breathing air filter. If leaks are detected, close the appropriate air-outlet valves on the compressors, relieve pressure on the lines by opening and then closing an appropriate downstream valve; then replace the hose, coupling, or coupling gasket. When the repair is completed, repeat the leak check until all leaks are repaired.

Step 7 Connect the respirator equipment to the breathing air filter and open the air-inlet valve to the filter. Check for proper pressure and airflow to the respirator equipment and airflow in the helmet. Adjust as required.

> **WARNING!**
> Breathing air must be monitored or periodically tested to prevent death from the inhalation of carbon monoxide.

Step 8 If used, turn on CO monitors or converters and check their operation in accordance with the manufacturer's instructions. If monitors or converters are not used, have the breathing air tested for the presence of CO. The test will have to be repeated periodically to make sure CO is not present in the breathing air.

Step 9 Put on the respirator equipment, communications equipment (if used), and gloves.

> **WARNING!**
> Before emptying waste or loading abrasives, make sure the proper respirator equipment is used and is functioning.

Step 10 If not automatically emptied, check the dust and waste storage hoppers on the dust collectors and recycling equipment and empty them if necessary.

Step 11 If used, start the loading/recovery/recycling equipment in accordance with the manufacturer's instructions.

Step 12 With the abrasive metering valve and air-inlet valve closed (shut off) and the pressure tank exhaust (blowdown) valve open, load the blast machine pressure tank with the desired abrasive manually or with the loading equipment.

Step 13 With the pressure tank exhaust valve closed, open the air-inlet valve to the blast machine and check that the pop-up valve seals and does not leak and that the tank pressurizes to approximately the correct pressure.

Step 14 Put on the blast suit along with the respirator equipment, communications equipment (if used), and gloves.

> **WARNING!**
> Make sure that you wear the proper blast suit and protective equipment. Do not point the nozzle at any body parts or another person; severe injury may occur from the jet of abrasive and/or air emerging from the nozzle. Always engage the safety latch on the control handle when blast cleaning is stopped, even for a moment.

Step 15 With the blast machine abrasive metering valve closed (shut off), check the operation of the blast-nozzle control handle. Make sure that the dead-man switch starts and stops airflow and that the safety latch locks the dead-man switch so that it cannot be operated. If not, find and correct the problem in the control handle or the blast machine control system.

Safety

Stress the importance of monitoring breathing air when starting up a blast system. Emphasize the use of appropriate personal protective equipment. Remind trainees not to point the nozzle at any body parts or another person.

Classroom

Have trainees explain the safety guidelines that must be followed when operating blast systems.

Step 16 Release the safety latch, trigger the control handle dead-man switch, and with air flowing from the nozzle, check for blast hose, coupling, and nozzle holder leaks. If leaks are detected, release the control handle dead-man switch, engage the safety latch, and replace the blast hoses, couplings, coupling gaskets, or nozzle holder and/or gasket as required. When the repair is completed, repeat the leak check until all leaks are repaired.

Step 17 After the leak check, make sure the blast hose is purged with air for several minutes to remove any moisture condensation. Then, direct the air jet from the nozzle onto a piece of white cloth for a few seconds. Check the cloth for traces of oil or moisture. If detected, find and correct the problem before continuing.

Step 18 Open the abrasive metering valve on the blast machine. Trigger the control handle. Check that the abrasive flows from the nozzle after a short period of time. Adjust the abrasive metering valve for the correct flow characteristics. With the abrasive flowing, check the nozzle pressure with a hypodermic needle pressure gauge for the correct operating pressure. Adjust the regulator at the air input to the blast machine for the correct nozzle pressure, but do not exceed the working pressure of the blast machine. Re-adjust the abrasive flow if necessary.

NOTE

If the pressure drop at the nozzle is excessive or the abrasive flow is abnormal, the blast machine or the blast hoses may be clogged, or the hose size may be too small. If the pressure at the blast machine drops as well, the nozzle may be too large for the capacity of the compressor or the blast machine.

Step 19 Load the abrasive into the blast machine storage hopper manually or with loading equipment.

4.5.3 Blast System Operation Guidelines

Before performing the following steps, review these safety guidelines:

- Any scaffolding used must be secure and working areas must be clear of debris, materials, and hoses. A proper fall protection system must be used. Grip the control handle and nozzle firmly and make sure footing is secure before triggering the control handle. Be prepared for the reactive force from the nozzle.
- Make sure that you wear the proper blast suit, air-supplied helmet, and hearing protection. Do not point the nozzle at any body parts or another person as severe injury may occur from the jet of abrasive and/or air emerging from the nozzle. Always engage the safety latch on the control handle when blast cleaning is stopped, even for a moment.
- Make sure that you wear any additional respiratory protection required and/or an air-supplied helmet before commencing blast cleaning operations.
- If using a wet blast attachment or injector, make sure the electricity to motors, electrical boxes, or devices is turned off and locked out or that the boxes have been water sealed with material resistant to the abrasive blast jet.
- Make sure that the blast machine and any conductive work surfaces are grounded and the blast hose is the static-suppressing or grounded type. When working in **confined spaces** such as tanks, make sure all volatile fumes and dust are removed by ventilation.
- Abrasive blast media can cause undesired damage to soft substrates. Check that the substrate can tolerate the abrasion, if any, that may be caused by the media that you will be using before you begin blast cleaning.

If not already accomplished, perform the blast system start-up guidelines and follow these steps to operate the blast systems:

Step 1 Check that the surfaces to be blasted have been cleaned of soluble salts, grease, and oil.

Step 2 Make sure that any required protection or containment has been implemented for the adjacent surfaces or areas.

Instructor's Notes:

Step 3 Check that any scaffolding used is secure and that the work area is clear.

Step 4 With the proper grip and stance, disengage the safety latch, trigger the dead-man switch on the control handle, and begin blast cleaning operations using the general air abrasive cleaning operating techniques outlined in this module.

Step 5 Grip the nozzle and keep it moving in slow, steady strokes with little **overlap** of the blast pattern. Keep the nozzle at the same distance from the surface throughout the stroke. Do not whip it around or move it in an arc. Do not dwell in one spot too long because undesired damage to the substrate may occur.

Step 6 Blast from the top down, side-to-side on flat, vertical surfaces and side-to-side, front-to-back on flat, horizontal surfaces, if possible.

Step 7 Blow clean (or blow wash) blasted surfaces periodically or before moving to the next section of the work surface to be blasted, if possible.

Step 8 Stop blast cleaning operations periodically, re-engage the dead-man switch safety latch, refill the blast machine, and conduct expended-abrasive cleanup and recovery as needed.

4.5.4 Air-Abrasive Blast Cleaning Operating Techniques

Regardless of which type of abrasive blast cleaning is used, always ensure that the surface has first been thoroughly washed with solvent or detergent to remove oil, grease, and soluble salts that could otherwise be spread or driven into the surface by the impact of the blasting.

Stand-off distance, angle of attack, and dwell time are three variables that affect the quality and effectiveness of the blasting operation.

Stand-off distance is the distance from the nozzle to the surface being blasted. It determines both the nozzle's cleaning power and the size of the blast pattern it produces. The closer the nozzle, the stronger the blast and the smaller the pattern. While old coating may be removed effectively at a distance of several feet, tight mill scale may require a stand-off of only 12 inches. The smaller the blast pattern, the longer it takes to blast the entire surface. Experiment with stand-off distance to find the right combination of power and productivity for each job.

The angle of attack, or blast angle, is the position of the abrasive/air stream from the nozzle with respect to the surface. The most effective angle of attack also depends on the needs of each job. For general cleaning, 60 degrees to 70 degrees is a good range for the angle of the nozzle to the surface. Rust, mill scale, and heavily pitted surfaces require a head-on attack with a nozzle angle of 80 degrees to 90 degrees.

Dwell time is how long the operator spends blasting any one spot of the surface. Loose contaminants may disappear immediately, while tightly adhered contaminants may take several seconds of blasting to remove.

To achieve a consistent cleaning and blast pattern, move the nozzle in a straight pass across the surface, maintaining the same stand-off distance, angle of attack, and dwell time. Slightly overlap each pass. It may be necessary to make adjustments if the surface is irregular or its conditions vary.

4.5.5 Blast System Shutdown Guidelines

At the end of blast cleaning operations or end-of-day, perform the following applicable operations before removing the blast suit and protective equipment. Before removing filters, emptying waste, or loading or cleaning up abrasives, make sure the proper respirator equipment is used and is functioning.

Step 1 If the abrasive left in the blast machine will not be used the following day or must be changed to a different abrasive, purge the blast machine and hopper by fully opening the abrasive metering valve and discharging the remaining abrasive into the abrasive loading/recovery/recycling equipment with the blast machine hopper feed disabled.

Step 2 Completely purge the abrasive from the blast hose. Close the abrasive metering valve.

Step 3 Clean up any other loose expended abrasive using the loading/recovery/recycling equipment.

Step 4 Remove the abrasive from the loading/recovery/recycling storage hoppers after all residual abrasive is processed. Cover all open hoppers.

Classroom

Have three trainees identify and explain the three variables that affect the quality and effectiveness of the blasting operation.

Teaching Tip

See the Teaching Tip for Sections 4.5.0–4.5.5 at the end of this module.

Performance Testing

Have each trainee complete Performance Task 2 to your satisfaction. Fill out a Performance Profile Sheet for each trainee.

Homework

Assign the reading of Sections 5.0.0–5.3.0 and the completion of the Review Questions for the next session.

Materials

Ensure that you have everything required for testing during this session.

Classroom

Have trainees name the coatings suitable for chemical stripping. Ask trainees to name coating materials that require specialty strippers.

Ask a trainee to explain what pickling is used for. Have a trainee explain how metal is pickled.

Ask a trainee to identify uses for high-pressure and ultra-high-pressure water jetting. Have a trainee identify situations in which water jetting is economical.

Step 5 If the abrasive in the blast machine will be used the following day or in the near future, close the abrasive metering valve and completely purge all abrasive from the blast hose. Clean up any loose expended abrasive using the loading/recovery/ recycling equipment. After all residual abrasive is processed, cover any open hoppers.

Step 6 If they are not emptied automatically, empty the dust and waste storage hoppers on the dust collectors and recycling equipment.

Step 7 Shut down the loading/recovery/recycling equipment and any auxiliary ventilation equipment. Remove the filters on the dust collectors, and check for any deterioration and for excessive accumulation of dust and dirt.

Step 8 Remove the blast suit and protective equipment once the air has been tested and shown to be clear of dust.

Step 9 Turn off the breathing air CO monitors or converters.

Step 10 Close the inlet-air valve on the blast machine and then shut down the blast and breathing air compressors. Vent the pressure in the receivers by opening the vent valves.

Step 11 Drain all oil and moisture traps and separators. Close the drain and vent valves.

Step 12 Open the pressure tank exhaust valve on the blast machine to relieve the pressure.

Step 13 Drain oil and moisture separators and traps. Close the exhaust and drain valves. Cover any open hoppers.

Step 14 Cover any stockpiled abrasive to prevent any contamination or water damage.

5.0.0 ◆ CHEMICAL STRIPPERS

In some situations where minimal substrate damage is desired, chemical stripping may be the only surface preparation method available to accomplish the task. Chemical strippers are suitable for removing most oil-based, thermoplastic, and water-based coatings. However, it is very difficult and labor intensive, and sometimes impossible, to remove thermoset materials such as epoxies, urethanes, polyesters, and other high-performance coating materials with conventional removers. In some cases, specialty strippers that will be effective in removing these types of coatings may be available.

There are three general classes of removers, or strippers, that are available in either thin or heavy-body viscosities. The classes are nonflammable, flammable, and alkali removers. Some removers are blends of different types. For example, some proprietary formulations combine flammable solvents with alkalies to enhance the stripping ability.

5.1.0 Pickling (*SSPC-SP8*)

Pickling is a method of preparing relatively new steel surfaces for coating by chemical reaction and/or electrolysis. Usually, it is used to remove mill scale from new hot-rolled plate, sheet, or structural steel stock and is accomplished by immersion of the steel in dilute acid baths. It is also used to remove oxides (light rust) from new steel.

A pickled surface is defined by *SSPC-SP8* as a surface that, when viewed without magnification, shall be free of all visible mill scale and rust. The surface shall also be etched to a degree suitable for the specified coating system. The color of the pickled surface may be affected by the grade, original surface condition, and configuration of the material being cleaned, as well as by discolorations from mill or fabrication marks and the shadowing from etching patterns.

5.2.0 High- and Ultra-High-Pressure Water Jetting (*SSPC-SP12/NACE 5*)

High-pressure water jetting (HP WJ) and ultra-high-pressure water jetting (UHP WJ), both also called *water blasting* or *hydro-blasting*, are becoming viable alternatives to dry abrasive blasting for the preparation of certain concrete and steel surfaces for recoating. This is possible because the newer equipment is available that can operate up to 40,000 psi, or 276 megaPascals (MPa) with reduced water volume, which results in efficient contaminant and coating removal rates with less water stream kick to the operator. Water jetting can also effectively remove soluble salt contaminants that dry abrasive blasting, for the most part, cannot remove. When removing hazardous coatings or when total coating removal or total profiling of a surface is not required, water jetting becomes an economical choice.

Instructor's Notes:

5.3.0 Environmental Humidity/ Temperature Control

Environmental conditions such as humidity and temperature have a substantial impact on cleaning and coating operations. In some cases, the cleaning or coating of a substrate, especially ferrous materials (steel or iron) or concrete, must be delayed because of excess moisture, humidity, or temperature. In other cases (especially outdoors), weather, humidity, and temperature are controlling factors in the cycle of ferrous or concrete surface preparation and coating. In nearly all instances, manufacturers' predictions of coating service life are based on optimum, ideal surface preparation and application conditions that are difficult, if not impossible, to attain in the field without some form of containment and environmental control.

Summary

Surface preparation before coating application is a critical step in ensuring the durability, quality, and effectiveness of the coating. Standards for proper preparation of surfaces or substrates for coating are covered by SSPC and NACE. The three major steps in the preparation process are cleaning, repairing, and pretreating the surface/substrate as necessary.

Blasting is generally used to thoroughly clean the surface, although other methods such as chemical stripping and high-pressure water jet cleaning are also used. Blasting equipment includes air compressors, pipes, hoses, couplings, and blast machines. The blasting process consists of using high-pressure air mixed with abrasives to scour the surface clean of dirt, grime, and other debris.The amount of air and abrasive flow is governed largely by the nozzle size and the air pressure used. The size and extent of the project determines the size of the equipment needed.

Blasting operators must wear appropriate personal protective equipment and follow company safety procedures. Blast cleaning can be a dangerous, even deadly, procedure if proper precautions are not taken.

Pressurized water is used in some systems instead of air. Centrifugal blast cleaning is similar to blast cleaning. The difference is that pressurized air or water is not used; abrasives are accelerated by centrifugal force and hurled against the surface at high speed.

A variety of abrasives may be used in these systems. Factors involved in abrasive selection include the type of finish desired, the type of surface to be blasted, the surface condition, the abrasive consumption rate and cost, and the environmental requirements. The finished surface is inspected and measured against standards for quality.

Review Questions

1. The most important part of a pipeline coating job is the _____.
   a. curing time
   b. finish application
   c. surface preparation
   d. material mixing

2. A _____ stores an adequate supply of abrasive and mixes it with high-volume compressed air.
   a. blast machine
   b. compressor
   c. mobile mixer
   d. tank

3. For compressed air blast systems, the _____ is the most critical component of the system.
   a. blast pipe
   b. air compressor
   c. blast machine
   d. air pump

4. To prevent pressure drop from friction, the inside diameter of hoses and couplings should be at least _____ times the size of the orifice diameter.
   a. 10
   b. 6
   c. 5
   d. 3

5. The bursting strength rating of hoses should be about _____ times the working pressure rating.
   a. five
   b. four
   c. three
   d. two

6. Nearly all of the machines used for commercial work are _____-feed.
   a. gravity
   b. siphon
   c. vacuum
   d. water

7. High production machines have _____ cubic feet of abrasive storage.
   a. 30 to 40
   b. 20 to 30
   c. 6 to 20
   d. 1 to 10

Performance Testing

Have each trainee complete Performance Task 3 to your satisfaction. Fill out a Perfomance Profile Sheet for each trainee.

Classroom

Go over the Review Questions for this module, and answer any questions trainees may have.

8. Blast hoses should be run _____.
   a. in large curved patterns
   b. on an incline
   c. in a looped pattern
   d. in a straight line

9. The outer covering of blast hoses is covered with _____.
   a. ground wires
   b. slits
   c. rubber sleeves
   d. pin holes

10. _____ are used to prevent the hoses from flying around if the joint breaks.
    a. Whip checks
    b. Couplings
    c. Rope supports
    d. Sleeves

11. The air/abrasive flow through a nozzle is determined by the _____.
    a. hose size
    b. compressor speed
    c. orifice size
    d. abrasive type

12. The air stream from the nozzle should have _____ discoloration from the abrasive.
    a. heavy
    b. slight
    c. substantial
    d. no

13. Blast cleaning propels abrasive media to speeds in excess of _____ mph.
    a. 100
    b. 200
    c. 300
    d. 400

14. Instead of pressurized air, centrifugal blast cleaning uses _____.
    a. water
    b. a motorized wheel
    c. steam
    d. a vacuum pump

15. Centrifugal blast cleaning is _____ than air abrasive blast cleaning.
    a. less expensive
    b. cleaner
    c. more difficult
    d. slower

16. Wet blast cleaning is a method used with a conventional blast machine to _____.
    a. suppress dust
    b. cool the surface
    c. speed up cleaning
    d. intensify blasting

17. Used with recyclable abrasive, _____ tools can provide environmentally safe blast cleaning.
    a. portable
    b. wet blast
    c. closed-circuit recovery
    d. centrifugal blast

18. There are four types of abrasives: _____, mineral, by-product, and special purpose.
    a. gravel
    b. chemical
    c. plastic
    d. metallic

19. From the beginning, _____ has been used for abrasive blast cleaning.
    a. sand
    b. plastic pellets
    c. water
    d. acid

20. Rounded or spherical abrasive particles clean by _____ the surface.
    a. cutting
    b. hitting
    c. gouging
    d. compressing

21. The two primary areas of concern in the surface preparation of steel are _____.
    a. profile and smoothness
    b. cleanliness and rust prevention
    c. cleanliness and profile
    d. oxidation and pitting

Instructor's Notes:

22. Once steel surfaces are cleaned and have metal exposed, _____ may begin almost immediately.
   a. streaking
   b. flash-rusting
   c. build-up
   d. pitting

23. To check the surface profile, use tools such as the _____.
   a. micrometer
   b. thickness gauge
   c. scale
   d. comparator

24. In some cases, the cleaning of a substrate, especially ferrous materials, must be delayed because of _____.
   a. low humidity
   b. low temperatures
   c. wind
   d. high humidity

25. Breathing-air compressors can, and do, produce _____.
   a. hydrogen chloride
   b. carbon monoxide
   c. carbon dioxide
   d. dust particles

26. Make sure the breathing-air compressor is located _____ of other engine-driven equipment.
   a. away from and upwind
   b. near but upwind
   c. away from and downwind
   d. anywhere, irrespective

27. _____ is the distance from the nozzle to the surface being blasted.
   a. Range
   b. Length
   c. Stand-off
   d. Spread

28. For general cleaning, _____ degrees is a good range for the angle of the nozzle to the surface.
   a. 60 to 70
   b. 45 to 50
   c. 20 to 30
   d. 10 to 20

29. Chemical strippers are most suitable for removing most _____ coatings.
   a. oil-based
   b. urethane
   c. polyester
   d. epoxy

30. _____ is a method of preparing relatively new steel surfaces for coating by chemical reaction and/or electrolysis.
   a. Water jetting
   b. Oxidation
   c. Pickling
   d. Stripping

Examination

Administer the Module Examination. Be sure to record the results of the Exam on Craft Training Report Form 200 and submit the results to the Training Program Sponsor.

Performance Testing

Ensure that all Performance Tests have been completed and Performance Profile Sheets for each trainee are filled out. Be sure to record the results of the testing on Craft Training Report Form 200 and submit the results to the Training Program Sponsor.

Instructor's Notes:

GLOSSARY

Glossary

This glossary will help you learn some terms specific to performing high-pressure blasting and surface preparation. Trade Terms introduced in this module appear in italic text.

***Abrasive blast cleaning:** Cleaning and roughening of a surface produced by the high-velocity impact of an abrasive that is propelled by the discharge of pressurized fluid from a blast nozzle or by a mechanical device such as a centrifugal blasting wheel. (Also referred to as Abrasive Blasting.)

Blast pots: Another name for blasting machines.

Bond/adhesion: The attachment of a coating to a substrate. Adhesion is the key to coating, effectively allowing the coating to become part of the substrate. The three types of adhesive bonds are chemical, polar, and mechanical.

***Brush-off blast cleaned surface:** A brush-off blast cleaned surface, when viewed without magnification, shall be free of all visible oil, grease, dirt, dust, loose mill scale, loose rust, and loose coating. Tightly adherent mill scale, rust, and coating may remain on the surface. Mill scale, rust, and coating are considered tightly adherent if they cannot be removed by lifting with a dull putty knife. [See *NACE No. 4/SSPC-SP 7*.]

Cigarette wrap: Application technique for cut lengths of sheet coating. The width of the sheet must exceed the circumference of the pipe to provide an overwrap. The sheet is placed longitudinally and wrapped no more than twice around the pipe. Ending seams should be alternated along the pipe as to not be continuous. Each sheet should overlap the previous. This method is not commonly used, with the exception of heat shrink sleeves.

Coal tar: A tar distilled from bituminous coal used to coat piping. Usually a paper wrap is used over coal tar.

***Commercial blast cleaned surface:** A commercial blast cleaned surface, when viewed without magnification, shall be free of all visible oil, grease, dust, dirt, mill scale, rust, coating, oxides, corrosion products and other foreign matter. Random staining shall be limited to no more than 33 percent of each unit area (approximately 58cm^2 [9.0 in.^{2}]) of surface and may consist of light shadows, slight streaks, or minor discolorations caused by stains of rust, stains of mill scale, or stains of previously applied coating. [See *NACE No. 3/SSPC-SP 6*.]

Confined space: A space that: 1) Is large enough and so configured that an employee can bodily enter and perform assigned work; and 2) Has limited or restricted means for entry and exit (e.g. tanks, vessels, vaults, etc.); and 3) Is not designed for continuous employee occupancy.

***Cracking (of coating):** Breaks in a coating that extend through to the substrate.

Fusion bond: Application of a powder coating to a heated pipe, performed by electrostatic methods. The powder gels, flows with the heat, then cures and hardens during cooling. The process creates a tight physical bond between the coating and metal.

Hand cleaning: The process of removing burrs, slag, rust, mill scale and other contaminants from steel pipe. Burrs or slag may be removed by grinding.

* Term published in NACE *Glossary of Corrosion-Related Terms*. Reproduced with permission from NACE International, Houston, TX. All rights reserved. © NACE International 2002. For a copy of the complete *Glossary of Corrosion-Related Terms*, contact NACE International, 1440 South Creek Dr., Houston, TX 77084. The Glossary is revised periodically, so information in an outdated version may not be accurate.

GLOSSARY (Continued)

***Hand tool cleaning:** Removal of loose rust, loose mill scale, and loose paint to degree specified, by hand chipping, scraping, sanding, and wire brushing. [See *SSPC-SP 2*.]

Hand wash: Removal of oil, dirt, soil, grease, and other contaminants by cleaning with solvents and/or detergents and/or similar products.

***Holiday:** A discontinuity in a protective coating that exposes unprotected surface to the environment.

Mastic: Asphalt-like mixture used to coat piping.

NACE: National Association of Corrosion Engineers.

***Near-white blast cleaned surface:** A near-white blast cleaned surface, when viewed without magnification, shall be free of all visible oil, grease, dust, dirt, mill scale, rust, coating, oxides, corrosion products and other foreign matter. Random staining shall be limited to no more than 5% of each unit area of surface (approximately 58 cm^2 [9.0 $in.^2$]) and may consist of light shadows, slight streaks, or minor discolorations caused by stains of rust, stains of mill scale, or stains of previously applied coating. [See *NACE No. 2/SSPC-SP 10*.]

Overlap: An amount of coating covering the previous row of coating in a row-to-row sequence.

Permit-required confined space: A confined space that has one or more of the following characteristics: 1) Contains or has a potential to contain a hazardous atmosphere; 2) Contains a material that has the potential for engulfing an entrant; 3) Has an internal configuration such that an entrant would be trapped or asphyxiated by inwardly converging walls or by a floor which slopes downward and tapers to a smaller cross-section; or 4) Contains any other recognized serious safety or health hazard.

***Power tool cleaning:** Removal of loose rust, loose mill scale, and loose paint to the degree specified by power tool chipping, descaling, sanding, wire brushing, and grinding. [See *SSPC-SP 3*.]

Pre-heat: Pre-heating of the pipeline metal prior to application of certain coatings. Heat shrink sleeves and several other coatings require the pipe to be pre-heated prior to applying the coating. Caution: Excessive heat can damage the coating.

**Primer:* A coating material intended to be applied as the first coat on an uncoated surface. The coating is specifically formulated to adhere to and protect the surface as well as to produce a suitable surface for subsequent coats. [Also referred to as Prime Coat.]

Shrink sleeve: A conformable, wrap-around sleeve designed to protect below-ground and submerged metal surfaces from corrosion and electrolysis. The sleeve consists of two overlapping pieces or a single tube surrounding the pipe. Heat is applied to the sleeves until an adhesive material in the sleeves bonds them to the pipe and joins the overlapping sections to form a tight bond.

SSPC: Society for Protective Coatings (formerly Steel Structures Painting Counsel).

Surface profile: Graphic representation of a surface condition.

Wax: The wax present in petroleum crudes primarily consists of paraffin hydrocarbons known as paraffin wax and naphthenic hydrocarbons.

Whip check: A safety cable used to connect hoses near the coupling to prevent hoses from flying around if the joint breaks.

***White metal blast cleaned surface:** A white metal blast cleaned surface, when viewed without magnification, shall be free of all visible oil, grease, dust, dirt, mill scale, rust, coating, oxides, corrosion products and other foreign matter. [See *NACE No. 1/SSPC-SP 5*.]

* Term published in NACE *Glossary of Corrosion-Related Terms*. Reproduced with permission from NACE International, Houston, TX. For a copy of the complete *Glossary of Corrosion-Related Terms*, contact NACE International, 1440 South Creek Dr., Houston, TX 77084. The Glossary is revised periodically, so information in an outdated version may not be accurate.

Instructor's Notes:

MODULE 61207-02 — TEACHING TIPS

The following are suggested activities or instructional methods to help you teach the material in this AIG.

General

When you call on someone to answer a question, the rest of the class relaxes or even tunes out because they expect that the question and answer will take place only between you and the trainee you called on. Instead, use this technique to involve more trainees in answering questions and to keep them on their toes.

1. Ask trainees to define a term or explain a concept.
2. After one trainee has answered, ask a trainee seated nearby if the answer is right. Then ask whether a trainee in the back of the room agrees.
3. Ask trainees to explain why they think an answer is right or wrong.
4. Use the session to clear up incorrect ideas and encourage trainees to learn from their mistakes.

Section 2.2.0 *Compressor Air Pipes/Hoses and Couplings Inspection*

Trainees will practice performing daily checks on compressor components. Trainees will need appropriate personal protective equipment, pencils, paper, and their Trainee Guides. You will need to provide access to a compressor. Allow 1 hour for this exercise.

1. Have trainees review the inspection procedure as explained in Section 2.2.0 of the Trainee Guide.
2. Ask a trainee to explain the importance of daily checks for damage or leaks. Discuss possible consequences of faulty components.
3. Supervise trainees as they inspect the whip checks, hoses, couplings, and coupling gaskets. Have trainees take notes on the components they check and their findings.
4. Ask trainees to present their findings to the class. Answer any questions trainees may have.

Section 4.4.0 *Safety Guidelines*

In this exercise, trainees will familiarize themselves with the safety guidelines for operating a blast system. Trainees will need pencils, paper, and their Trainee Guides. You will need to describe scenarios in which safety guidelines are violated during operation of a blast system. Allow 1 hour for this exercise.

1. Have trainees take turns simulating or explaining the safety guidelines for operating a blast machine, as listed in Section 4.4.0 of their Trainee Guides.
2. Present scenarios in which safety guidelines are not followed during operation of a blast system. Have trainees explain which safety guidelines were violated and the potential results of violating those guidelines.
3. Ask trainees if they have any questions.

Sections 4.5.0–4.5.5

Blast System Start-Up, Operation, and Shutdown Guidelines

In this exercise, trainees will practice the procedures for operating a blast system. Trainees will need appropriate personal protective equipment and their Trainee Guides. You will need to arrange access to a blast system and provide necessary materials and equipment, including the manufacturer's operation and instruction manual. Alternatively, you may arrange for trainees to observe an expert operate a blast system or have trainees simulate the use of the blast system. Allow 2 to 3 hours for this exercise.

1. Have trainees review the manufacturer's procedures for starting up, operating, and shutting down a blast system.
2. Have trainees start up, operate, and shut down the blast system. Emphasize safety procedures. Either during or after the exercise, ask trainees to explain the reasons for various steps in the procedure.
3. Ask trainees if they have any questions.

MODULE 61207-02 — ANSWERS TO REVIEW QUESTIONS

Answer	Section
1. c	1.0.0
2. a	2.0.0
3. b	2.1.0
4. d	2.2.0
5. b	2.2.0
6. a	2.3.0
7. c	2.3.0
8. d	2.4.0
9. d	2.4.0
10. a	2.4.0
11. c	2.5.0
12. b	2.5.0
13. d	2.6.0
14. b	2.7.0
15. b	2.7.0
16. a	2.8.1
17. c	2.8.4
18. d	3.1.0
19. a	3.1.0
20. b	3.1.0
21. c	4.0.0
22. b	4.1.0
23. d	4.2.0
24. d	4.3.0
25. b	4.4.0
26. a	4.5.1
27. c	4.5.4
28. a	4.5.4
29. a	5.0.0
30. c	5.1.0

CONTREN™ LEARNING SERIES — USER UPDATES

The NCCER makes every effort to keep these textbooks up-to-date and free of technical errors. We appreciate your help in this process. If you have an idea for improving this textbook, or if you find an error, a typographical mistake, or an inaccuracy in NCCER's Contren™ textbooks, please write us, using this form or a photocopy. Be sure to include the exact module number, page number, a detailed description, and the correction, if applicable. Your input will be brought to the attention of the Technical Review Committee. Thank you for your assistance.

Instructors – If you found that additional materials were necessary in order to teach this module effectively, please let us know so that we may include them in the Equipment/Materials list in the Instructor's Guide.

Write: Curriculum Revision and Development Department
National Center for Construction Education and Research
P.O. Box 141104, Gainesville, FL 32614-1104

Fax: 352-334-0932

E-mail: curriculum@nccer.org

Craft ________ Module Name ________

Copyright Date ________ Module Number ________ Page Number(s) ________

Description ________

____

____

____

(Optional) Correction ________

____

____

(Optional) Your Name and Address ________

____

____

Module 61208-02

Apply Coatings Using Spray Applications (CT 7.6 and 13.5)

Apply Coatings Using Spray Applications
Annotated Instructor's Guide

Module 61208-02

MODULE OVERVIEW

This module introduces the basic ingredients and film-forming processes common to all coatings. It also describes coating systems and the functional categories of coatings with an emphasis on industrial high-performance coatings. Trainees will learn to apply coatings according to manufacturer specifications.

PREREQUISITES

Please refer to the Course Map in the Trainee Module. Prior to training with this module, it is recommended that the trainee shall have successfully completed the following:

Core Curriculum; Pipeline Core, Modules 66101-02 and 66102-02; Pipeline Corrosion Control Level One; Pipeline Corrosion Control Level Two, Modules 61201-02 through 61207-02

OBJECTIVES

Upon completion of this module, the trainee will be able to:

1. Identify and explain different coating types (CT 13.5).
2. Determine whether or not the coating is compatible with environmental conditions, the substrate to be coated, any related job-site conditions, or any other requirements of the job (CT 7.6).
3. Prepare the surface for coating application according to manufacturer specifications (CT 7.6 and CT 13.5).
4. Identify and explain the different spray application systems used for coating.
5. Apply coating according to manufacturer specifications (CT 7.6 and CT 13.5).

PERFORMANCE TASKS

Under the supervision of the instructor, the trainee should be able to:

1. Identify the correct coating repair method/procedure for a given type of coating (CT 7.6).
2. Correctly prepare a pipeline surface for coating application (CT 7.6 and 13.5).
3. Properly apply coating to a pipeline exposed to atmospheric conditions (CT 7.6 and 13.5).
4. Visually inspect and/or repair coating defects (CT 13.5).

NCCER STANDARDIZED CRAFT TRAINING PROGRAM

The National Center for Construction Education and Research (NCCER) provides a standardized national program of accredited craft training. Key features of the program include instructor certification, competency-based training, and performance testing. The program provides trainees, instructors, and companies with a standard form of recognition through a National Craft Training Registry. The program is described in full in the *Guidelines for Accreditation*, published by the NCCER. For more information on standardized craft training, contact the NCCER by writing us at P.O. Box 141104, Gainesville, FL 32614-1104; calling 352-334-0911; or e-mailing info@nccer.org. More information may be found at our Web site, www.nccer.org.

HOW TO USE THIS ANNOTATED INSTRUCTOR'S GUIDE

Each page presents two sections of information. The larger section displays each page exactly as it appears in the Trainee Module. The narrow column ties suggested trainee and instructor actions to each page and provides icons to call your attention to material, safety, audiovisual, or testing requirements. The bottom of each page includes space for your notes.

If you see the Teaching Tip icon, that means there is a teaching tip associated with this section. Also refer to any suggested teaching tips at the end of the module.

SAFETY CONSIDERATIONS

Ensure that the trainees are equipped with appropriate personal protective equipment. Emphasize the importance of following all safety precautions and procedures when working with power tools.

PREPARATION

Before teaching this module, you should review the Module Outline, Objectives, Performance Tasks, and the Materials and Equipment List. Be sure to allow ample time to prepare your own training or lesson plan and gather all required equipment and materials.

MATERIALS AND EQUIPMENT LIST

Overhead projector and screen

Whiteboard/chalkboard

Transparencies

Markers/chalk

Blank acetate sheets

Transparency pens

Pencils and scratch paper

Module Examinations*

Performance Profile Sheets*

Copies of Quick Quiz**

Copies of the Performance Verifications for Covered Tasks 7 and 13†

Appropriate personal protective equipment

Copies of your company's policies and procedures manual

Sample manufacturer's coating system data sheets

Respirator

Necessary equipment and materials for spraying

Necessary equipment and materials for cleaning and maintaining conventional spray equipment

Necessary equipment and materials for using, cleaning, and maintaining HVLP spray equipment

Samples of various types of spray coating problems

Viscosity cup and coating material

Wet-film thickness gauge, dry-film thickness gauge, and coated substrates

* Located in the Test Booklet.
**Located at the end of this module.
† Available for download from www.nccer.org.

TEACHING TIME FOR THIS MODULE

An outline for use in developing your lesson plan is presented below. Note that each Roman numeral in the outline equates to one session of instruction. Each session has a suggested time period of 3 hours. This includes 10 minutes at the beginning of each session for administrative tasks and one 10-minute break during the session. Approximately 18 hours are suggested to cover *Apply Coatings Using Spray Applications (CT 7.6 and 13.5)*. You will need to adjust the time required for hands-on activity and testing based on your class size and resources.

Topic	Planned Time
Session I. Coating Materials, Part One	
A. Pigments	____________
B. Resins	____________
C. Solvents	____________
D. Additives	____________
E. Film-Forming Mechanisms and Generic Coating Types	____________
Session II. Coating Materials, Part Two	
A. Coating Systems and Functional Classifications of Coatings	____________
B. Powder Coatings	____________
C. Selecting a Coating	____________
D. Surface Preparation	____________
1. Performance Testing	____________
E. Coating Preparation and Application	____________
F. Testing and Inspection	____________
Session III. Air Spray Systems, Part One	
A. Conventional Air Spray Systems and Components	____________
B. HVLP Spray Systems	____________
C. Spraying with a Conventional Spray System	____________
1. Laboratory	____________
Session IV. Air Spray Systems, Part Two	
A. Spraying with an Airless System	____________
B. Spraying with an HVLP System	____________
C. Common Spray Coating Problems	____________
D. Measuring the Viscosity of a Coating Material	____________
1. Laboratory	____________
E. Measuring the Thickness of a Coating Material	____________
1. Laboratory	____________
Session V. Electrostatic Spray Systems	
A. Types of Electrostatic Systems	____________
B. Principles of Electrostatic Sprayer Operation	____________
C. Key Equipment Characteristics	____________
D. Coating Materials	____________
Session VI. Thermal Spray Systems	
A. Types of Processes	____________
B. Flame and Wire Arc Spraying System Equipment	____________
C. Thermal Spray Coating Materials	____________
1. Laboratory	____________
D. Thermal Spraying Guidelines for Zinc and Aluminum Coatings	____________
E. Dry-Powder Coating Application Processes	____________

F. Summary
   1. Summarize module ____________
   2. Answer questions ____________

G. Module Examination ____________
   1. Trainees must score 70% or higher to receive recognition from the NCCER.
   2. Record the testing results on Craft Training Report Form 200 and submit the results to the Training Program Sponsor.

H. Performance Testing ____________
   1. Trainees must perform each task to the satisfaction of the instructor to receive recognition from the NCCER.
   2. Record the testing results on Craft Training Report Form 200 and submit the results to the Training Program Sponsor.

Module 61208-02

Apply Coatings Using Spray Applications

(CT 7.6 and 13.5)

Instructor's Notes:

Instructor's Notes:

Introduction to the Pipeline Operator Qualification Rule, API, and NCCER

It could be said that petroleum products keep America running. Petroleum products like oil and natural gas are the principal sources of energy in the United States and are used in many manufacturing and chemical processes. Petroleum products are primarily transported by pipeline. There are more than a million miles of liquid and natural gas pipeline in this country, ranging in diameter from two inches to four feet. These pipelines bring crude oil and natural gas from production fields to processing plants and refineries—and ultimately to the consumer.

Pipeline Operator Qualification

Petroleum products are considered hazardous, so pipeline personnel must take extra precautions and follow specific procedures when working with them. To help ensure the safe handling and transportation of petroleum products, the United States Department of Transportation (DOT) mandates safe practices as federal regulations.

One such regulation, the Operator Qualification (OQ) rule, requires pipeline operators to develop a written qualification plan to evaluate their personnel and their contract personnel (see *49 CFR Part 192 Subpart N—Gas*, and *49 CFR Part 195 Subpart G—Liquid*). The evaluation hinges on the operator's ability to perform certain covered tasks and to be able to recognize and respond to abnormal operating conditions that may be encountered while performing these covered tasks.

Covered Tasks

A pipeline performance task is deemed *covered* if the activity passes the following four-part test:

1. Is it performed on a pipeline facility?
2. Is it an operations or maintenance task?
3. Is it performed as a requirement of *49 CFR Part 195 (Liquid)* or *49 CFR Part 192 (Gas)*?
4. Does it affect the operation or integrity of the pipeline facility?

If the answer to all four questions is *yes*, then the activity is considered *covered* and must figure into the evaluation process of pipeline personnel.

American Petroleum Institute

The OQ rule and its related covered tasks provided the catalyst to develop a unified pipeline training and assessment program—so that pipeline operators and owners have ready access to the assessment and training tools they need. This effort was the direct result of the vision and leadership provided by The American Petroleum Institute (API) and API members.

Efforts to unify the industry are nothing new to API. Indeed, the impetus for forming the Institute in 1919 was the need to standardize engineering specifications for drilling and production equipment. Today, the most pressing issues revolve around public perceptions and government policies toward the industry. Speaking with one voice on these issues has become as essential as having interchangeable parts in the field. API's members determine the petroleum industry's positions on public policy as well as the standards that govern its day-to-day operations worldwide.

Consortium on Operator Qualification

API assembled a focus group whose primary purpose was to identify covered tasks and provide compliance recommendations to pipeline operators.

The Consortium on Operator Qualification (COOQ) became the guiding force for the identification and development of covered tasks for the liquid pipeline industry. Their work product, API publication 1161, *Guidance Document for the Qualification of Liquid Pipeline Personnel*, formed the basis for the resulting Pipeliner Training and Assessment Program.

The covered task list created in API publication 1161 has now evolved to include gas covered tasks and is maintained by NCCER. The most current covered task list can be downloaded at www.nccer.org.

National Center for Construction Education and Research

With the guidance and support of API, its members, and the Consortium on Operator Qualification, the National Center for Construction Education and Research (NCCER) was able to design and implement the Pipeliner Training and Assessment Program.

NCCER is a not-for-profit 501(c)(3) education foundation established in 1995 by the world's largest and most progressive construction and maintenance contractors and national contractor associations. NCCER was founded to address the

severe workforce shortage facing the construction and maintenance industries and to develop a standardized training process and curricula. Today, NCCER is supported by hundreds of leading construction and maintenance companies, manufacturers, and national associations and developed more than 35 construction and maintenance titles.

NCCER also maintains a National Registry, which contains the records of trainees and instructors who have received NCCER training. All completion information for every trainee is sent to the NCCER and filed within the Registry. The Registry can then confirm training and skills for workers as they move from company to company, state to state, or even within their own company.

Standardized Curriculum (NCCER's Contren™ Learning Series)

NCCER curricula are developed using the expertise of industry subject matter experts (SMEs). The use of SMEs as an integral part of every new development or revision ensures that the curriculum is current, industry-driven, and indicative of industry practices. The development and/or revision process is facilitated from start to finish by NCCER project managers, who work closely with the SMEs and a select team of technical writer(s) on the development and presentation of the curriculum.

Pipeline: Five Training Paths

The pipeline curriculum focuses on essential skills and competencies within five skill areas:

- Corrosion Control
- Electrical and Instrumentation
- Maintenance
- Mechanical
- Operations (Field and Control Center)

SMEs from the pipeline industry have determined the skills and competencies needed based on API's covered tasks and the consensus of the group/industry. Specific company information is generally not incorporated into the curriculum unless the SMEs determine that the information is the industry standard.

Training Modules and Covered Tasks

The smallest unit of instruction in the curriculum is a *module*. Each curriculum is composed of modules (the equivalent of stand-alone chapters) that are logically sequenced into complete *levels* of instruction. Every module is task-oriented and requires the successful completion of a written (knowledge-based) examination and a performance (hands-on) test. Modules may be individually selected and/or combined for efficient task training. Modules in the pipeline curricula, to the extent possible, will align with API-developed covered tasks. While not all modules will relate to covered tasks, all covered tasks will relate to at least one module.

Benefits of Training to the Pipeline Marketplace

Applying NCCER's standardized training process allows pipeline owners and contractors in the United States to establish the benchmark for acceptable skill and competency levels within the pipeline industry. This process will assist training organizations in developing and/or improving standardized pipeline training programs that effectively match their resources and operating conditions while responding to regulatory demands and to the needs of the industry. The process will also recognize training organizations that meet the minimum standards of competence, warranting regulatory, public, and professional confidence. This in turn will foster national unity within the pipeline industry for the mutual benefit of individual technicians, owners, contractors, and ultimately pipeline-users and society.

Additional Resources

To review the Pipeliner Training and Assessment Program Implementation and Resource Guide and the latest version of pipeline covered tasks, visit NCCER's Web site at www.nccer.org, or contact:

National Center for Construction Education and Research
PO Box 141104
Gainesville, FL 32614-1104
(888) 622-3720
Web site: www.nccer.org

For more information on API, please contact:

American Petroleum Institute
1220 L Street, Northwest
Washington, D.C. 20005
(202) 682-8000
Web site: www.api.org

To order *API Pub 1161*, contact Global Engineering Documents by calling 1-800-854-7149 or visiting their Web site at www.global.his.com.

Additional sources of information on the pipeline operator qualification regulation include:

DOT – Office of Pipeline Safety
www.ops.dot.gov

Transportation Safety Institute
www.tsi.dot.gov

Instructor's Notes:

COURSE MAP

This course map shows all of the modules in the second level of the Pipeline Corrosion Control curriculum. The suggested training order begins at the bottom and proceeds up. Skill levels increase as you advance on the course map. The local Training Program Sponsor may adjust the training order.

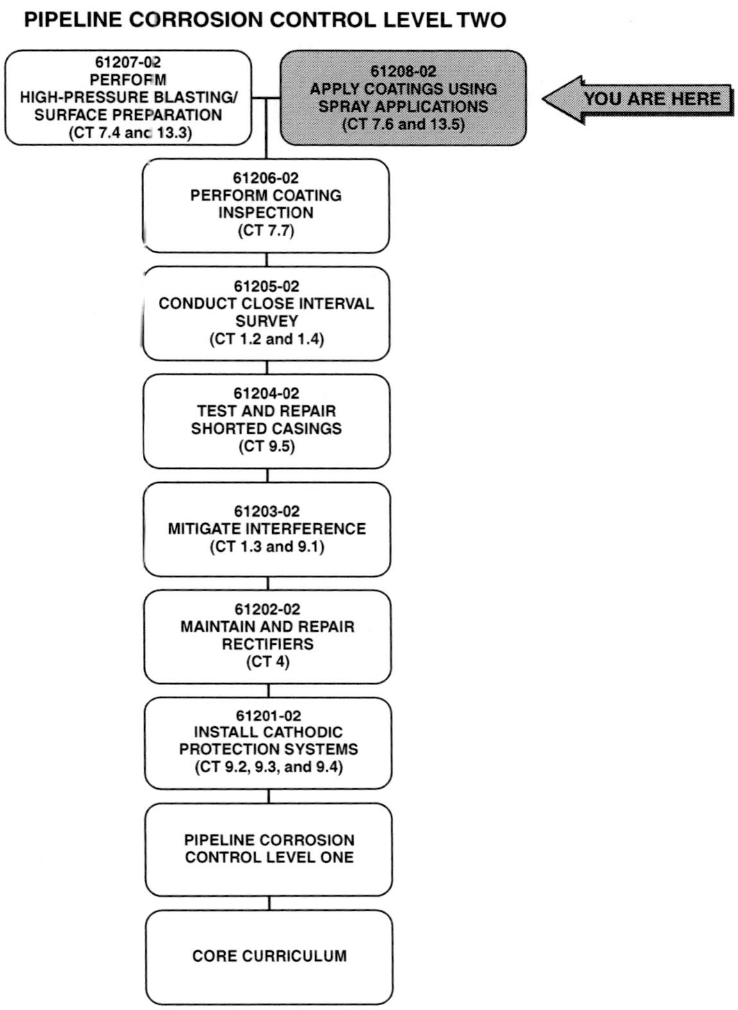

208CMAP.EPS

Homework

Assign reading of Module 61208-02.

Instructor's Notes:

MODULE 61208-02 CONTENTS

Instructor's Notes:

MODULE 61208-02 CONTENTS (Continued)

Figures

Instructor's Notes:

Tables

Instructor's Notes:

MODULE 61208-02

Apply Coatings Using Spray Applications (CT 7.6 and 13.5)

Objectives

When you have completed this module, you will be able to do the following:

1. Identify and explain different coating types (CT 13.5)
2. Determine whether or not the coating is compatible with environmental conditions, the substrate to be coated, any related job-site conditions, or any other requirements of the job (CT 7.6)
3. Prepare the surface for coating application according to manufacturer specification (CT 7.6 and CT 13.5)
4. Identify and explain the different spray application systems used for coating.
5. Apply coating according to manufacturer specifications. (CT 7.6 and CT 13.5)

Prerequisites

Before you begin this module, it is recommended that you successfully complete the following: Core Curriculum; Pipeline Core, Modules 66101-02 and 66102-02; Pipeline Corrosion Control Level One; Pipeline Corrosion Control Level Two, Modules 61201-02 through 61207-02.

Required Trainee Materials

1. Pencil and paper
2. Appropriate personal protective equipment

1.0.0 ◆ INTRODUCTION

As a result of the Clean Air Act and many other laws and regulations, coating manufacturers have stepped up research and development for both water-based and oil-based coatings. Coatings have been re-engineered to eliminate or reduce the use of **volatile** organic compounds (VOCs) and lead. Today's coatings are quite different in their composition and performance from the best coatings of just a few years ago. Improved synthetic resins have yielded better performance, especially in water-based/water-borne coatings. As a result of these efforts to meet VOC regulations, many new water-based coatings have been developed that have properties equal to or better than oil-based coatings.

The first part of this module introduces the basic ingredients and film-forming processes common to all coatings. The remainder of the module describes coating systems and the functional categories of coatings with an emphasis placed on industrial high-performance coatings and their use.

2.0.0 ◆ COATING MATERIALS

Coatings are applied to substrates for the purpose of protection, decoration, and/or to serve a functional purpose such as weatherproofing. All coatings contain some or all of the four basic ingredients listed here and shown in *Figure 1*:

- Solvents
- Resins
- Pigments
- Additives

2.1.0 Pigments

Pigments consist of fine, solid particles dispersed in the liquid portion of the coating. They may be white or colored, opaque or translucent. Pigments give a coating its color, level of sheen or gloss, and **opacity**.

Materials

Ensure that you have everything required to teach the course. Check the Materials and Equipment List at the front of this module.

Classroom

Explain that terms shown in bold are defined in the Glossary at the back of this module.

Audiovisual

Show Transparency 1, Course Objectives.

Show Transparency 2, Performance Tasks.

Teaching Tip

See the general Teaching Tip at the end of this module.

Classroom

Ask a trainee to explain why coating manufacturers are continually improving water-based and oil-based coatings.

Audiovisual

Show Transparency 3 (Figure 1). Ask a trainee to name the components of oil-based coating. Have a trainee name the components of water-based coating.

Audiovisual

Show Transparency 4 (Figure 2). Ask a trainee to explain the main reasons pigments are added to coatings.

Classroom

Ask a trainee to identify the most popular inorganic white pigment. Have trainees name characteristics of organic pigments.

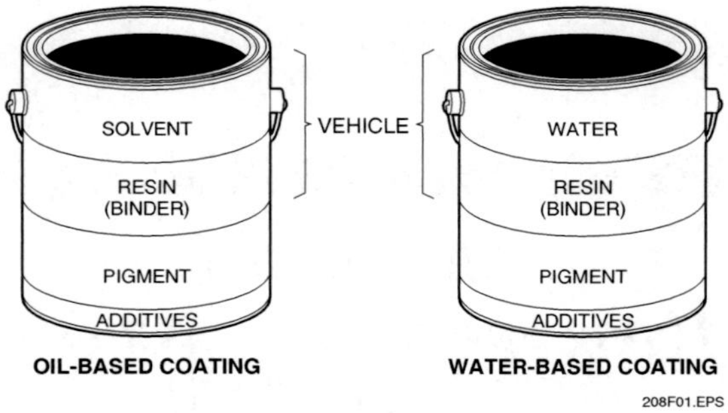

Figure 1 ◆ Basic components of coatings.

Opacity is a coating's ability to cover or hide the underlying surface. Also, within limits, the finer the pigment is ground, the better the hiding power of the coating. Fine grinding is also used to provide smoothness in coating films. Most pigment particles used in commercial coatings range in diameter from 0.0001 to 0.006 millimeters (mm). The addition of special pigments to coatings can provide extra protection against ultraviolet (UV) rays, moisture, and airborne contaminants. They also help the coating resist heat, abrasion, acid, or alkali.

Most pigments consist of two or more elements combined to form an organic or inorganic compound. Basically, inorganic pigments are chemical compounds made from minerals that contain no carbon. They can be made of simple oxides of metal such as iron, zinc, or chromium, or they can be made of a complex combination of elements. Most white and translucent pigments are examples of inorganic pigments. Titanium dioxide is the most popular inorganic white pigment because of its good hiding power and ability to reflect light (**refractive index**). Rutile titanium, another inorganic pigment, has the highest refractive index and the best coloring power (tinting strength) of all pigments.

Organic pigments are chemical compounds based on the element carbon. They are typically derived from **coal tar** and petroleum. When elements such as copper, sodium, barium, or calcium are added to organic pigment compounds, they provide a wide range of colors. Compared to inorganic pigments, most organic pigments are generally brighter in color. However, they are usually not as permanent because they are more susceptible to attack from the sun's UV rays, heat, and airborne contaminants.

Pigments may be classified by their primary function (*Figure 2*). These include:

- Color pigments
- Hiding pigments
- Extender pigments
- Barrier pigments
- Inhibitive pigments
- Sacrificial pigments

2.2.0 Resins

Resins or binders are fusible materials that bind the solid, nonevaporating (nonvolatile) particles of pigment and additives together and cause them to adhere to the substrate, creating a strong and durable **bond**. The type of resin used contributes to the useful life of the coating. It also determines most of the properties of the coating. By varying the resins, many different types of coating can be produced.

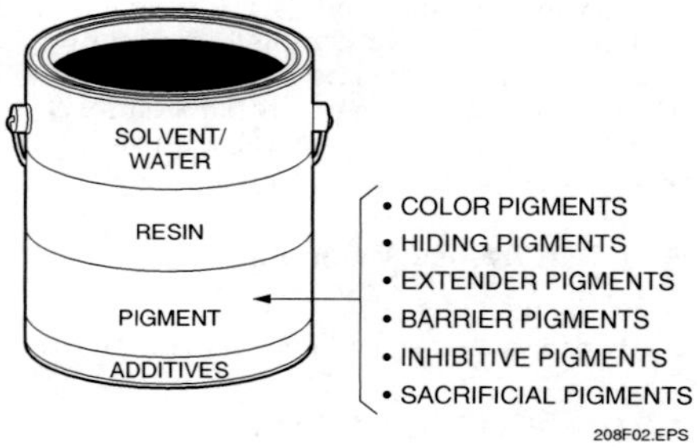

Figure 2 ◆ Functional categories of pigments used in coatings.

Instructor's Notes:

Resins are typically classified by their origin: natural or synthetic. Natural resins are **thermoplastic**, meaning that they become soft when heated and hard when cooled. They are flammable and can be dissolved in some organic solvents. They are not dissolvable in water. Before the development of synthetic resins, natural resins were the only source of material for making coatings with high gloss and exceptional hardness. Some common natural resins are: copal, lac, rosin, dammon, and manila. Use of natural resins has decreased as a result of the development of improved synthetic resins.

Synthetic resins are man-made resins usually derived from by-products of chemical refining or manufacturing processes. Synthetic resins can be used in both water-based and oil-based coatings. When compared to natural resins, they have better resistance to moisture, chemicals, and UV light. For this reason, they are widely used in coatings made to protect substrates in harsh environments. Resins can be used alone or in combination (*Figure 3*).

2.3.0 Solvents

Solvents are easily evaporated (volatile) liquids used to dissolve the solid coating pigment and additive particles in liquid oil-based coating. Note that the water used in water-based coatings (emulsion coatings) acts not to dissolve but to disperse the pigments and additive particles in the coating. Most coatings use a blend of solvents to achieve the desired properties. Solvents reduce the viscosity of oil-based coating so that it can be readily applied using a brush or roller. Depending on the particular solvents, they can also affect a coating's film formation, consistency, leveling, drying, **adhesion**, and durability.

Most coating solvents are VOCs. After application of a coating to a substrate, the solvents evaporate from the drying coating film. Upon evaporation into the atmosphere, VOCs react with nitrous oxides (combustion compounds from automotive emissions and the burning of fuels) and sunlight to form ozone and air pollutants. Ozone is an unstable form of oxygen that is highly reactive. When trapped within the Earth's troposphere, it contributes to smog that can be hazardous to humans, animals, and plant life. VOCs have also been identified as one of many sources contributing to poor indoor air quality in buildings. To be compliant with increasingly restrictive VOC legislation, coating manufacturers are working to replace high-VOC solvents with low-VOC solvents and/or are reducing or eliminating the amount of VOC solvents in their products.

Solvents can be grouped into five broad classes (*Figure 4*):

- Hydrocarbon solvents
- Turpentines
- Ketones
- Esters
- Alcohols

2.4.0 Additives

Additives are special-purpose ingredients added in small quantities to a coating to improve or enhance its performance. Additives are classified by the function they perform (*Figure 5*). These functions are:

- Driers
- Antiskinning agents
- Fungicides
- Thixotropes
- Flow agents
- Emulsion aids
- UV absorbers
- Plasticizers

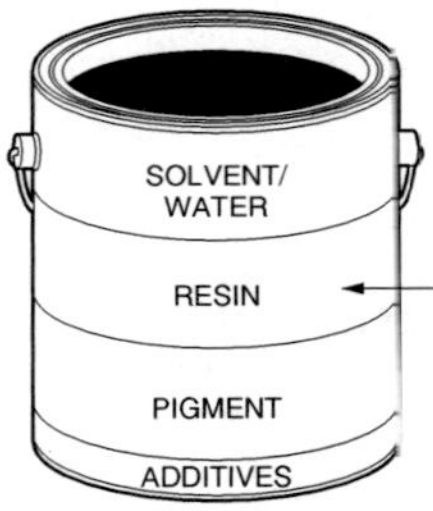

- ACRYLIC RESINS
- ALKYD RESINS
- CHLORINATED RUBBER RESINS
- EPOXY RESINS
- INORGANIC RESINS
- LATEX
- OIL RESINS
- PHENOLIC RESINS
- RUBBER RESINS
- SILICONE RESINS
- URETHANE RESINS
- VINYL RESINS

208F03.EPS

Figure 3 ◈ Common resins used in coatings.

Audiovisual

Show Transparency 5 (Figure 3). Have a trainee compare and contrast natural and synthetic resins.

Show Transparency 6 (Figure 4). Ask trainees to list facts about solvents on a blank sheet of paper. Discuss trainees' answers.

Show Transparency 7 (Figure 5). Have a trainee define additives. Ask trainees if they have any questions.

Classroom

Have trainees name and explain the film-forming mechanisms of industrial coatings.

Audiovisual

Show Transparencies 8 and 9 (Figures 6 and 7). Ask a trainee to explain how thermoset coatings dry and cross-link.

2.5.0 Film-Forming Mechanisms and Generic Coating Types

The basic film-forming mechanisms of industrial coatings are oxidation, evaporation, and chemical curing or cross-linking. Coatings can also be classified as thermoplastic or **thermoset**, depending on the way they cure. The two terms *drying* and *curing* are sometimes used interchangeably to describe this process, and although their meanings are similar, there are subtle differences. In simplest terms, drying means that the coating no longer is wet, that the solvent has volatilized. Curing means that the coating material has undergone a change in property that is characterized by the wet pigmented binder becoming a dry protective and/or decorative film. The dry film is a result of a chemical or oxidative reaction.

Coatings are called thermoplastic coatings when they form a film by solvent evaporation because they become plastic and can soften and deform when heated. Oxidized and chemically cross-linked coatings set and harden when cured. These are referred to as thermoset coatings and do not soften or deform upon heating. The binder, solvent, and additives control the type of film-forming mechanisms.

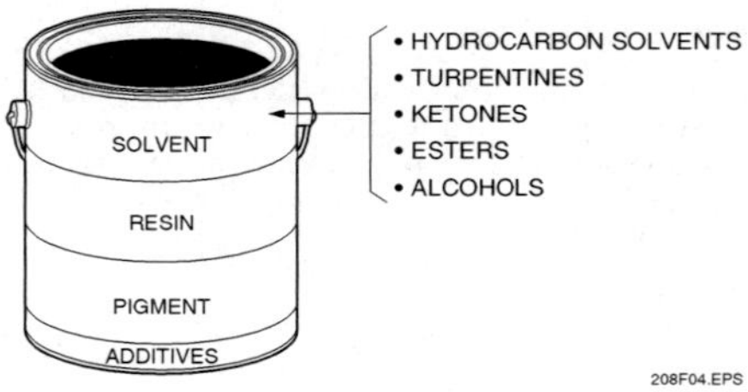

Figure 4 ◈ Classes of solvents used in coatings.

2.5.1 Thermoset Coatings (Films Formed by Oxidation)

These coating types dry and cross-link by reaction with oxygen from the atmosphere (*Figure 6*). All coatings in this class contain drying oils that consist mainly of polyunsaturated fatty acids. The drying oil is combined with a resin, usually by cooking or heating, to improve water and chemical resistance. The drying reaction is accelerated by the presence of metallic salts as driers.

After application, the coating is initially dried by solvent evaporation (*Figure 7*). However, to attain maximum chemical and moisture resistance, the oil must react with oxygen from the air to cross-link, cure, and harden further. This process, called *oxidation*, occurs at a relatively fast rate shortly after application of the wet coating, and continues throughout the life of the coating, although at a much slower rate.

For most oil-based coatings, suitable moisture and chemical resistance occurs within a few days after application, although maximum resistance may require months or years to obtain. Eventually (in some cases 20 or 30 years after application of the coating), oxidation and drying of the oleoresinous binder system goes beyond the final curing stage and leads to **cracking**, embrittlement, and deterioration of the coating film. Drying oil, oil-based alkyd, and modified alkyd coatings are thermoset-type coatings that dry and cross-link by reaction with oxygen.

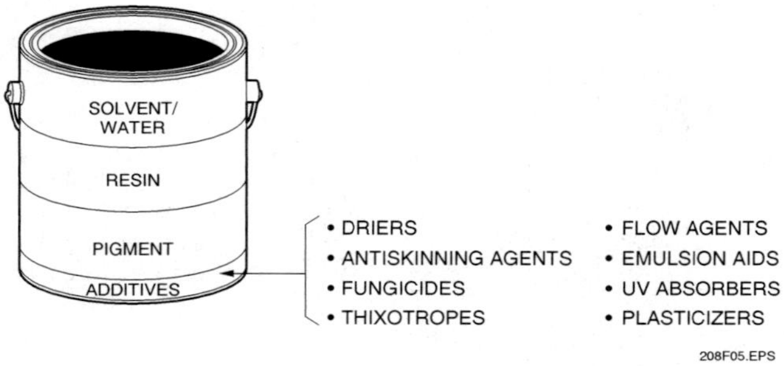

Figure 5 ◈ Functional categories of additives used in coatings.

Instructor's Notes:

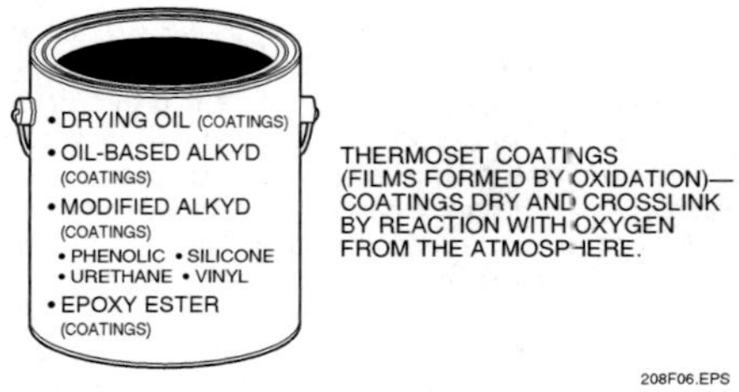

Figure 6 ◆ Thermoset coatings – films formed by oxidation.

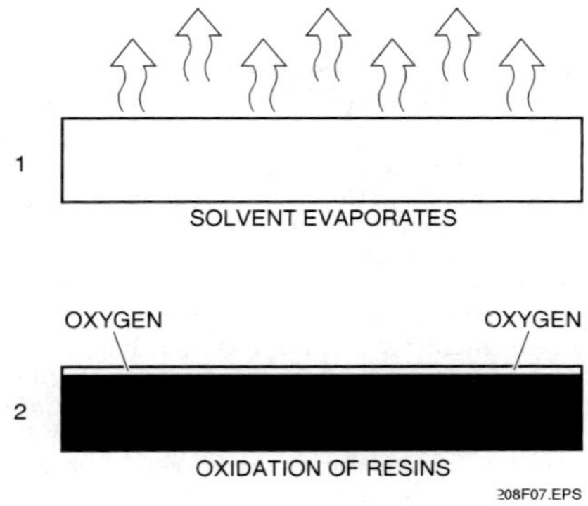

Figure 7 ◆ Film forming by oxidation.

2.5.2 Thermoplastic Coatings (Films Formed by Evaporation)

Another film-forming mechanism is evaporation. Coatings in this category are created by dissolving a resin in solvent or emulsifying it in water, then mixing in the pigments and other constituents. The coating is converted to a solid material by evaporation of the solvent or water used to dissolve or disperse the film-forming components (*Figure 8*). Unlike oxidized and chemically cross-linked coating materials, no chemical change occurs to the resin in a coating that cures by evaporation.

For solvent-borne coatings that cure by this method, the process is called solvent evaporation. For waterborne coatings that cure in this manner, the process is called coalescence (*Figure 9*).

A solvent-borne coating cured by evaporation can be redissolved in the solvents in which it originally was dissolved. This is generally not a disadvantage because even though the solvent resistance of these coatings may be poor, they rarely, if ever, are used in a solvent environment. Instead, they are used to withstand weathering, acid or alkali exposures, and moisture, to which they usually show excellent resistance. Also, these coatings often feature excellent recoatability, even after extended periods of time, because when a new topcoat is applied over an old topcoat of the same type of system, the solvent in the new coating partially dissolves the existing coating, providing adhesion between the two.

With the exception of certain resins and insect lacquers, all resins used in coatings that dry by solvent (or water) evaporation are manufactured.

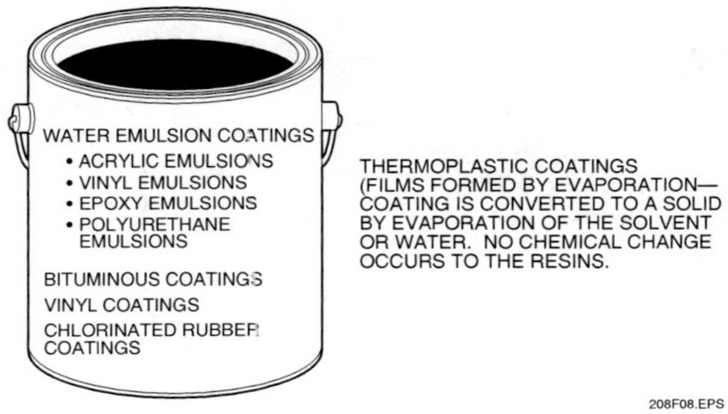

Figure 8 ◆ Thermoplastic coatings–films formed by evaporation.

Classroom

Have trainees name types of thermoplastic coatings. Ask a trainee to explain how thermoplastic coatings are created.

Audiovisual

Show Transparency 10 (Figure 9). Have a trainee explain the process of evaporation. Ask trainees to name advantages and disadvantages of thermoplastic coatings.

Classroom

Have trainees name some coatings that chemically cross-link by copolymerization and some coatings that chemically cure by reacting with moisture.

They are either highly purified or refined natural materials such as bitumens; or synthetic organic resins such as vinyls; chlorinated rubbers; or acrylics commonly used in water emulsions.

Waterborne coatings consist of pigment and resin dissolved or emulsified in water. Most commercial and industrial water-based systems are emulsions.

An emulsion is essentially a dispersion of droplets or particles of a material in a liquid. An example of an emulsion is milk, which consists of fat globules dispersed in water. The dispersed droplets or particles are not soluble in the liquid. They remain distinct entities. In coating technology, emulsion coatings consist of pigmented synthetic resin particles dispersed and emulsified in water. These coatings dry by evaporation of the water (and perhaps 1% or less of a coalescing organic solvent).

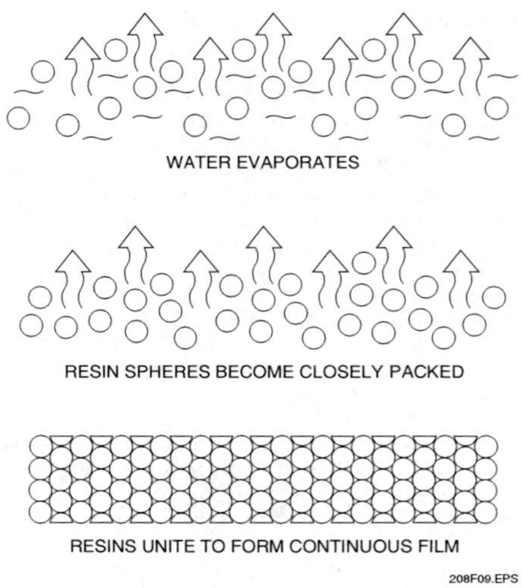

Figure 9 ◆ Water-based coating film formation.

Except for water emulsion coatings, the use of most coatings that dry solely by solvent evaporation, such as vinyls and chlorinated rubbers, is now severely restricted due to VOC regulations. The reason is that these resins must be dissolved in relatively high amounts of strong solvents to be formulated into a coating. Bituminous coatings, which include coal tars and asphalts, can be formulated as solvent cutbacks, but because they also use high amounts of solvent, their use is also limited.

2.5.3 Thermoset Coatings (Films Formed by Chemical Cross-Linking)

Chemically cured coatings harden or cure and attain their final resistance properties as a result of a chemical reaction that combines small molecules (mers) to create large molecules (polymers). These reactions can consist of the same type of molecules (homopolymers) or two or three kinds of molecules (copolymers or terpolymers). Generally, however, they are all polymers, and the process by which molecules join together, or cross-link, is called polymerization (*Figure 10*).

Coatings that chemically cross-link by copolymerization include epoxies, polyesters, urethanes, silicones, and phenolics. In addition, some coatings chemically cure by reacting with moisture. Examples of those are moisture-cured polyurethanes and some types of inorganic zinc-rich coatings.

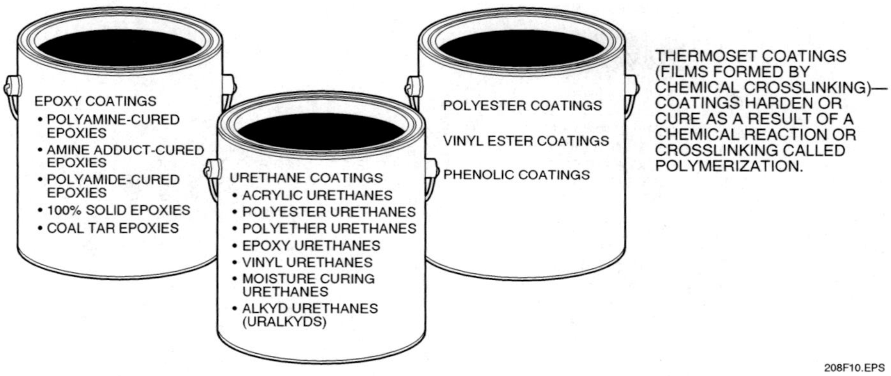

Figure 10 ◆ High-performance thermoset coatings – films formed by chemical cross-linking.

Instructor's Notes:

The basic principle of chemical cross-linking in coating formulation is to use a resin or binder of low molecular weight. Small molecules have low molecular weight; large molecules have high molecular weight. The smaller the molecule, the greater its ability to be dissolved in a solvent. Chemical cross-linking begins with small molecules of low molecular weight being dissolved in a solvent. After mixing and application of the coating material, the cross-linking reaction occurs to form large, three-dimensional molecules. When applying a cross-linked coating system to a large surface, it is possible that the film may cure into one large macromolecule, covering the entire area. If suitably formulated, such a coating will form a tough, flexible, and highly chemical-resistant corrosion-protection barrier over the substrate. As a general rule, large molecules cannot be redissolved in even strong solvents, which is why chemical cross-linking is used to obtain solvent-, moisture-, and chemical-resistant coatings.

Chemically cured coatings can be formulated to have excellent resistance to acids, alkalis, and moisture, and to resist abrasion, UV degradation, and thermal degradation. Generally, toughness and flexibility increase with the length of the molecule, and chemical and moisture resistance increase with the cross-linking density.

Cross-linking occurs in one of three ways: oxidative polymerization, addition polymerization, or condensation polymerization. Oxidative polymerization involves the addition of oxygen to a drying oil. Addition polymerization links or adds simple molecules (monomers) together without forming any by-products, while condensation polymerization produces a new, large molecule with a small molecule (usually water or carbon dioxide) as a byproduct at each cross-link reaction site during polymer formation.

To illustrate how cross-linking works to produce a coating, here is an example of addition polymerization. Vinyl acetate molecules can be polymerized to form a large molecule of polyvinyl acetate, which does not have very good moisture or water resistance and is unsuitable for exterior use as a protective coating resin. Likewise, vinyl chloride molecules can be polymerized to form polyvinyl chloride, and while it has excellent moisture resistance, it cannot be suitably dissolved with most common solvents to form a liquid suitable for a protective coating. However, when vinyl acetate is joined with vinyl chloride in the proper ratio, the resulting copolymer can be dissolved in solvents to make a coating with excellent moisture resistance.

Polymerization cross-linking reactions are becoming increasingly important in the formulation of modern VOC-compliant coating systems. Liquid molecules with low-molecular-weight can be combined with other low molecular weight liquid molecules to form a polymer with higher molecular weight. This will harden to form a solid protective film with the addition of little or no organic solvent. The most successful low-VOC coatings are two-pack, chemically cured coatings such as the epoxy, polyurethane, polyester, and vinyl ester systems.

While thermoset materials that react with moisture are generally single-pack systems, most thermoset coatings are multicomponent coatings consisting of two or more containers that must be mixed prior to application. The curing and hardening of some chemically cross-linked systems, such as most polyesters and vinyl esters, is so rapid that plural-component spray application is required. Such systems mix the co-reactant materials at the time of spray application, and the chemical cross-linking reaction occurs so fast that by the time the coating has deposited on the surface, it is beginning to set. Such coatings are hard-to-the-touch within minutes, but solvent volatilization and additional cross-linking may continue for days before the coating is ready for service in a severe environment.

2.5.4 Zinc-Rich Coatings

Zinc-rich **primers** are unique in providing galvanic protection to steel. Galvanic protection is the process of electrically connecting two dissimilar metals and inducing the corrosion of the more active (electrically negative) metal in order to protect the other less active metal.

As the name suggests, the binder of zinc-rich coatings is highly loaded with metallic zinc dust pigment. After the coating is applied to a thoroughly cleaned steel substrate, the binder holds the zinc particles together and in contact with the steel. This metal-to-metal contact of the two dissimilar metals results in a galvanic cell in which the zinc becomes the anode and sacrifices itself to protect the underlying steel cathode.

Properly applied zinc-rich coatings eliminate pitting and underfilm corrosion even at voids, pinholes, scratches, and abrasions in the coating system. This cannot be said of any other type of coating, and it is this capability that makes zinc-rich coatings unique and widely used.

The binders of zinc-rich coatings can be organic or inorganic. Organic zinc-rich binders use organic resins that contain carbon, such as epoxies, urethanes, and vinyls. Inorganic zinc-rich binders are always silicates, an inorganic material, or one that does not contain the carbon atom.

Classroom

Ask a trainee to explain the basic principle of chemical cross-linking. Have a trainee identify the most successful low-VOC coatings.

Have a trainee explain how zinc-rich primers provide galvanic protection to steel. Ask trainees to name examples of organic and inorganic zinc-rich binders.

Homework

Assign the reading of Sections 2.6.0–2.11.0 for the next session.

Materials

Ensure that you have everything required for the demonstration and testing during this session.

Demonstration

Hand out sample manufacturer's coating system data sheets. Ask trainees specific questions about the data on each sheet.

Classroom

Have a trainee name the three coating layers in a multicoat system. Ask trainees to explain the purpose of each coating layer.

2.6.0 Coating Systems and Functional Classifications of Coatings

Coating manufacturers group their products into systems. A coating system is defined as a protective film consisting of one or more coats, applied in a specific order by recommended methods. Manufacturers' coating systems are based on the type of substrates on which a particular group of coatings can be applied and under what conditions. The coatings grouped into a particular coating system can have different properties and serve different functions, but they are all designed to be compatible when used with one another.

Figure 11 shows an example of the coating system data given by one major manufacturer. The method or layout used by different manufacturers will vary, but they all contain the same basic information. As shown, this information specifies:

- Types of coatings for use with different substrates (metal, concrete, other) and surrounding conditions
- Recommended surface preparation method that should be used prior to application of the coating
- The sequence and number of coatings recommended for application of the coating

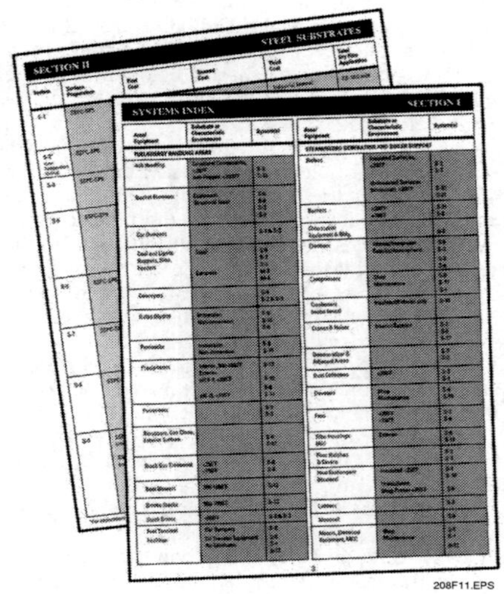

208F11.EPS

Figure 11 ◆ Example of one manufacturer's coating system data.

Manufacturers' coating system data and product literature often describe their coating products by the function or end use they serve. These classifications include:

- Primers
- Tie coats
- Finish coats
- High-performance coatings

2.6.1 Primers

Primers are the first coat in a multicoat system. They are formulated differently than tie coats or finish coats because they serve a different purpose or function. Their purpose is to provide adhesion and protect the substrate. Primers are made to be compatible with and aid in the adhesion of the coatings applied on top of them. The properties of various primers are different and are determined by the type of substrate over which they are to be applied. Different types of primers are made for use on metal, masonry, and concrete.

2.6.2 Tie, Intermediate, Build, and Guide Coats

Tie coats are coatings made specifically to provide a transition from a primer or undercoat to a finish coat. Examples of their use is to seal the surface of a zinc-rich primer, to bond different types of coatings, or to improve the adhesion of the next coating. Intermediate and build coats add thickness to the coating film. Guide coats are coatings that are similar to the finish coat but of a different color, making it easier to assure good coverage when applying the finish coat.

2.6.3 Finish Coats

A finish coat, also called the topcoat, is the last coat applied in a multicoat system. Topcoats are made to protect a properly prepared and primed substrate from conditions such as abrasion, heat, cold, sunlight, moisture, water immersion, or damaging chemicals. Finished coats also give the finished surface the desired finished appearance.

Water-based acrylics, vinyl emulsions, and oleoresinous binders are commonly used for exterior finish coats. When chemical resistance or corrosion protection is needed, or in areas of immersion or high moisture, more chemically resistant resin systems are used, such as epoxies, chlorinated rubbers, polyurethanes, and other high-performance materials. For improved corrosion protection and weatherability, most exterior finish coats are formulated with a gloss or

Instructor's Notes:

semigloss appearance. This is to make sure that a sufficient amount of resin or binder is available to seal the surface of the finish coat and to minimize or prevent moisture or chemicals from entering into the coating. For interior use, coatings using less resistant resins, such as polyvinyl butyral or certain acrylics, and a higher pigment content are often used. These formulations can result in finishes having different sheens, from flat to gloss.

The sheen is the amount of luster the finish coating has when fully cured. The degree of sheen is determined by the proportion of resin in the coating. The resin determines the degree to which the coating is absorbed into the coating surface and how much pigment is left to form a film on the surface. In short, the more resin, the less absorption, thus the more sheen or glossier the coating (*Figure 12*). Terms commonly used by manufacturers to give the sheen of their coating, from lowest to highest, are flat or matte, eggshell and satin, semigloss, and gloss. Note that there is no industry standard for these terms. Therefore, semigloss or gloss coatings made by one manufacturer very often have a different level of sheen than others. Enamel is another term used by manufacturers that indicates the glossiness of a coating. It usually means that the coating is an eggshell or higher gloss coating. Again, one manufacturer's enamel may be glossier than others.

2.6.4 High-Performance Coatings

There are a wide variety of high-performance coatings designed for meeting special coating needs and/or substrate application conditions or problems (*Figure 13*).

Some commonly used types of high-performance coatings are:

- Heat-resistant coatings
- Fire-retardant coatings
- Encapsulants
- Linings
- Anti-corrosion coatings
- Low temperature
- Dry fall/fog coatings
- Elastomeric coatings
- Roof coatings
- Floor coatings
- Surface-tolerant coatings
- Waterproof/damp-proof coatings
- Fillers
- Secondary containment coatings

2.7.0 Powder Coatings

Up to this point, we have described coatings that rely on solvents or water to convey the resinous binder over a surface. A powder coating is a dry coating material that does not need a volatile carrier to accomplish coverage or flow. The powder coating material, produced as a finely ground powder, is spread upon the surface to be coated as a layer of powder. It is then subjected to heat, which melts the powder layer, allowing it to flow and fuse into a continuous coating.

Two methods of application are used: fluidized bed and electrostatic spray. In the fluidized bed method, the powder is placed in a suitable container and fluidized (bubbled with air so that it

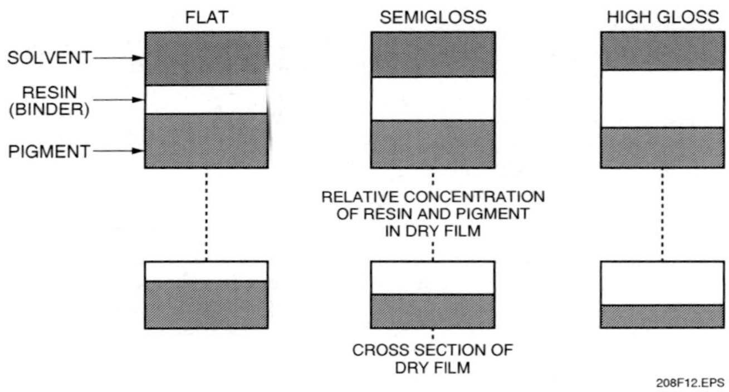

Figure 12 ◆ Example of how the proportion of resin in a coating determines the sheen.

Audiovisual

Show Transparency 11 (Figure 12). Ask a trainee to explain how sheen is determined.

Show Transparency 12 (Figure 13). Discuss some sample applications for each of the high-performance coatings.

Classroom

Ask a trainee to describe powder coatings. Have two trainees describe the two methods of powder coating application.

Teaching Tip

See the Teaching Tip for Section 2.7.0 at the end of this module.

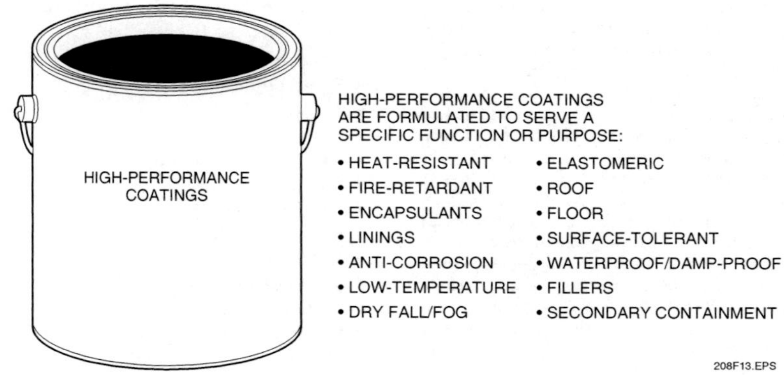

Figure 13 ◈ High-performance coatings classified by function.

resembles a boiling liquid). A preheated part is then dipped into the bed of fluidized powder and the powder, upon contacting the hot part, melts and adheres. In the electrostatic spray method, the powder is dispersed in an air stream and each particle is given an electrical charge. The charge cloud is then directed at the part and each particle is attracted by the positive-negative electrical attraction. This attraction holds the powder on the surface during the subsequent handling and heating of the part. In both methods, the part with the powder layer in place is heated enough to completely convert the powder layer to the final desired coating. Detailed information about the application methods and equipment used with powder coatings is provided in another module.

Powder coatings are used in many automotive, appliance, and similar manufacturing applications. Highly protective powder coatings are also being used to line the inside of oil drilling pipes that are subject to severe pressures, high temperatures, and corrosives. Reinforcing steel bars for highway and bridge decks are powder coated to drastically cut the formation of corrosion products that cause concrete cracking and high repair costs.

Powder coatings have many desirable features:

- Their use greatly reduces solid waste disposal problems. Typically, there is no sludge, fouled spray booth filters, solvent, clean-up waste, or hazardous wastes involved with their use. Most powder coatings used today are formulated to be nontoxic, allowing waste powders to be disposed of in an ordinary landfill.
- No solvents are required in mixing, cleaning, or maintenance, making it easy to comply with government VOC emission and other EPA regulations.
- The fire hazards associated with liquid solvents are eliminated.
- Up to 99% of powder **overspray** can be recovered and reused.

Powder coatings are available in both thermoplastic and thermoset formulations. Thermoplastic powder coatings are more suitable for the coating of items requiring thicker film for the best performance under extreme conditions. Examples of thermoplastic powder coatings include:

- *Polyethylene* – Has excellent chemical resistance or toughness with outstanding electrical insulation properties. It has many uses in the coating of laboratory equipment.
- *Polypropylene* – Has the same useful properties as plastic material.
- *Nylon* – A tough coating with excellent abrasion, wear, and impact resistance. It has a low coefficient of friction and good lubricity. It is typically used to coat sliding and rotating bearing mechanisms in appliances, farm equipment, and textile machinery.
- *Polyvinyl chloride* – Has good exterior durability and provides coatings with a medium soft glossy finish. Bonds well to most metals and can withstand the stress of metal fabrication operations, such as bending, embossing, and drawing.
- *Thermoplastic polyester* – Has good adhesion to most metals without the requirement of a primer and also has good exterior weatherability. It is typically used on such items as outdoor metal furniture.

Instructor's Notes:

Thermoset powder coatings can produce thin (1 to 3 mils), coating-like surfaces with properties equal to (or sometimes better than) those produced by liquid coating formulations. Examples of thermoset powder coatings include:

- *Epoxy* – Tough and chemically resistant. They provide low-cost, low-maintenance, and long-lasting protection against most chemically aggressive environments such as those found in gas and oil field distribution piping systems. They are also used to coat or line the inside of oil drilling pipes and to coat reinforcing steel bars used in highway and bridge decks.
- *Epoxy/polyester hybrids* – Have a decorative film performance similar to epoxies with somewhat better ultraviolet resistance and weatherability.
- *Polyester urethane* – Exterior durable, thin-film coating. Chemically comparable to the exterior-quality urethane coatings used on aircraft, and buses.
- *Polyester TGIC* – Exterior durable, thicker-film coating than polyester urethane. Provides a combination of good sharp-edge coverage with good exterior durability.
- *Acrylic urethane* – Exterior durable, thin-film coating with trouble-free application characteristics. Used extensively in the appliance industry.

2.8.0 Selecting a Coating

Coatings or coating systems are selected for special reasons and unique job requirements. Therefore, it is extremely important to make sure that the coating system selected is compatible with the substrate to be coated, any related job-site conditions, and the contractual requirements of the job. In general, the selection of a coating is based on the following:

- *Type of substrate and exposure* – The type of substrate affects the chemical compatibility, thermal compatibility, flexibility, and adhesion of the coating applied to it. A reaction between the coating and the substrate could be harmful if it lessens adhesion or leads to bleeding. Exposure is also a highly important consideration. Coatings should always be selected with regard to the service environment.
- *Operating conditions/upset conditions* – The specific operating conditions to which a coating is subjected are a key factor in coating selection. For example, if a coating is to be applied to a structure operating at high temperature and exposed to acid fumes, the coating selected is limited to one that can withstand high temperatures and also has good resistance to acid. Upset conditions refers to a change in the operation that could damage or destroy a coating. For example, a vat properly protected on its interior and containing an acid or alkali might be coated on its exterior with a mildly resistant coating. As long as the contents stay in the vat, no harm may come to the coating. If the tank were to overflow, the exterior coating could fail. If this type of upset condition could occur, then the coating selected for the exterior of the vat should be one that is highly resistant to the vat's contents.
- *Ambient conditions during application* – The anticipated ambient conditions that exist during the application of a coating must be considered. Temperature, relative humidity, and dew point can all affect how well a coating cures. The coating selected must be one that can be successfully applied in the anticipated ambient conditions.
- *Government and environmental regulations* – In all cases, the coating selected must comply with the prevailing federal, state, and local laws. In many locations, government regulations may restrict the use of certain techniques and/or materials such as abrasive blasting or solvent emission. There are also U.S. Department of Agriculture (USDA) and Food and Drug Administration (FDA) regulations regarding coatings to be used in contact with potable water or foodstuffs.
- *Budget* – The amount of money allocated for the purchase of coating materials for the job can influence the coating selected.
- *During operation or shutdown* – If the coating must be applied when a unit is in operation, various factors may influence the coating selection. Some of these factors include metal temperature during operation, application methods permissible during application, personnel who will be present during operation, and safety.
- *Time constraints* – The time allowed to work on and/or to complete the job may affect the coating selection.
- *Surface preparation* – Generally, new construction projects specify the degree of surface preparation and allow for, or require, a complete coating system. With maintenance work, surface preparation efforts usually tend to be minimal, and the coating may be only touched up. Coatings compatible with those to be maintained must be chosen.

Classroom

Discuss the factors to consider when selecting a coating.

Describe some scenarios in which coating is to be applied. Provide the necessary information so that trainees can list the coating characteristics that would be required.

Classroom

Ask trainees specific questions about the information contained in the data sheets and MSDS in Figures 14 and 15.

Ask a trainee to explain the basic steps in surface preparation of a steel substrate. Have a trainee explain the meaning of SSPC and NACE numbers.

In order to determine the individual coating to use, *always* check the manufacturer's recommendations and the job requirements. Job requirements are usually expressed as one or more American Society for Testing and Materials (ASTM) standards. The ASTM standards used to select these types of coatings usually refer to coating thickness, moisture resistance, fire resistance, specific chemical resistance, and resistance to abrasion. Always check these and resolve any coating requirement issues *before* starting the job. For example, if the exposure level and substrate require a coating 20 mils thick, but the intended epoxy coating is supposed to be no more than 7 mils thick, an alternative coating or system must be selected. Coating failures can be life-threatening in these situations, rather than just resulting in poor quality decoration.

Selection of a correct coating is also aided by referring to and comparing the manufacturer's product information data sheets for each coating product being considered. *Figure 14* shows an example of the typical product data sheets used by one manufacturer. The method or layout used by different manufacturers will vary, but they all contain basically the same information. As shown, it includes the following kinds of information:

- Product description
- Recommended uses
- Product characteristics
- Performance characteristics
- Recommended coating systems
- Surface preparation
- Color availability and tinting
- Application
- Ordering information
- Safety precautions

The use of industrial high-performance coatings may involve hazards to personnel or the environment. In some instances, these hazards may prohibit a particular product from being used for a specific application. For this reason, the Material Safety Data Sheets (MSDSs) for each product being considered for a job should always be reviewed to identify any such hazards. MSDSs will also identify any special safety equipment and/or precautions that must be taken into consideration when using the product.

Figure 15 shows an example of an MSDS used by one manufacturer. The example shown is for the same coating product as presented in the product data sheets shown in *Figure 14*.

Note that the example shows multiple product lines presented on a single MSDS; this is only because they share similar hazards. MSDSs should not be used for product evaluations or comparisons of performance, quality, or specific composition.

2.9.0 Surface Preparation

Compatibility of the substrate and existing coating system with the intended coating is very important. With industrial high-performance coatings, special surface preparation steps are normally required, such as heat blasting, air or chemical etching, or complete moisture removal. Since steel is a common substrate for these coatings, acceptable surface preparation procedures are often referred to by The Society for Protective Coatings (SSPC) number. When applicable, the National Association of Corrosion Engineers (NACE) number is also used. *Table 1* is a comparison of general surface preparation methods with some additional steps that may be necessary.

Table 1 Surface Preparation Comparison

Common Steps in Surface Preparation	Additional/Alternative Steps Required for Specialty Coatings (Steel Substrate Used as an Example)
Clean (solvent or soap and water)	Strip old coatings and profile the surface:
Strip old coatings	Solvent cleaning (SSPC-SP1)
Sand	Acid etching
Brush off foreign materials	Hand-tool cleaning (SSPC-SP2)
Dry	Power-tool cleaning (SSPC-SP3 and SSPC-SP11)
Apply primer/sealer	White metal blast cleaning (SSPC-SP5/NACE 1)
	Commercial blast cleaning (SSPC-SP6/NACE 3)
	Sweep blasting
	Brush-off blast cleaning (SSPC-SP7/NACE 4)
	Near-white blast cleaning (SSPC-SP10/NACE 2)
	High- and ultra-high-pressure jetting (SSPC-SP12/NACE 5)
	Apply a job-, substrate-, and coating system-specific primer

208T01.EPS

Instructor's Notes:

Industrial and Marine Coatings

4.66

POLAR-POXY™

LOW TEMPERATURE CURE EPOXY

PART A	B62W700	WHITE
PART A	B62S701	ALUMINUM
PART B	B62V702	HARDENER

PRODUCT INFORMATION

PRODUCT DESCRIPTION

POLAR-POXY LOW TEMPERATURE EPOXY is a high performance, interior/exterior, self-priming, high build, VOC compliant, epoxy amine adduct epoxy, formulated to be applied at temperatures as low as 20°F. Specifically designed as a one coat primer or finish coat where conventional coatings cannot be used due to cold weather conditions.

- Will wet and penetrate firm rust and tight crevices
- Chemical and moisture resistant
- Suitable for use in USDA inspected facilities

RECOMMENDED USES

For use over prepared steel surfaces such as:

- Structural steel
- Industrial equipment
- Railings
- Tank exteriors
- Bridges
- Catwalks
- Piping
- Water towers

Do not use in hot weather applications over 80°F.
Not intended for architectural applications.

PRODUCT CHARACTERISTICS

Finish: Low Sheen

Color: White and Aluminum

Volume Solids: 67% ± 2%, mixed

Weight Solids: 80% ± 2%, mixed

VOC (EPA Method 24):

White, mixed	Unreduced:	280 g/L; 2.33 lb/gal
Aluminum, mixed	Unreduced:	316 g/L; 2.63 lb/gal

Mix Ratio: 2 components, 1:1 by volume

Recommended Spreading Rate per coat:

Wet mils:	5.75 - 8.75
Dry mils:	5.0 - 7.0
Coverage:	160 - 220 sq ft/gal approximate

NOTE: Brush or roll application may require multiple coats to achieve maximum film thickness and uniformity of appearance.

Drying Schedule @ 7 mils wet @ 50% RH:

	@ 35°F	@ 50°F	@ 75°F
To touch:	12 hours	6 hours	4 hours
To handle:	48 hours	12 hours	6 hours
To recoat:			
minimum:	24 hours	24 hours	18 hours
maximum:	6 months	6 months	6 months
Pot Life:	8 hours	5 hours	3 hours

If maximum recoat time is exceeded, abrade surface before recoating. Drying time is temperature, humidity, and film thickness dependent.

Sweat-in-time: 15 minutes, @ 50°F, 50% RH

Shelf Life: 24 months, unopened, at 77°F

Flash Point: mixed — White: 87°F Seta Flash; Aluminum: 81°F Seta Flash

Reducer/Clean Up: Reducer #54, R7K54

PERFORMANCE CHARACTERISTICS

System Tested: (unless otherwise indicated)
Substrate: Steel
Surface Preparation: SSPC-SP6
Finish: 1 ct. Polar-Poxy @ 6 mils dft

Abrasion Resistance:
Method: ASTM D4060, CS17 wheel, 1000 cycles, 1 kg load
Result: 80 mg loss

Adhesion:
Method: ASTM D3359 Method B
Result: 5B

Direct Impact Resistance: (on cold rolled steel)
Method: Fed. Standard 141, Method 6226
Result: 24 in. lbs. minimum

Dry Heat Resistance:
Method: ASTM D2485
Result: 175°F, intermittent 200°F

Exterior Durability:
Method: 1 year at 45° south
Result: Chalks, discolors (untopcoated)

Flame Spread Rating:
Method: ASTM-E84 Tunnel Test
Result: Class A on noncombustible surfaces

Flexibility:
Method: ASTM D522, 180° bend, 1/8" mandrel
Result: Passes

Moisture Condensation Resistance:
Method: ASTM D4585, 100°F, 2000 hours
Result: Passes, no evidence of rusting, blistering, cracking, or adhesion loss

Pencil Hardness:
Method: ASTM D3363
Result: 7H

Prohesion:
Method: ASTM G85, Annex A5, 2500 hours
Result: Rating 10 per ASTM D610 for rusting
Rating 10 per ASTM D714 for blistering

Salt Fog Resistance:
Method: ASTM B117, 1000 hours
Result: No surface or pinpoint rusting, softening, or delamination; no more than 1/8" creepage at scribe.

Epoxy coatings may darken or yellow following application and curing.

208F14A.EPS

Figure 14 ◆ Example of a manufacturer's coating product data sheet. (1 of 2)

4.66

POLAR-POXY™

LOW TEMPERATURE CURE EPOXY

Sherwin Williams — Industrial and Marine Coatings

PART A	B62W700	WHITE
PART A	B62S701	ALUMINUM
PART B	B62V702	HARDENER

PRODUCT INFORMATION

RECOMMENDED SYSTEMS

Steel:
1-2 cts. Polar-Poxy @ 5.0 - 7.0 mils dft/ct

Steel, acrylic topcoats:
1-2 cts. Polar-Poxy @ 5.0 - 7.0 mils dft/ct
1-2 cts. DTM Acrylic Finish @ 2.5 - 4.0 mils dft/ct
or Metalatex Semi-Gloss

Steel, epoxy topcoats:
1-2 cts. Polar-Poxy @ 5.0 - 7.0 mils dft/ct
1-2 cts. Sher-Tile Epoxy @ 4.0 - 8.0 mils dft/ct
or Heavy Duty Epoxy
or Macropoxy
or Water Based Catalyzed Epoxy

Steel, polyester epoxy:
1-2 cts. Polar-Poxy @ 5.0 - 7.0 mils dft/ct
1-2 cts. Armor-Tile HS @ 2.5 - 4.0 mils dft/ct

Steel, polyurethane topcoats:
1-2 cts. Polar-Poxy @ 5.0 - 7.0 mils dft/ct
1-2 cts. Hi-Solids Polyurethane @ 3.0 - 4.0 mils dft/ct
or Corothane II Polyurethane

Concrete:
2 cts. Polar-Poxy @ 5.0 - 7.0 mils dft/ct

Check minimum application temperatures of topcoats prior to use.

The systems listed above are representative of the product's use. Other systems may be appropriate.

SURFACE PREPARATION

Surface must be clean, dry, and in sound condition. Remove all oil, dust, grease, dirt, loose rust, and other foreign material to ensure adequate adhesion.

Refer to product Application Bulletin for detailed surface preparation information.

Minimum recommended surface preparation:

Iron & Steel:	SSPC-SP2
Concrete:	Cured, clean, sound, dry

COLOR AVAILABILITY / TINTING

Do not tint. Not controlled for tint strength.

Colors: White and Aluminum

APPLICATION

Temperature:	20°F minimum, 80°F maximum (air and surface; material must be mixed at 50°F minimum) At least 5° above dew point
Relative humidity:	85% maximum

Refer to product Application Bulletin for detailed application information.

ORDERING INFORMATION

Packaging:

Parts A & B:	5 gallon containers

Weight per gallon:

White:	12.78 ± 0.3 lb, mixed
Aluminum:	11.88 ± 0.3 lb, mixed

SAFETY PRECAUTIONS

Refer to the MSDS sheet before use.

Published technical data and instructions are subject to change without notice. Contact your Sherwin-Williams representative for additional technical data and instructions.

208F14B.EPS

Figure 14 ◆ Example of a manufacturer's coating product data sheet. (2 of 2)

Instructor's Notes:

97 —SECTION I— PRODUCT IDENTIFICATION

MATERIAL SAFETY DATA SHEET

THE SHERWIN-WILLIAMS CO.
101 PROSPECT AVE. N.W.
CLEVELAND, OH 44115

EMERGENCY TELEPHONE NO. (216) 566-2917
INFORMATION TELEPHONE NO. (216) 566-2902
DATE OF PREPARATION 7-JAN-97

Epoxy Coatings - 5

B62/5

Numbers In Boxes Are Percent By Weight

—SECTION II— CAS No.	HAZARDOUS INGREDIENT (percent by weight)	ACGIH TLV <STEL>	OSHA PEL <STEL>	Units	Vapor Pressure (mm Hg)	POLAR-POXY™ LT Epoxy B62W700 Part A White	B62S701 Part A Aluminum	B62V702 Part B Hardener	TANK-O-LON® High Solids NSF Epoxy B62W600 Part A Sanitary White	B62L601 Part A Sanitary Blue	B62V602 Part B Hardener
64742-88-7	Mineral Spirits.	100	100	PPM	2.0		2				
100-41-4 §	Ethylbenzene	100 <125>	100 <125>	PPM	7.1			2			
1330-20-7 §	Xylene.	100 <150>	100 <150>	PPM	5.9			9			
67-63-0	2-Propanol	100 <500>	400 <500>	PPM	33.0			6			
71-36-3 §	1-Butanol	C 50	C 50	PPM (Skin)	5.5		4	7	2	1	
107-98-2	1-Methoxy-2-propanol	Not Established			10.9	8	8		2	2	
79-24-3	Nitroethane	100	100	PPM	15.6	1					
108-03-2	1-Nitropropane	25		PPM	7.5	2					
107-15-3	Ethylenediamine.	10		PPM (Skin)	10.7						3
90-72-2	Tri(dimethylaminomethy)phenol	Not Established									1
Unknown	Epoxy Polymer.	Not Established				39	45		52	52	
Unknown	Polyamine.	Not Established						22			96
14807-96-6	Talc	2	2	Mg/M3	as Resp. Dust	8	24	53	10	10	
12001-26-2	Mica	3	3	Mg/M3	as Resp. Dust	8	7				
7727-43-7	Barium Sulfate	10	10[5]	Mg/M3 [Resp.	as Dust Fraction]				17	19	
13463-67-7	Titanium Dioxide.	10	10[5]	Mg/M3 [Resp.	as Dust Fraction]	30			14	12	
	[% Barium]								[9.7]	[11.1]	
	Weight per Gallon (lbs.)					13.24	11.44	12.33	13.21	13.22	7.1
	Solids by Weight (%)					87.0	83.9	75.8	95.7	96.0	97.0
	Solids by Volume (%)					77.7	74.3	56.5	92.0	92.6	97.2
	VOC (Volatile Organic Compounds) - lbs./gal.					1.72	1.83	2.98	0.57	0.52	0.21
	Photochemically Reactive					No	Yes	Yes	No	No	No
	Flash Point (°F) / DOL Storage Category					105 / 2	98 / 1C	65 / 1B	89 / 1C	89 / 1C	89 / 1C
	Flammability Classification / HMIS (NFPA) Rating					Comb / 2* 2 2	Flam / 2* 3 1	Flam / 2 3 0	Flam / 2* 3 0	Flam / 2* 3 0	Flam / 3* 3 0

§ Ingredient subject to the reporting requirements of the Superfund Amendments and Reauthorization Act (SARA) Section 313, 40 CFR 372.65 C

208F15A.EPS

Figure 15 ◆ Example of a manufacturer's material safety data sheet (MSDS). (1 of 2)

Epoxy Coatings

B62/5

Section III — PHYSICAL DATA

PRODUCT WEIGHT – See TABLE
SPECIFIC GRAVITY – 0.86-1.59
BOILING RANGE – 178-395 °F
VOLATILE VOLUME – 2-43 %
EVAPORATION RATE – Slower than Ether
VAPOR DENSITY – Heavier than Air
MELTING POINT – N.A.
SOLUBILITY IN WATER – N.A.

Section IV — FIRE AND EXPLOSION HAZARD DATA

FLAMMABILITY CLASSIFICATION *FLASH POINT* See TABLE *LEL* 1.0 *UEL* 16.6
See TABLE
EXTINGUISHING MEDIA
Carbon Dioxide, Dry Chemical, Foam
UNUSUAL FIRE AND EXPLOSION HAZARDS
Keep containers tightly closed. Isolate from heat, electrical equipment, sparks, and open flame. Closed containers may explode when exposed to extreme heat. Application to hot surfaces requires special precautions. During emergency conditions overexposure to decomposition products may cause a health hazard. Symptoms may not be immediately apparent. Obtain medical attention.
SPECIAL FIRE FIGHTING PROCEDURES
Full protective equipment including self-contained breathing apparatus should be used. Water spray may be ineffective. If water is used, fog nozzles are preferable. Water may be used to cool closed containers to prevent pressure build-up and possible autoignition or explosion when exposed to extreme heat.

Section V — HEALTH HAZARD DATA

ROUTES OF EXPOSURE
Exposure may be by INHALATION and/or SKIN or EYE contact, depending on conditions of use. To minimize exposure, follow recommendations for proper use, ventilation, and personal protective equipment.
ACUTE Health Hazards
EFFECTS OF OVEREXPOSURE
Irritation of eyes, skin and respiratory system. May cause nervous system depression. Extreme overexposure may result in unconsciousness and possibly death.
SIGNS AND SYMPTOMS OF OVEREXPOSURE
Headache, dizziness, nausea, and loss of coordination are indications of excessive exposure to vapors or spray mists.
Redness and itching or burning sensation may indicate eye or excessive skin exposure.
MEDICAL CONDITIONS AGGRAVATED BY EXPOSURE
May cause allergic skin reaction in susceptible persons.
EMERGENCY AND FIRST AID PROCEDURES
If INHALED: If affected, remove from exposure. Restore breathing. Keep warm and quiet.
If on SKIN: Wash affected area thoroughly with soap and water. Remove contaminated clothing and launder before re-use.
If in EYES: Flush eyes with large amounts of water for 15 minutes. Get medical attention.
If SWALLOWED: Get medical attention.
CHRONIC Health Hazards
No ingredient in these products is an IARC, NTP, or OSHA listed carcinogen.
Prolonged overexposure to solvent ingredients in Section II may cause adverse effects to the liver, urinary, and reproductive systems.
Rats exposed to titanium dioxide dust at 250 mg./m3 developed lung cancer, however, such exposure levels are not attainable in the workplace.
Reports have associated repeated and prolonged overexposure to solvents with permanent brain and nervous system damage.

Section VI — REACTIVITY DATA

STABILITY – Stable
CONDITIONS TO AVOID
None known.
INCOMPATIBILITY
None known.
HAZARDOUS DECOMPOSITION PRODUCTS
By fire: Carbon Dioxide, Carbon Monoxide, Oxides of Metals in Section II
HAZARDOUS POLYMERIZATION – Will Not Occur

Section VII — SPILL OR LEAK PROCEDURES

STEPS TO BE TAKEN IN CASE MATERIAL IS RELEASED OR SPILLED
Remove all sources of ignition. Ventilate and remove with inert absorbent.
WASTE DISPOSAL METHOD
Waste from these products may be hazardous as defined under the Resource Conservation and Recovery Act (RCRA) 40 CFR 261. Waste must be tested for ignitability to determine the applicable EPA hazardous waste numbers.
Incinerate in approved facility. Do not incinerate closed container. Dispose of in accordance with Federal, State, and Local regulations regarding pollution.

Section VIII — PROTECTION INFORMATION

PRECAUTIONS TO BE TAKEN IN USE
Use only with adequate ventilation. Avoid breathing vapor and spray mist. Prevent contact with skin and eyes. Wash hands after using.
These coatings may contain materials classified as nuisance particulates (listed "as Dust" in Section II) which may be present at hazardous levels only during sanding or abrading of the dried film. If no specific dusts are listed in Section II, the applicable limits for nuisance dusts are ACGIH TLV 10 mg./m3 (total dust), 3 mg./m3 (respirable fraction), OSHA PEL 15 mg./m3 (total dust), 5 mg./m3 (respirable fraction).
VENTILATION
Local exhaust preferable. General exhaust acceptable if the exposure to materials in Section II is maintained below applicable exposure limits. Refer to OSHA Standards 1910.94, 1910.107, 1910.108.
RESPIRATORY PROTECTION
If personal exposure cannot be controlled below applicable limits by ventilation, wear a properly fitted organic vapor/particulate respirator approved by NIOSH/MSHA for protection against materials in Section II.
When sanding, wirebrushing, abrading, burning or welding the dried film, wear a particulate respirator approved by NIOSH/MSHA for protection against non-volatile materials in Section II.
PROTECTIVE GLOVES
Wear gloves which are recommended by glove supplier for protection against materials in Section II.
EYE PROTECTION
Wear safety spectacles with unperforated sideshields.
OTHER PROTECTIVE EQUIPMENT
Use of barrier cream on exposed skin is recommended.

Section IX — PRECAUTIONS

DOL STORAGE CATEGORY – See TABLE
PRECAUTIONS TO BE TAKEN IN HANDLING AND STORING
Keep away from heat, sparks, and open flame. During use and until all vapors are gone: Keep area ventilated - Do not smoke - Extinguish all flames, pilot lights, and heaters - Turn off stoves, electric tools and appliances, and any other sources of ignition.
Consult NFPA Code. Use approved Bonding and Grounding procedures.
Keep container closed when not in use. Transfer only to approved containers with complete and appropriate labeling. Do not take internally. Keep out of the reach of children.
OTHER PRECAUTIONS
These products must be mixed with other components before use. Before opening the packages, READ AND FOLLOW WARNING LABELS ON ALL COMPONENTS.
Intentional misuse by deliberately concentrating and inhaling the contents can be harmful or fatal.

Section X — OTHER REGULATORY INFORMATION

CALIFORNIA PROPOSITION 65
WARNING: These products, except for B62V602, contain chemicals known to the State of California to cause cancer and birth defects or other reproductive harm. B62V602 contains a chemical(s) known to the State of California to cause cancer.

TSCA CERTIFICATION
All chemicals in these products are listed, or are exempt from listing, on the TSCA Inventory.

The above information pertains to these product as currently formulated, and is based on the information available at this time. Addition of reducers or other additives to these products may substantially alter the composition and hazards of the products. Since conditions of use are outside our control, we make no warranties, express or implied, and assume no liability in connection with any use of this information.

covers MSDS page B62/5

208F15B.EPS

Figure 15 ◈ Example of a manufacturer's material safety data sheet (MSDS). (2 of 2)

Instructor's Notes:

Consider coatings used as lining in tanks. Surface preparation commonly calls for blasting of the steel to create an anchor pattern, or profile, for the coating to adhere to. Insufficient cleaning or profiling can mean that the coating does not adhere and bond to the surface properly. Also, consider the temperature and humidity limits for the coating during storage, shipping, and application, and the curing requirements afterward.

Detailed information about the recommended surface preparation required before using a manufacturer's coating product is also available on application sheets or bulletins published by the manufacturer. *Figure 16* shows an example of one manufacturer's application bulletin. The example shown is for the same coating product as presented in the product data sheets shown in *Figure 14*.

2.10.0 Coating Preparation and Application

Mixing of coatings should be done as directed in the mixing instructions section of the coating manufacturer's application bulletin. Thinning a coating is sometimes necessary to obtain the required wet-film thickness or viscosity for application equipment. Thinning must always be done according to the manufacturer's instructions and within current VOC restrictions.

Induction time (also known as sweat-in time), pot life, and drying requirements vary by product; be sure to read the manufacturer's literature for each product used. Some two-part epoxies or urethanes have a short pot life once mixed. Plan ahead so they are not mixed too soon and begin curing too early. Mixing sequence and pot life vary from product to product. Mixing of two multicomponent coatings with long pot lives can be done in large containers. Multicomponent coatings with very short pot lives must sometimes be mixed in special spray equipment as they are applied.

Epoxies have some of the most involved coating preparation requirements, including induction or sweat-in times and procedures. The induction, or sweat-in, time is the time you must wait after mixing the components of a multicomponent coating before you can begin application. This wait is required to allow the components to begin the chemical reaction that creates a hard, dry film. Induction times depend on the temperature and relative humidity (RH) of the area where the coating is mixed and applied. *Table 2* shows typical induction, pot life, and drying times for several epoxy coatings.

Application procedures for coatings can be more forgiving than the exacting application procedures for high-performance coatings. Application procedures are coating-specific, even within a single product line for a single manufacturer. Most industrial high-performance coatings are applied with spray equipment, although some may require a roller, squeegee, or spreading device.

Before using a coating, a good practice is to always refer to the coating manufacturer's application bulletin for detailed information on how to properly apply the coating. As shown in *Figure 16*, this information typically includes:

- Application conditions
- Application equipment
- Application procedures
- Cleanup instructions
- Performance tips
- Safety precautions

WARNING!
Some jobs may be in confined spaces. Ensure that personnel have proper training and follow all safety precautions for confined space entry to prevent injury or asphyxiation.

2.11.0 Testing and Inspection

Industrial coatings are chosen and applied in jobs with special requirements. Coating failures in these situations usually result in more than just staining or discoloration of the substrate. Coating failures in these situations can result in loss of structure, loss of contained resources, spills, accidents, other hazards, and even loss of life. It is especially important to ensure proper surface preparations and coating application. One way to verify the protective qualities of a coating is with inspections and tests. Each inspection or test verifies a certain property. Minimum chemical, corrosion, and fire resistance/retardance are determined when the coating is selected. Control and checking begins with determining the atmospheric conditions of the job, especially if multicomponent coatings are used, since relative humidity and temperature affect their application. Determining the appropriate profiling of blasted substrates and the moisture content of cement substrates is also necessary. Once applied, wet-film thickness can be measured. Once dry, the dry-film thickness, film continuity, and gloss can be checked, since they contribute to the protective qualities of a coating. The inspection and test results are often expressed as some measure of SSPC or ASTM standards. The coating is often required to meet certain ASTM standards per the specification requirements.

Performance Testing

Have each trainee complete Performance Task 2 to your satisfaction. Fill out a Performance Profile Sheet for each trainee.

Classroom

Ask trainees to point out similarities and differences among the product data sheet, MSDS, and application bulletin.

Have trainees explain induction time, pot life, and drying requirements.

Safety

Stress the importance of following safety precautions for confined space entry.

Classroom

Ask a trainee to explain the possible consequences of coating failures. Have a trainee briefly explain the inspection process.

Homework

Have trainees review Sections 3.0.0–3.4.5 for the next session.

Sherwin Williams
Industrial and Marine Coatings

4.66A
POLAR-POXY™
LOW TEMPERATURE CURE EPOXY

PART A	B62W700	WHITE
PART A	B62S701	ALUMINUM
PART B	B62V702	HARDENER

APPLICATION BULLETIN

Revised 1/97

SURFACE PREPARATION

Surface must be clean, dry, and in sound condition. Remove all oil, dust, grease, dirt, loose rust, and other foreign material to ensure adequate adhesion.

Iron & Steel
Minimum surface preparation is Hand Tool Clean per SSPC-SP2. Remove all oil and grease from surface by Solvent Cleaning per SSPC-SP1. For better performance, use Commercial Blast Cleaning per SSPC-SP6, blast clean all surfaces using a sharp, angular abrasive for optimum surface profile (2 mils). Remove all weld spatter and round all sharp edges by grinding to a minimum 1/4" radius. Prime any bare steel the same day as it is cleaned or before flash rusting occurs.

Poured Concrete
New
Surface must be clean, dry, sound, and offer sufficient profile to achieve adequate adhesion. Minimum substrate cure is 28 days at 75°F. Remove all form release agents, curing compounds, salts, efflorescence, laitance, and other foreign matter by sandblasting, shotblasting, mechanical scarification, or suitable chemical means, such as muriatic acid etch, refer to ASTM D4260. Rinse thoroughly to achieve a final pH between 6.0 and 10.0. Allow to dry thoroughly prior to coating.

Old
Surface preparation is done in much the same manner as new concrete; however, if the concrete is contaminated with oils, grease, chemicals, etc., they must be removed by cleaning with a strong detergent. Refer to ASTM D4258. Form release agents, hardeners, etc. must be removed by sandblasting, shotblasting, mechanical scarification, or suitable chemical means. If surface deterioration presents an unacceptably rough surface, Kem Cati-Coat Epoxy Filler/Sealer is recommended to patch and resurface damaged concrete.

Previously Painted Surfaces
If in sound condition, clean the surface of all foreign material. Smooth, hard or glossy coatings and surfaces should be dulled by abrading the surface. Apply a test area, allowing paint to dry one week before testing adhesion. If adhesion is poor, or if this products attacks the previous finish, removal of the previous coating may be necessary. If paint is peeling or badly weathered, clean surface to sound substrate and treat as a new surface as above.

APPLICATION CONDITIONS

Temperature:	20°F minimum, 80°F maximum (air and surface; material must be mixed at 50°F minimum) At least 5° above dew point
Relative humidity:	85% maximum

APPLICATION EQUIPMENT

The following is a guide. Changes in pressures and tip sizes may be needed for proper spray characteristics. Always purge spray equipment before use with listed reducer. Any reduction must be compatible with the existing environmental and application conditions.

Reducer/Clean Up .... Reducer #54, R7K54

Airless Spray
Pump 30:1
Hose .... 3/8" ID
Tip .... .019" - .023"
Filter .... 60 mesh
Reduction .... as needed up to 6% by volume

Brush
Brush .... Nylon/Polyester or Natural Bristle
Reduction .... not recommended

Roller
Cover .... 3/8"-1/2" woven with phenolic core
Reduction .... not recommended

If specific application equipment is listed above, equivalent equipment may be substituted.

208F16A.EPS

Figure 16 ◆ Example of a manufacturer's application bulletin. (1 of 2)

Instructor's Notes:

Sherwin-Williams

Industrial and Marine Coatings

4.66A

POLAR-POXY™

LOW TEMPERATURE CURE EPOXY

PART A	B62W700	WHITE
PART A	B62S701	ALUMINUM
PART B	B62V702	HARDENER

APPLICATION BULLETIN

APPLICATION PROCEDURES

Surface preparation must be completed as indicated.

Mixing Instructions: All material must be above 50°F before use. Mix both components thoroughly with power agitation. Make certain pigment does not remain on the bottom of the container. Under agitation, mix equal parts, by volume, of paint and hardener until thoroughly mixed. Avoid over agitation. Always mix a complete unit. Allow to sweat-in as indicated. Re-stir before using.

If reducer solvent is used, add only after both components have been thoroughly mixed, after sweat-in.

Apply paint at the recommended film thickness and spreading rate as indicated below:

Recommended Spreading Rate per coat:

Wet mils:	5.75 - 8.75
Dry mils:	5.0 - 7.0
Coverage:	160 - 220 sq ft/gal approximate

NOTE: Brush or roll application may require multiple coats to achieve maximum film thickness and uniformity of appearance.

Drying Schedule @ 7 mils wet @ 50% RH:

To touch:	12 hours	6 hours	4 hours
To handle:	48 hours	12 hours	6 hours
To recoat:			
minimum:	24 hours	24 hours	18 hours
maximum:	6 months	6 months	6 months
Pot Life:	8 hours	5 hours	3 hours

If maximum recoat time is exceeded, abrade surface before recoating. Drying time is temperature, humidity, and film thickness dependent.

Sweat-in-time: 15 minutes, @ 50°F, 50% RH

Application of coating above maximum or below minimum recommended spreading rate may adversely affect coating performance.

PERFORMANCE TIPS

Stripe coat all crevices, welds, and sharp angles to prevent early failure in these areas.

When using spray application, use a 50% overlap with each pass of the gun to avoid holidays, bare areas, and pinholes. If necessary, cross spray at a right angle.

Spreading rates are calculated on volume solids and do not include an application loss factor due to surface profile, roughness or porosity of the surface, skill and technique of the applicator, method of application, various surface irregularities, material lost during mixing, spillage, overthinning, climatic conditions, and excessive film build.

Excessive reduction of material can affect film build, appearance and adhesion.

In order to avoid blockage of spray equipment, clean equipment before use or before periods of extended downtime with Reducer #54, R7K54.

Do not apply the material beyond recommended pot life.

Do not mix previously catalyzed material with new.

Do not use in hot weather applications over 80°F.

Not intended for architectural applications.

Coating will chalk and discolor in atmospheric service untopcoated.

Refer to Product Information sheet for additional performance characteristics and properties.

CLEAN UP INSTRUCTIONS

Clean spills and spatters immediately with Reducer #54, R7K54. Clean tools immediately after use with Reducer #54, R7K54. Follow manufacturer's safety recommendations when using any solvent.

SAFETY PRECAUTIONS

Refer to the MSDS sheet before use.

Published technical data and instructions are subject to change without notice. Contact your Sherwin-Williams representative for additional technical data and instructions.

208F16B.EPS

Figure 16 ◆ Example of a manufacturer's application bulletin. (2 of 2)

Materials

Ensure that you have everything required for the laboratory during this session.

Audiovisual

Show Transparency 13 (Figure 17). Ask a trainee to explain how a conventional air spray system operates. Have trainees name the advantages and disadvantages of the conventional air spray system.

Table 2 Typical Induction and Drying Times for Several Epoxy Coatings

Coating	Application Temperatures (Min./Max.)	Pot Life (@ 77°F, 50% RH)	Induction Time (@ 77°F, 50% RH)	Drying Time (@ 77°F, 50% RH)
Epoxy/polyamide @ 4 to 8 mils wet	55°–100°F	8 hours	1 hour	To touch: 3 to 5 hours To recoat: 24 hours To cure: 14 days
Epoxy/polyamine @ 6 mils wet	50°–120°F	40 minutes	N/A	To touch: 2 to 3 hours To recoat: 4 to 6 hours To cure: 10 days
Epoxy/acrylic @ 8 mils wet	55°–120°F	24 hours	30 minutes	To touch: 30 minutes To recoat: 6 hours To cure: 10 days
Zinc-rich epoxy @ 5 mils wet	40°–120°F	6 hours	30 minutes	To touch: 30 minutes To recoat: 6 hours min.; 1 year max. To cure: 14 days
Recoatable epoxy primer @ 6 mils wet	35°–120°F	8 hours	30 minutes	To touch: 15 minutes To recoat: 2 hours min.; 1 year max. To cure: 14 days
Surface-tolerant epoxy @ 7 mils wet	55°–120°F	4 hours	15 minutes	To touch: 3 hours To recoat: 18 hours min.; 10 days max. To cure: 10 days
Epoxy mastic @ 7 mils wet	55°–120°F	2½ hours	15 minutes	To touch: 8 hours To recoat: 16 hours min.; 72 hours max. To cure: 10 days
Polyester epoxy @ 4 mils wet	50°–100°F	8 hours	30 minutes	To touch: 2 to 4 hours To recoat: 16 hours min.; 72 hours max. To cure: 7 days
Coal tar epoxy @ 29 mils wet	50°–100°F	4 hours	30 minutes	To touch: 8 to 10 hours To recoat: 16 hours min.; 48 hours max. To cure: 7 to 10 days
100% solids epoxy @ 20 mils wet	50°–100°F	30 minutes	None required	To touch: 4 hours To recoat: 18 hours min.; 4 days max. To cure: 7 days
Phenolic epoxy @ 9 mils wet	50°–100°F	6 hours	15 minutes	To touch: 4 hours To recoat: 18 hours min.; 2 days max. To cure: 7 days

208T02.EPS

3.0.0 ◆ AIR SPRAY SYSTEMS

There are many types of air spray systems available. This section discusses conventional spray systems, airless systems, and high-volume, low-pressure (HVLP) systems.

3.1.0 Conventional Air Spray Systems and Components

The oldest of all spraying systems is the conventional air spray system. It can be used to apply a wide range of coatings from fine lacquers to heavy **mastics**. Conventional air spray systems use compressed air at pressures between 40 and 70 pounds per square inch (psi) or higher to atomize (break) a stream of liquid coating material into a fine mist of small particles as it leaves the spray gun nozzle (air cap). Atomizing the coating is done by using a high-velocity compressed airstream to hit a relatively slow-moving coating stream as both leave the spray gun air cap (*Figure 17*). In addition to atomizing the coating, the compressed air acts to propel the atomized coating to the surface in a pattern determined by the spray gun nozzle.

The advantage of using conventional air spraying is its flexibility and control. Atomization of the coating produces a very high-quality finish. The spray pattern from the gun can be adjusted to

Instructor's Notes:

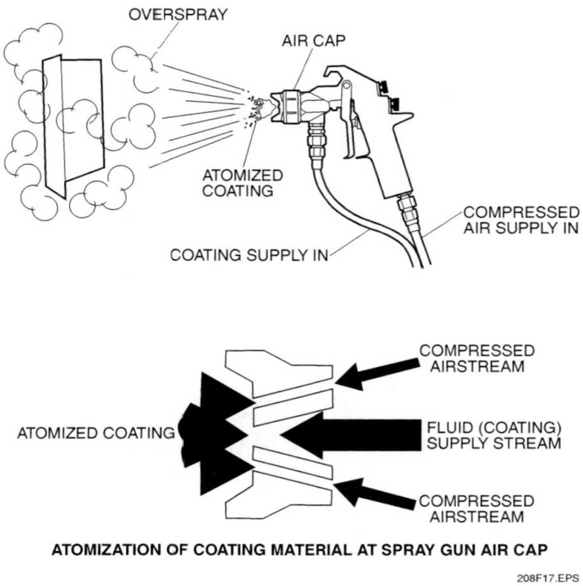

Figure 17 ◆ Conventional air spray.

make it narrower or wider, or to spray coatings at different rates. Also, the spray gun design allows the operator to blow air without the coating, which is useful to clean or dry a surface. The type of coating can be changed with minimal cleanup or loss of time and materials.

The major disadvantage of conventional spraying is that there is a substantial amount of overspray. Overspray, also called **bounceback**, is the atomized coating particles that rebound or deflect from the surface, as well as the particles that miss the surface being sprayed.

As a result of overspray, more effort is required to mask or otherwise protect adjacent surfaces from coating damage than is required with airless, HVLP, or electrostatic spray systems (discussed later). Because of overspray, conventional spray systems have a low transfer efficiency (typically between 30% and 40%). Transfer efficiency is the amount of coating sprayed that actually goes onto the part or surface as compared to the total amount used.

Conventional air sprayers can be used for the following applications:

- Textured coatings
- Multicolored specialty coatings
- Protective coatings

There are two types of conventional spray systems: pressure-feed and siphon-feed systems.

3.1.1 Pressure-Feed Systems

Pressure-feed systems are used for various finishing jobs when large amounts of coating need to be delivered to the finish area, when fast application is required, and when the coating material is too heavy to be siphoned from a container. Pressure-feed flow is dependent upon the system air or pump pressure. One common type of pressure-feed system is the pressure-feed tank system (*Figure 18*). In this system, the coating is placed in an airtight tank called a pot. Air supplied from an air

Audiovisual

Show Transparency 14 (Figure 18). Ask a trainee to explain the uses of pressure-feed systems. Have a trainee explain how a pressure-feed tank system works.

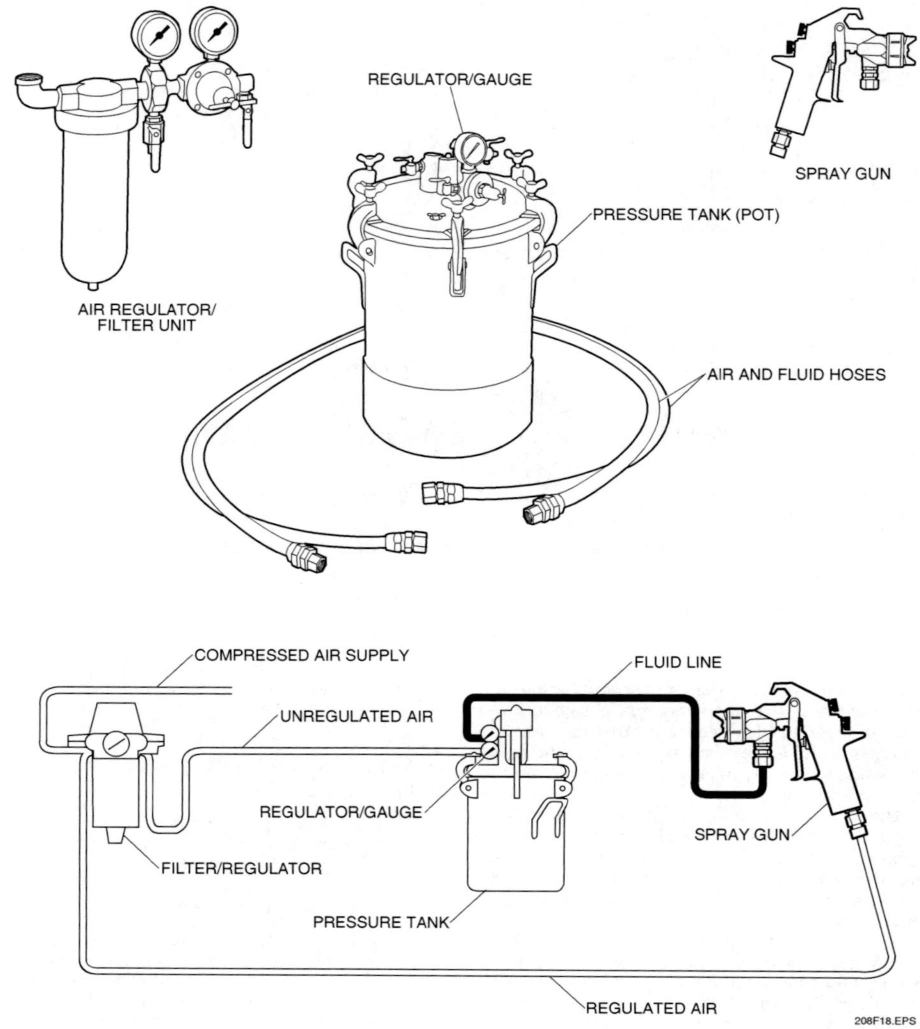

Figure 18 ◆ Components of a typical pressure-feed air spray system (compressor not shown).

Instructor's Notes:

compressor is used to pressurize the tank and force the coating through the fluid hose to the fluid nozzle within the spray gun. Pressurized air from the air compressor is also supplied through a separate air hose to the spray gun air cap. Outlets on the spray gun air cap direct the jets of air into the coating stream to atomize the coating and shape the spray pattern. The air system contains filters to clean the compressed air and regulators to adjust the system air pressures.

Some pressure-feed systems use a pneumatic pump (*Figure 19*) instead of a pressurized tank to supply the coating to the spray gun. This type of pump is operated by air pressure. Another type of pressure-feed system, called a pressure cup system, uses a quart-sized cup attached to the spray gun (*Figure 20*) or a one-quart or two-quart cup (called a remote cup) carried by the operator to supply pressurized material to the spray gun. Because these pressure cups have a small capacity, use of this system is generally limited to fine finishing jobs with a limited amount of spraying.

The air-coating balance in a pressure-feed tank spray system is controlled by adjusting the viscosity of the coating material, the air pressure applied to the coating material, and the volume of atomizing air. The spray gun has a fluid adjustment knob for controlling material flow and a knob for adjusting the width and shape of the spray fan.

Audiovisual

Show Transparencies 15 and 16 (Figures 19 and 20). Have trainees point out the differences among the pressure-feed tank system, pressurized pump system, and pressure cup system.

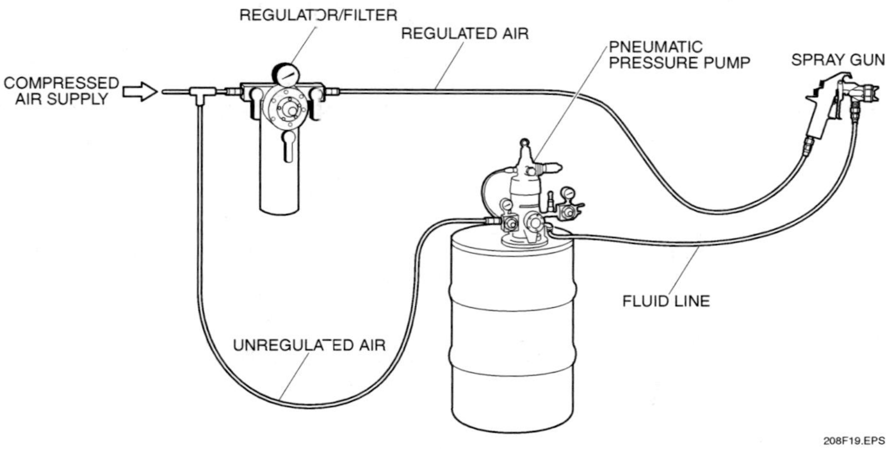

Figure 19 ◆ Pressurized pump system.

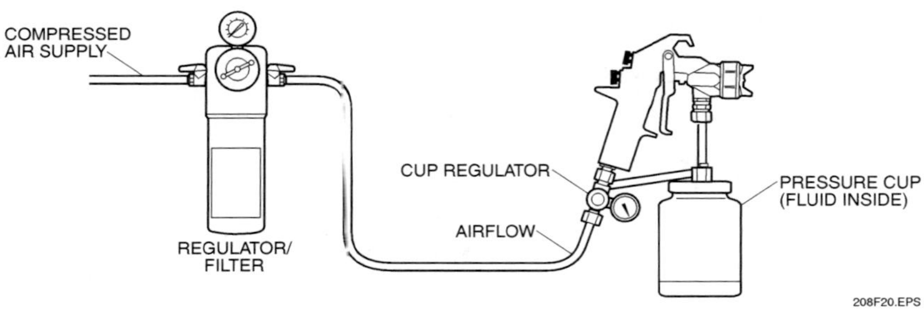

Figure 20 ◆ Pressure cup system.

Audiovisual

Show Transparency 17 (Figure 21). Ask a trainee to explain the uses for siphon-feed systems. Have a trainee explain how the siphon-feed system works.

Show Transparency 18 (Figure 22). Have trainees identify the advantages and disadvantages of airless spray systems.

Show Transparency 19 (Figure 23). Have trainees label the components of both types of airless spray systems.

3.1.2 Siphon-Feed Systems

Siphon-feed systems are typically used for jobs that involve low-viscosity fluids, frequent color changes, and finishing jobs that require only small amounts of material. In the siphon-feed system (*Figure 21*), the spray gun nozzle draws coating material contained in a spray gun cup by a vacuum created at the spray gun cap. Air pressure applied from the compressor to the spray gun for atomization of the coating is controlled by adjustment of the air control regulator. The amount of fluid is adjusted by a fluid control screw on the gun, the viscosity of the coating, and the air pressure.

3.1.3 Conventional Spray System Components

The basic components used in typical pressure-feed and siphon-feed conventional spray systems include:

- Spray guns
- Air compressors
- Air pressure regulators and air-cleaning devices
- Tanks and cups
- Fluid (coating) pumps
- Hoses, couplings, and fittings

3.2.0 Airless Spray Systems

Airless spray systems are so named because air is not used to atomize the coating, as it is with conventional air spray systems. Instead, fluid is pumped under high pressure through a small, specially shaped opening or hole called an **orifice** in the spray gun tip (*Figure 22*). The tip controls the flow of the fluid and creates the spray pattern.

As the fluid passes through the orifice at a high pressure (typically 2,000 to 3,000 psi), it separates into small droplets, resulting in a very fine spray mist. The process is similar to the way a garden hose works. The velocity with which the spray leaves the nozzle gives it enough momentum to reach the surface. The higher the viscosity of the coating material, the higher the pressure required to atomize it.

Airless systems (especially the heavy-duty, air-powered units) can be used to spray almost any type of coating. The heavy-duty, air-powered units can readily support multiple guns operating simultaneously. Airless systems can provide complete and uniform coverage because they apply only the coating material, not air, to the surface. They can coat corners, edges, and porous surfaces evenly because there is little air turbulence to deflect the coating material or cause it to rebound. Airless spraying is best suited for jobs where medium to large areas must be covered quickly. Its strong points are its simplicity and speed. Airless systems can achieve up to 50% more coverage per day on large, regular surfaces than conventional or HVLP air spray systems. Because there is little overspray, typical airless systems have a transfer efficiency that ranges between 50% and 60%. The disadvantages associated with airless spray systems include safety hazards due to its high operating pressures, a greater chance of clogging problems, moderate overspray, and limited methods of controlling the spray pattern.

There are many types of airless spray systems. However, all consist of the same basic components (*Figure 23*):

- Spray gun
- Fluid (coating) pump and pump power source
- Fluid hose, couplings, and fittings

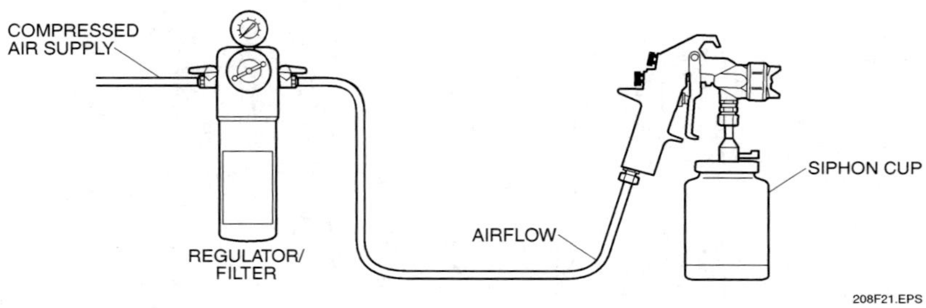

Figure 21 ◆ Siphon-feed system.

Instructor's Notes:

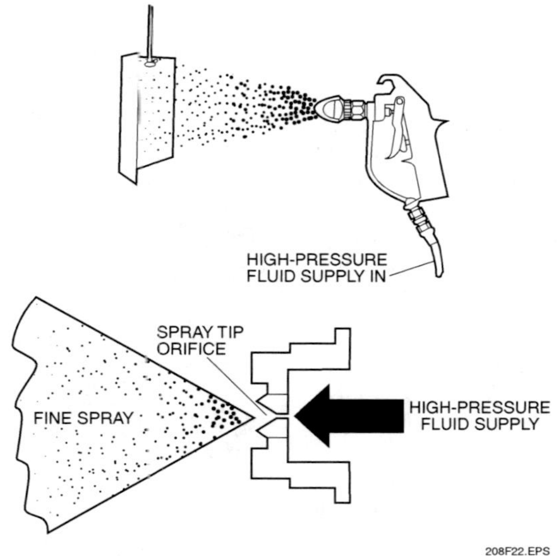

Figure 22 ◆ Airless spray gun and orifice.

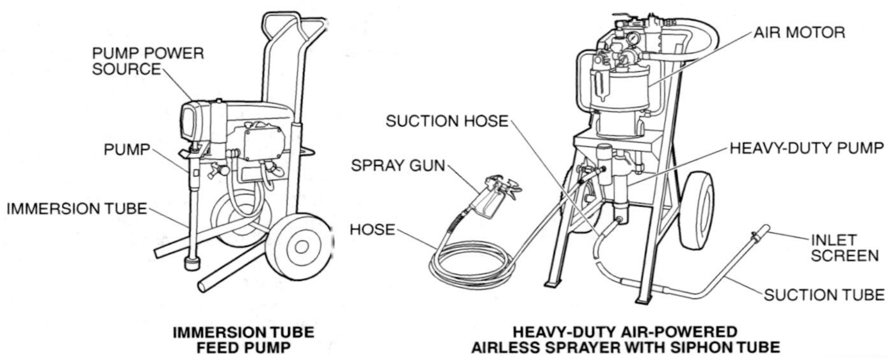

Figure 23 ◆ Components of a typical airless spray system.

Classroom

Ask a trainee to describe an air-assisted airless spray system. Answer any questions trainees may have.

Audiovisual

Show Transparency 20 (Figure 24). Ask a trainee to explain the difference between conventional and HVLP spray systems. Have a trainee name the two spray methods that meet the 65% transfer efficiency criteria.

Show Transparency 21 (Figure 25). Have trainees describe the two main types of HVLP systems.

3.2.1 Air-Assisted Airless Spray Systems

A variation of the airless spray system is the air-assisted airless system. In this system, the guns use an airless spray tip that shapes the coating material into a fan pattern at somewhat lower fluid pressure, while a special air cap atomizes the fluid. This eliminates tails and controls the width of the fan pattern. The result is a high-quality finish that is better than that of an airless spray system but not quite as fine as a conventional air spray system. An air-assisted airless system consists of an air-assisted airless spray gun and spray tip, a pump, air and fluid hoses, a regulator, and an air compressor.

3.3.0 HVLP Spray Systems

In an HVLP system, a high volume of air (typically 50 to 130 cubic feet per minute [cfm] or more), delivered at a low pressure (10 psi or less, measured at the spray gun air cap), is used to atomize a coating into a soft, low-velocity pattern. This reduction in airstream velocity over the 40 to 70 psi typically delivered by conventional spray systems results in a more controlled spray pattern with less overspray (*Figure 24*). Consequently, HVLP systems have a much higher transfer efficiency (65% to 90%) than conventional or airless systems (normally less than 60%). Because there is little overspray, HVLP spraying can easily be performed indoors in closed areas. Also, there is less coating material consumption per job and lower VOC emissions because of the higher transfer efficiency. This is especially important in Southern California and other areas of the country where VOC legislation requires that all spray equipment have a transfer efficiency of at least 65%. In establishing its VOC requirements, the California South Coast Air Quality Management District (SCAQMD) has specified HVLP as one of two spray methods that meets the 65% transfer efficiency criteria. Electrostatic spraying is the other method. It is described later in this module.

HVLP systems are not designed to spray large, flat surfaces. They are best suited for smaller jobs that require a very fine finish. When compared to an airless system for coating high-volume jobs, production with an HVLP system is much slower because it operates at such low pressures.

HVLP systems can be used to spray a wide variety of conventional and low-VOC coatings. However, some coatings must be thinned to sprayable viscosities for use with HVLP.

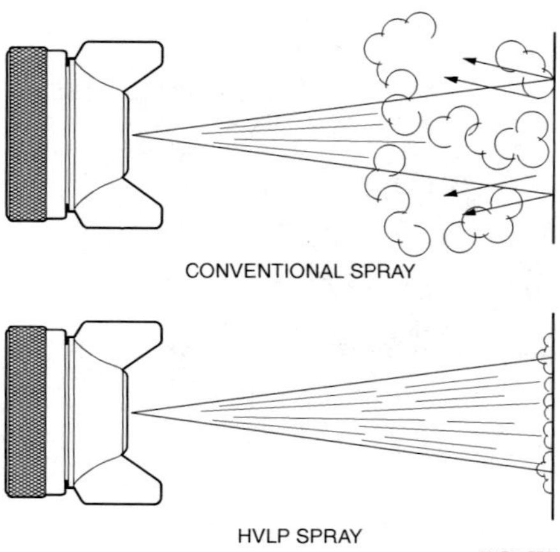

Figure 24 ◆ Conventional spray versus HVLP air spray.

3.3.1 HVLP Spray System Components

Depending on the type of system, components of a typical HVLP system include an HVLP-type spray gun, air and fluid hoses, a pressure pot or cup with pressure regulators, and a compressor or turbine. There are two main types of HVLP systems: compressed air and turbine. In the compressed air system, the high-volume air supply is provided by a standard air compressor. In one common compressed air system, the higher pressure air supplied from the air compressor is reduced to the required pressure of 10 psi or less within the spray gun. In another common system, the higher pressure air supplied by the compressor is passed through an air-regulating device in the compressor output air line to reduce the air pressure before the air reaches the spray gun. Most systems need at least 3 horsepower (hp) for an electric unit or 5hp for a gasoline-driven unit to satisfy the air volume demand of a single spray gun nozzle. The air lines and hoses must also be capable of delivering a sufficient volume of clean, dry, oil-free air for the gun to work efficiently and produce a flawless finish.

In the turbine-type HVLP system, air is supplied by a turbine (*Figure 25*). This type of system is portable and is used for smaller jobs.

Except for the HVLP spray gun or turbine, the remaining components, such as compressors and pressure pots or cups with pressure regulators, used in compressed-air HVLP spray systems are basically the same as those described earlier for use in conventional pressure-feed spray systems. Only the air pressures and airflow involved are different.

Instructor's Notes:

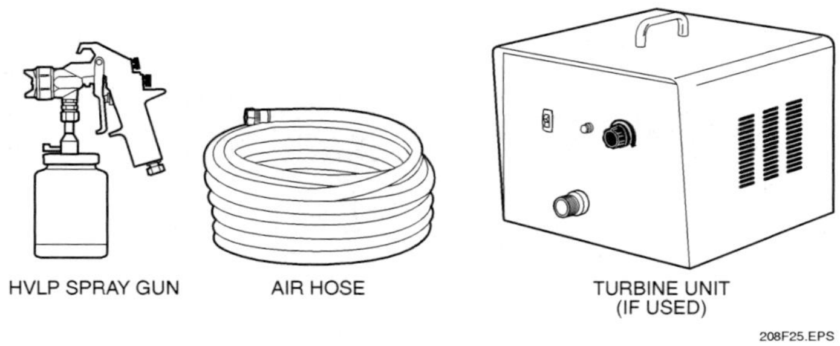

Figure 25 ◆ Components of a typical turbine HVLP spray system.

3.4.0 Spraying with a Conventional Spray System

There are many different kinds of conventional spray equipment and spray system configurations. Regardless of the type, all should be operated and maintained as directed in the manufacturer's operation and service manuals supplied with the equipment and/or system. General guidelines that apply to the operation and maintenance of most conventional spray systems are described in the following paragraphs.

3.4.1 Personal Protective Equipment/Safety Guidelines

When using any type of spray system, always use the proper respirator (*Figure 26*). Make sure that the respirator is approved by the National Institute for Occupational Safety and Health (NIOSH) and certified for use with the material you are spraying. Also, before attempting any spraying, make sure that the respirator has a good face seal by performing both positive and negative fit checks. Use of protective hand cream and/or gloves and eye protection is also important. Goggles offer good protection because of their wraparound design. Because overspray will obscure your goggles while spraying, you should consider keeping one or two extra pairs of goggles handy to exchange as necessary. Clean them often using nonabrasive cleaner. If in doubt about which respirator, gloves, or face/eye protection to use, refer to the Material Safety Data Sheet (MSDS) for the material being sprayed.

Whenever operating or maintaining a conventional spray system, you should also comply with the general safety guidelines that follow.

Spray Equipment:

- Ensure that there is adequate ventilation before starting. Mechanical ventilation must be provided when spraying in enclosed areas. Ventilation equipment must be located so that it will not circulate contaminated air back into the spraying area.
- The spray area must be free of open flames, sparks, and other ignition sources.
- Locate appropriate fire extinguishers in or near the spraying area. Make sure all personnel know where they are located.
- Use explosion-proof portable lamps when required. Check with your supervisor.
- Appropriate warning signs must be posted in hazardous areas.
- The spray area must be free of combustible materials and residue.
- OSHA requires that an operator's instruction manual be at the job site whenever spray equipment is being used. The operator's instruction manual is available from your supervisor.
- Check the equipment carefully before use. Ensure that the electrical cords, plugs, air and fluid hoses, couplings, and spray tips are in good condition. Remove any damaged equipment from service.
- Check the spray equipment's safety shutoff valve and pressure gauge to be sure that they are in good condition prior to each use. Equipment with a damaged or missing safety shutoff valve or pressure gauge must be removed from service.
- Close the safety shutoff valve when spraying operations are not being done.
- Do not exceed the equipment manufacturer's recommended safe working pressure. If the equipment is faulty and the safe working pressure cannot be maintained, immediately remove it from service for repairs or replacement.
- Before repairing or servicing the equipment, disconnect the power source and completely relieve the pressure from the system.
- Clean up overspray because it can be flammable.

Audiovisual

Show Transparency 22 (Figure 26). Discuss how to choose the proper respirator for a job.

Laboratory

Have trainees take turns practicing proper use of a respirator. Ensure that they perform fit checks and check for NIOSH certification.

Classroom

Ask a trainee to explain the importance of choosing the correct nozzle combination.

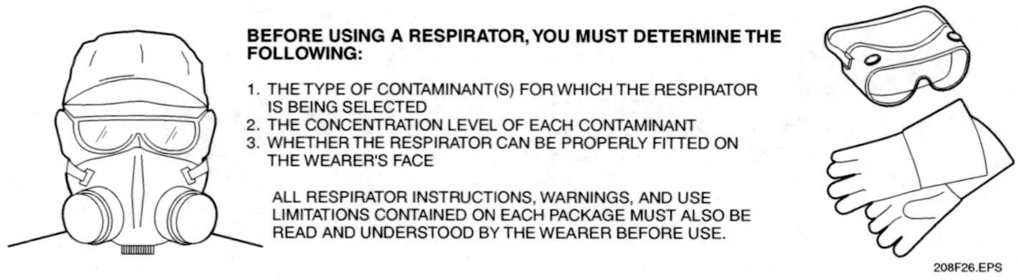

Figure 26 ◆ Use proper personal safety protection for the spraying job at hand.

Halogenated hydrocarbon solvents or fluids containing such solvents can cause an explosion when used in a pressurized spray system containing aluminum or galvanized components. Cleaning agents, coatings, and other chemicals may contain halogenated hydrocarbon solvents. If in doubt about the material you intend to use, always check the MSDS for information concerning halogenated hydrocarbon solvents.

Compressors:

- Never operate a compressor unless it is equipped with a pressure relief valve and pressure gauges.
- Operate and lubricate compressors in accordance with manufacturer's recommendations.
- Inspect all safety devices to ensure that they are in proper working order before using the compressor.
- Before any repair work is done on the pressure system of the compressor, bleed off the pressure and lock out the system.
- Obey any signs posted on compressors. They are there to warn you of important safety hazards, including the automatic starting feature of the compressor, if so equipped.
- *Do not* aim compressor air at yourself or toward others.
- Use safety cables, also known as *whip checks*, on couplings of high-pressure hose lines where a connection failure could occur.
- The compressed air pressure of spray pots must be checked and must not exceed the safe working pressure of the container.

Gasoline Engine-Driven Compressors:

- Shut the compressor down while filling the fuel tank. Do not refuel a hot engine.
- Always ground the fuel nozzle against the filler neck to avoid sparks.
- Handle all flammable material according to the procedures on the container.
- Refueling operations should be done outdoors. If you must refuel the compressor inside, make sure that the area is well ventilated.
- Never smoke while handling fuel. The fumes in an empty fuel container are explosive.
- Avoid spilling fuel. If a spill occurs, clean it up immediately.

3.4.2 Selecting the Conventional Spray Gun Nozzle Combination and Fan Spray

To obtain the required results, the spray gun air cap, fluid tip, and needle valve (needle) must be correctly selected for use with the material being sprayed and to achieve the desired fan pattern and application rate. Because they all affect the spray pattern, the air cap, fluid tip, and needle must be selected as a matched set, commonly called the *nozzle combination*. Generally, the larger the fluid tip orifice, the more coating is discharged and the faster the application. However, if the fluid tip orifice is too large, control over the material discharge can be lost. If the fluid tip orifice is too small, the discharge velocity will be too high. This means that the air coming from the air cap will not atomize the coating properly and the pattern will be distorted.

To select the proper nozzle combination, refer to the gun manufacturer's specifications for the particular spray gun being used and choose a nozzle combination recommended by the manufacturer for use with the type of material being sprayed. Most coating manufacturers also give cap/tip recommendations for use with each of their products. For example, to conventionally spray almost all interior and exterior latex and alkyd coatings, one coating manufacturer's data recommends using a general-purpose cap/tip combination that has an orifice inside diameter (ID) of 0.042" (flow rate up to 20 oz./min.) and a typical pattern size of 10" to 12".

Instructor's Notes:

Once the correct cap/tip combination is installed in the spray gun, the fan spray is adjusted using the pattern control on the gun. Typically, turning it clockwise until it is fully closed produces a round spray; turning it counterclockwise widens the spray into a fan shape. The fan spray pattern can be oriented horizontally or vertically (*Figure 27*) or anywhere through 360° by loosening the spray gun retaining ring and rotating the nozzle or air cap to the required position.

3.4.3 Operation and Adjustment of Conventional Sprayer Air Supply and Fluid Feed Equipment

Always prepare the spray equipment in accordance with the manufacturer's instructions. Make sure to follow all warnings and cautions marked on the equipment and/or given in the instructions.

If using a portable air compressor, locate it:

- As close as possible to the spray site to minimize pressure drop
- On as level a surface as possible
- Away from walls or obstacles to aid cooling
- In an area that allows easy access to the operating controls
- Upwind of any source of airborne dirt or contaminants

Consult the coating manufacturer's data for the type of coating being sprayed to find the recommended air and fluid pressures for use with conventional spraying. For example, to spray almost all interior and exterior latex and alkyd coatings, one major manufacturer recommends using air pressures between 40 and 60 psi and a fluid pressure of 20 psi.

After the system compressor is turned on and sufficient air pressure has built up, adjust the system fluid and air pressure controls to those recommended by the manufacturer, then fine-tune your adjustments while testing to find the desired spray pattern. A faulty spray pattern is a sign of a problem with the spray gun, usually a clog in the air cap or the fluid tip. Generally, the greater the fluid pressure, the higher the atomizing pressure and the greater the overspray. Use the lowest fluid and air pressures that will produce a quality finish with good coating material flow at the desired rate of application.

3.4.4 Conventional Spraying Techniques

Expertise in spraying can only be achieved with practice. Practice spraying should never be done on the customer's surfaces or objects. Typically, it is done using cardboard cartons or on sheets of newspaper stapled to a box or carton.

The most common mistake is moving your arm in an arcing motion while spraying (*Figure 28*). This creates an uneven coating that is heavy in the middle and light on the ends. Instead, at the beginning and end of each stroke, bend your wrist as necessary to always keep the sprayer at right angles (90°) to the surface and always at an equal distance of approximately 6" to 8". Holding the gun closer to the surface deposits more coating material on the surface and produces a narrower spray pattern. Holding it farther from the surface produces a thinner coat and wider spray pattern. Start your stroke before you pull the trigger, and continue your stroke after releasing the trigger. If you do not, there will be heavy spots of coating material where you start and stop. Reverse your stroke direction and apply the return stroke in the same manner as

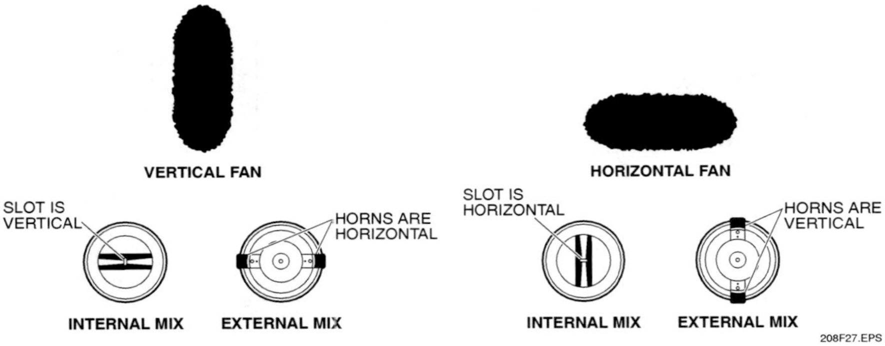

Figure 27 ◆ Selection of conventional sprayer pattern.

Classroom

Have a trainee explain the guidelines for locating a portable air compressor. Ask a trainee to explain what a faulty spray pattern may signal.

Audiovisual

Show Transparencies 23 and 24 (Figures 28 and 29). Discuss proper spraying techniques.

Teaching Tip

See the Teaching Tip for Sections 3.4.0–3.4.4 at the end of this module.

previously described. Move the gun either up or down the surface at a steady rate. Moving the gun at a constant speed results in even coverage. The correct spraying speed allows for a full wet coat of material without runs or **sags**. **Overlap** the edges of each spray stroke about 50% to get a uniform coating without streaks or thin spots.

Corners, edges, and other difficult areas should be sprayed first. For outside corners and edges, aim directly at the area and split the center of the spray pattern on the corner or edge and spray vertically or horizontally as required so that both adjoining sections receive about the same amount of coating (*Figure 29*). Spraying directly into an

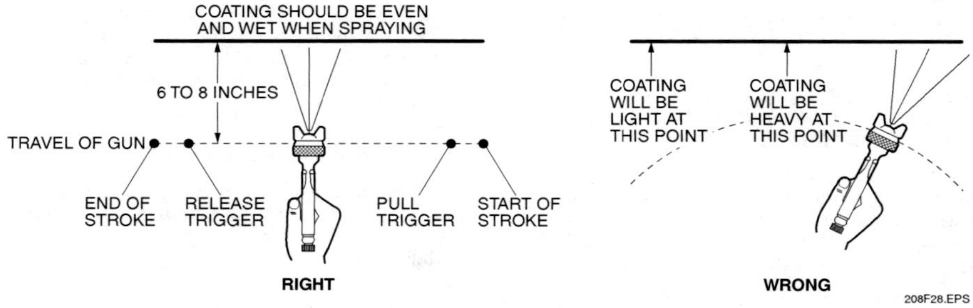

Figure 28 ◈ Right and wrong techniques for conventional air spraying.

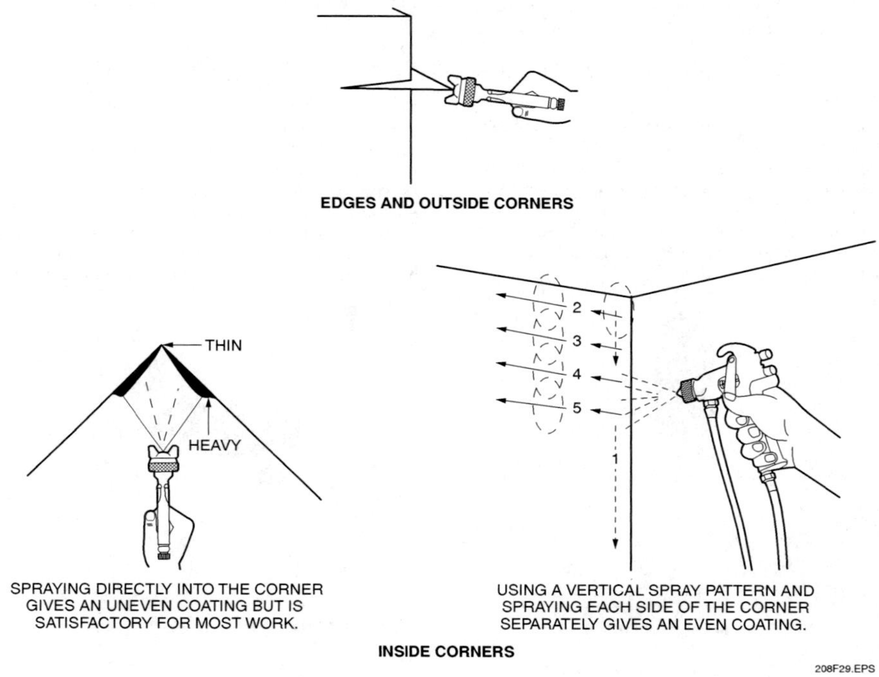

Figure 29 ◈ Spraying edges and corners.

Instructor's Notes:

inside corner is okay for many jobs, but it normally gives an uneven coating with a heavier buildup of material at the edges of the pattern. For even, smooth coverage on an inside corner, turn the spray gun 90 degrees or adjust the spray nozzle so that the spray pattern is applied vertically. Then, spray each side of the corner separately. Spray long, flat panels and/or surfaces in sections of convenient length. They may be sprayed with vertical strokes; however, most users prefer horizontal strokes. Spray each section with 18" to 36" horizontal strokes, overlapping the strokes as described earlier. Each completed section should be overlapped about 4" (*Figure 30*).

3.4.5 Cleaning and Maintenance of Conventional Spray Equipment

There are many different kinds of conventional spray equipment and spray system configurations. Regardless of the type, all should be cleaned and maintained as directed in the manufacturer's operation and service manuals supplied with the equipment and/or system. General guidelines that apply to the cleaning and maintenance of most conventional spray systems are described in the following paragraphs.

While cleaning the equipment, make sure to collect used solvent waste in a container with a lid. Once waste solvent is no longer usable, pour it into a suitable waste solvent barrel. Dispose of dirty solvent and any dry waste in accordance with all federal, state, and local laws.

In some states and/or localities, it is illegal to spray solvents containing VOCs into the atmosphere when cleaning a spray gun or equipment. In these areas, special cleaning equipment such as a closed system cleaner or a washer-type cleaner must be used. In a closed system cleaner, an enclosed unit or spray gun cleaning station is used to condense solvent vapors back into liquid form. This prevents the escape of VOCs into the atmosphere. In washer-type cleaners, the spray gun, cups, nozzles, and other parts are totally enclosed during the washing, rinsing, and draining cycles. This flushes solvent through the gun without releasing any VOCs into the atmosphere. If you are required to use these types of cleaners, operate them according to the manufacturer's instructions.

General guidelines for cleaning and maintaining conventional spray equipment after each use are outlined below. The maintenance is similar whether the filters are separate units or part of a combined filter-regulator unit.

Follow these steps to clean a gun and pressure container or pressure cup:

Step 1 Shut off the air supply to the container, and release the pressure on the container (or cup).

Step 2 Hold a wadded piece of cloth over the gun nozzle and pull the trigger. The air will back up through the nozzle and force the material out of the hose into the container.

Step 3 Empty the container, then pour an appropriate solvent or thinner into the container and spray the solvent through the gun and hose until they are clean.

CAUTION
Never soak the gun in solvent; it may cause the needle packing to dry out.

Step 4 Attach the material hose to an air line, then blow out the hose to dry it and remove all traces of the material.

Step 5 Thoroughly clean the inside and outside of all cups, tanks, and tank inserts using rags and the appropriate solvents after each use.

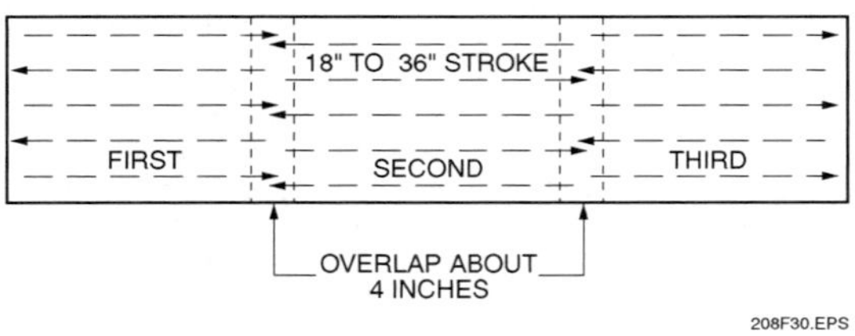

Figure 30 ◆ Spraying long panels or surfaces.

Safety

Ask a trainee to explain the danger of soaking the spray gun in solvent.

Teaching Tip

See the Teaching Tip for Section 3.4.5 at the end of this module.

Homework

Have trainees review Sections 3.5.0–3.9.0 for the next session.

Materials

Ensure that you have everything required for laboratories during this session.

Classroom

Have trainees discuss the safety guidelines for using an airless spray system. Describe some scenarios and problems and have trainees identify the possible causes/responses.

> **NOTE**
> Cleaning is most easily done while the coating residue is still wet.

Step 6 Inspect all sealing surfaces, gaskets, and clamps as you clean to ensure that they have been properly cleaned and are not damaged.

Step 7 Clean the outside of air and fluid hoses with a solvent-soaked rag and make sure that they are undamaged. Coil and hang up unused hoses for storage.

Step 8 Lubricate the pivot points and the packing around the fluid needle of the spray gun occasionally with several drops of light lubricating oil.

Step 9 Drain system air-cleaning filters daily or more often, as necessary. Clean or replace their elements regularly.

Step 10 Clean and perform maintenance on the system compressor per the manufacturer's instructions and maintenance schedule.

Follow these steps to clean a gun used with a siphon cup:

Step 1 Unscrew the cup, empty it, and rinse it carefully with an appropriate solvent or thinner.

Step 2 Place clean solvent or thinner in the cup, spray the solvent through the gun until it is clean, and then blow air through the gun to dry it.

3.5.0 Spraying with an Airless System

An airless spray gun operates at pressures many times higher than that of conventional spray equipment. The spray is powerful enough to easily penetrate the skin. Keep hands and feet away from the spray gun discharge and never aim the gun at anyone. The spray gun must have a tip guard to prevent fingers from getting into the way of the spray.

Use of a respirator and all the other general safety guidelines described earlier for use with conventional spraying also apply to airless spraying. Because of the noise levels involved with some gasoline engine-driven sprayers, hearing protection may also be required.

Always follow company safety rules and guidelines when spraying with airless spray guns. For your safety and that of others in the area, it is imperative that you be familiar with the following safety-related information.

Injection hazards exist when using airless spray equipment. Coatings and other fluids under high pressure can penetrate skin and cause serious injury that can lead to amputation. Injection into the skin is a serious traumatic injury. It is important to surgically treat the injury as soon as possible.

If you are injured (or even suspect an injury), go to an emergency room immediately. Tell the doctor that you suspect an injection injury. Do not treat it as a simple cut. Show the doctor the MSDS for the kind of material you were spraying. For treatment instructions, have the doctor call the local Poison Control Center.

To prevent fires and/or explosions, be sure the system and the object being sprayed are properly grounded before operating the spray gun. If there is any static sparking or you feel an electric shock while using the equipment, stop spraying immediately. Do not use the equipment until you identify the problem.

Halogenated hydrocarbon solvents or fluids containing such solvents can cause an explosion when used in a pressurized airless spray system containing aluminum or galvanized components. Cleaning agents and coatings may contain halogenated hydrocarbon solvents. If in doubt about the material you intend to use, always check the product's MSDS.

In addition, you must adhere to the following safety precautions:

- Get in the habit of using the spray gun trigger lock and turning off the power any time the system is not in use. Check the operation of the trigger lock before using the gun. Never remove or modify the trigger lock.
- Check for worn hoses, loose or leaking fittings, and improper couplings before using the system. Use only equipment designed to withstand the high pressure of airless spraying.
- Never exceed the equipment's maximum rated pressure.
- Never remove the fluid hose or nozzle tip without first shutting off the pump and relieving the pressure from the system. Failure to do so can cause serious injury from the high-pressure coating material.
- When working with a pressure hose, do not pull the hose over sharp edges or straighten out kinks by whipping the hose.
- Do not work on the spray gun while the hose is pressurized.

Instructor's Notes:

- Make sure the pump, spray gun, and object being sprayed are all properly grounded to prevent static sparks and possible explosions. Any static sparking or slight shock is reason to immediately stop spraying. Check the entire system for grounding, and do not resume spraying until the problem has been fixed. The fluid hose for airless equipment should come equipped with a ground wire. The electrical supply circuit should also be properly grounded.
- Never spray flammable or combustible material near a flame, heater, or any possible source of ignition, including a lit cigarette.
- Never remove the nozzle safety tip, which is fastened over the spray nozzle to prevent hands or other objects from getting too close to the spray orifice.

3.5.1 Selecting the Spray Gun Tip

As stated earlier, the airless spray tip creates a back pressure in the airless spray system to increase the velocity of the material being sprayed to atomize the material. The elliptical shape of the tip determines the angle and shape of the fan or the fan width. The size of the tip orifice determines the rate of fluid flow. Both the orifice size and fan width size determine the characteristics of the tip. For example, if you select increasingly larger orifice sizes while maintaining the same fan width, the result is a thicker coating buildup on the same area (*Figure 31*). If the same orifice size is maintained while increasing the fan width, the same amount of coating is applied over a greater surface area, resulting in a thinner coat.

Spray equipment manufacturers identify their tips in different ways. One common method used by some manufacturers involves a six-digit number stamped on the tip. For example, the tip may be marked 641-519. The first three digits (641) identify the specific type or style of tip. The second three digits (519) can be used to find the actual tip size. In the second set of numbers, double the first digit (5) to determine the minimum fan width when spraying 12" from the surface. In our example, the approximate fan width would be about 10" to 12", where the maximum fan width is (5 × 2) + 2 (for maximum) = 12. The last two digits (19) show the size of the tip orifice in thousandths of an inch (0.019").

Tip selection is based on the type of coating to be sprayed and its viscosity. On multigun systems, the maximum tip size depends on how many guns are being used with the system. The more guns, the smaller the tip size that can be used with each gun. Refer to the coating manufacturer's specifications and choose a tip recommended for use with the type of material being sprayed. Also, most airless spray equipment manufacturers recommend tip sizes for use with different categories of coating products. *Table 3* shows some typical tip sizes for use with different materials when spraying with a single gun.

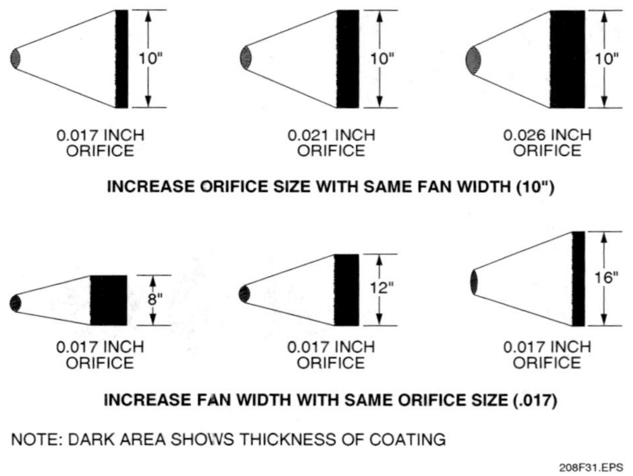

Figure 31 ◆ Examples of airless tip orifice and fan width variations.

Audiovisual

Show Transparency 25 (Figure 31). Provide some sample six-digit tip numbers and have trainees determine the minimum and maximum fan width and the size of the tip orifice.

Audiovisual

Show Transparency 26 (Figure 32). Ask a trainee to explain how to choose the correct tip. Have a trainee explain the problems associated with tip wear.

Show Transparencies 27 and 28 (Figures 33 and 34). Have trainees identify the components of each airless spray system.

Classroom

Ask a trainee to explain airless spraying techniques. Ask a trainee to state whether vertical or horizontal strokes are more productive on flat surfaces.

The tip used should be in good condition. If, as a result of wear, the tip orifice size is enlarged, the resulting thickness of the coating will be too heavy, and the finish will be poor. As shown in *Figure 32*, the pattern size decreases and the orifice size increases with increased wear. The result is lower productivity because additional strokes are required to complete a job. Also, as the size of the tip orifice increases, more coating material is sprayed over a smaller area, requiring the use of more material to complete the job. To maintain quality and productivity, most tip manufacturers recommend that the tip be replaced after spraying about 100 gallons of coating material.

Table 3 Typical Tip Orifice Size Used with Various Materials

Material to be Sprayed	Range of Tip Sizes (Inch)
Acrylics	0.015–0.017
Acrylic block filler	0.025–0.027
Alkyd high solids	0.015–0.019
Alkyd corrosion protection	0.015–0.019
Epoxy block fillers	0.019–0.023
Epoxy masonry surfacer	0.033–0.035
Epoxy high solids	0.015–0.019
Epoxy mastics	0.017–0.021
Coal tar epoxies	0.017–0.031
Polyurethane mastics	0.015–0.019
Coal tar polyurethanes	0.017–0.021
Catalyzed high-performance polyurethanes	0.015–0.017
Catalyzed high-solids polyurethanes	0.013–0.017
Organic or inorganic zinc-rich	0.015–0.072
Asphaltic, fibered or nonfibered	0.060–0.081

208T03.EPS

3.5.2 Operation and Adjustment of Airless Equipment

Always prepare the spray equipment in accordance with the manufacturer's operating instructions. Make sure to follow all warnings and cautions marked on the equipment and/or given in the instructions. For the purpose of an example, the procedure for the setup, start-up, and shutdown of a typical gasoline engine-driven airless sprayer (*Figure 33*) or a heavy-duty, air-powered airless sprayer (*Figure 34*) is given here. With the exception of the power source, the procedures used with other types of airless sprayers are similar.

3.5.3 Airless Spraying Techniques

The spraying techniques described earlier for conventional spraying apply to airless spraying with the following exceptions. Because of the higher pressures involved, you should maintain a distance of about 12" to 14" from the spray gun tip to the surface. If the gun is held too close, coating buildup will cause sags and runs. If it is held too far, a thinner coat and more overspray will result.

Unlike conventional air spraying, an airless system has only two settings, on and off, which makes it more difficult to control the coating material flow. Move the gun at a speed that is comfortable and provides proper coverage. If you have to move it too fast to prevent excess coating buildup, lower the pressure or use a tip with a smaller orifice. If you must move too slowly in order to get a good coating, raise the unit pressure or use a tip with a larger orifice.

The direction of the spray pattern is determined by job needs and the position of the spray tip on the gun. With the spray tip adjusted, you can use either a vertical or horizontal stroke. When spraying flat surfaces, a vertical stroke is usually more productive, as horizontal strokes may create some unnecessary overlapping.

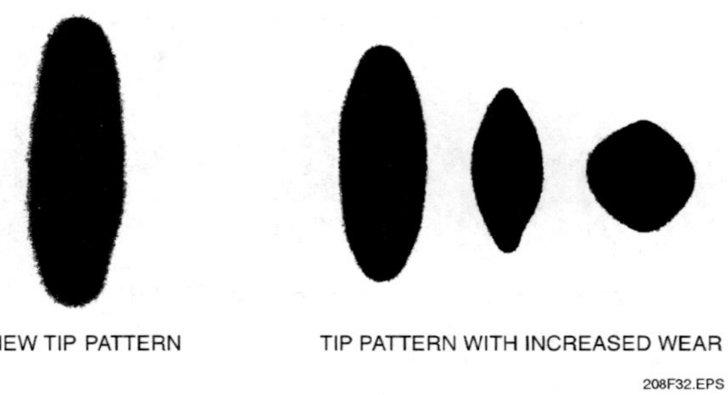

Figure 32 ◆ Tip wear causes changes in the spray pattern.

Instructor's Notes:

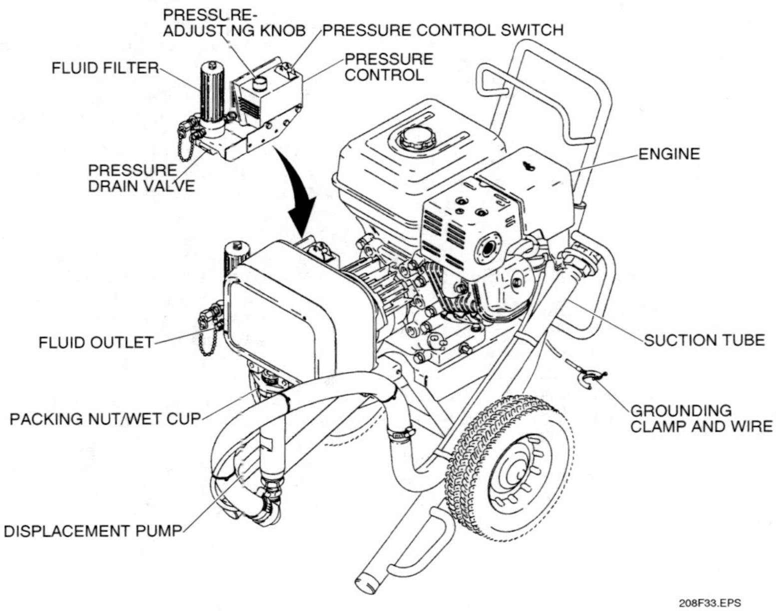

Figure 33 ◆ Typical gasoline-powered airless spray system components.

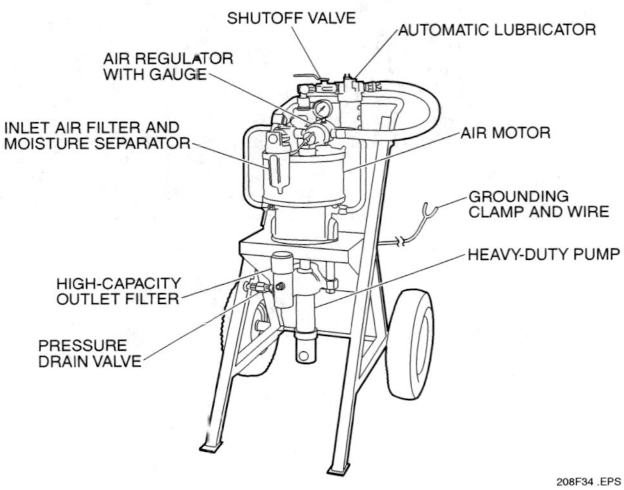

Figure 34 ◆ Typical heavy-duty, air-powered airless spray system components.

Classroom

Ask a trainee to explain the possible danger associated with halogenated hydrocarbon solvents or fluids.

Audiovisual

Show Transparency 29 (Figure 35). Have a trainee explain the purpose of an air nozzle test gauge.

3.5.4 Cleaning and Maintenance of Equipment

All airless spray systems should be cleaned and maintained as instructed in the manufacturer's operation and service manuals supplied with the equipment and/or system. Also, follow all laws and regulations for your location concerning the release of VOCs into the atmosphere and the proper disposal of waste solvents. Environmental and VOC considerations were discussed earlier in this module.

3.6.0 Spraying with an HVLP System

All the safety guidelines described earlier for use with conventional spraying also apply to HVLP spraying. In addition, never attempt to assemble or clean the exterior of a turbine unit while it is plugged in. If using compressed air and a pressure pot, never exceed 50 psi in the pressure pot.

Always be aware that halogenated hydrocarbon solvents or fluids containing such solvents can cause an explosion when used in a pressurized HVLP spray system containing aluminum or galvanized components. Cleaning agents and coatings may contain halogenated hydrocarbon solvents. If in doubt about the material you intend to use, always check the product's MSDS.

3.6.1 Selecting the Spray Gun Nozzle Combination and Fan Spray

Selection of an HVLP spray gun nozzle combination is done in the same way as described earlier for a conventional sprayer. However, because of the lower air pressures involved, correct selection of the air cap and fluid tip is more critical in HVLP spraying. Always follow the gun and/or coating manufacturer's recommendations.

Table 4 shows some typical fluid/needle and air cap sizes used with different materials. The sizes listed are typical only. A test should be performed to determine which works best with the coating being sprayed.

Depending on location, some regulatory agencies may require that HVLP spray gun operators have an air nozzle test gauge like the one shown in *Figure 35*. It is used to measure equipment compliance with the mandated maximum HVLP atomizing pressure of 10 psi.

Table 4 Typical HVLP Fluid/Nozzle and Air Cap Sizes Used with Various Materials

Uses	Typical Applications	Characteristics
Protective coatings	• Gas and oil pipe lining • Underground petroleum tanks • Chemical processing equipment	• Chemical resistance • Durability
Primer coatings	• Plastic and metal finishing	• Promotes adhesion
Sealants and adhesives	• Electrical and electronic components • Sporting equipment	• Corrosion resistance • Excellent adhesive strength • Moisture, solvent resistance • Little shrinkage • Electrical resistance

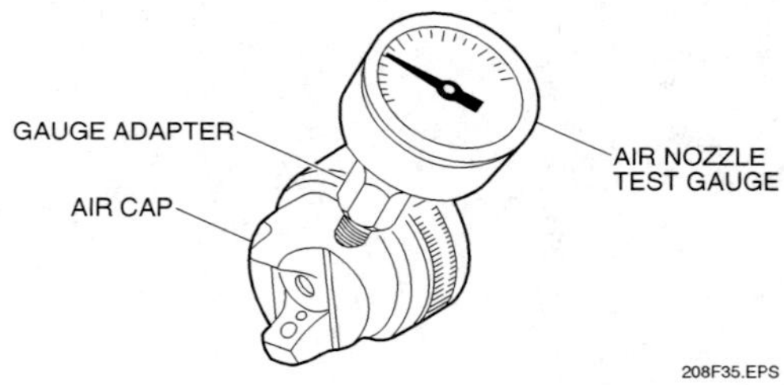

Figure 35 ◆ Air nozzle test gauge.

Instructor's Notes:

3.6.2 Operation and Adjustment of HVLP Spray Equipment

Always prepare the spray equipment in accordance with the manufacturer's operating instructions. Be sure to follow all warnings and cautions marked on the equipment and/or given in the instructions. Except for the pressures involved, the setup and adjustment of an HVLP system is basically the same as described earlier for a conventional spray system. The general procedure for the start-up of a typical HVLP system is given here.

Step 1 If a turbine is being used, set the turbine power switch to OFF, then plug the turbine into a grounded outlet that is at least 20' from the area of spray activity.

Step 2 Attach an air hose to the turbine (if used) or to an air pressure regulator on an air compressor if an HVLP conversion gun is being used.

Step 3 Prepare the coating material according to the manufacturer's instructions.

NOTE
For HVLP spraying, most materials need to be thinned to obtain the proper spraying consistency. If required, thin the material with the recommended solvent. When thinning any coating, comply with all VOC regulations for your location.

Step 4 Pour the prepared coating material into the quart cup, and attach the quart cup to the gun.

Step 5 Attach the tube from the gun to the cup, and attach the air hose to the inlet port on the gun. Make sure that all connections are tight.

Step 6 If a turbine is being used, turn the turbine power switch to ON.

Step 7 Adjust the spray gun fan size and fluid controls to get the desired pattern. A horizontal or vertical pattern can be achieved by loosening the retainer and rotating the air cap.

NOTE
Most turbines have an airflow adjustment (speed control) similar to an air pressure regulator, which can be used to reduce the pressure and air volume when spraying light-body coatings. When used, less air pressure will give a more effective spray pattern, resulting in less overspray.

Step 8 Spray the surface to be coated. When finished spraying, shut off the system and trigger the spray gun to relieve any system pressure.

3.6.3 HVLP Spraying Techniques

The spraying techniques described earlier for conventional and airless spraying apply to HVLP spraying with the following exceptions. When compared to working with an airless system, the user will have to slow down because the output is not as high as an airless sprayer. Maximum delivery of a typical HVLP system is about 0.10 gallon per minute (gpm) while typical airless sprayers provide maximum deliveries from 0.25 gpm to 7 gpm.

To adjust the speed of application, the nozzle size can be changed. The smaller the nozzle, the slower the application; the larger the nozzle, the faster the application. To adjust the degree of finish, the air cap size can be changed. Many times, the users of compressed air HVLP systems are tempted to increase air or fluid pressures in an effort to increase production or material output. Never do this. HVLP equipment is designed to be most efficient at compliant pressures. Increasing fluid or air pressures will reduce finish quality and transfer efficiency, defeating the purpose of using HVLP equipment.

3.6.4 Cleaning and Maintenance of Equipment

All HVLP spray systems should be cleaned and maintained as instructed in the manufacturer's operation and service manuals supplied with the equipment and/or system. While cleaning the equipment, remember to follow all laws and regulations for your location concerning the release of VOCs into the atmosphere and the proper disposal of waste solvents.

Classroom

Have trainees review the procedure for the operation of HVLP spray equipment. Ask a trainee to explain circumstances when less air pressure is desirable.

Ask a trainee to explain how to adjust the speed of application. Have a trainee explain how to adjust the degree of finish.

Teaching Tip

See the Teaching Tip for Sections 3.6.0–3.6.4 at the end of this module.

Safety

Caution trainees not to use a wire brush or hard tools when cleaning spray system components. Stress the importance of not using damaged valves or gaskets.

Audiovisual

Show Transparency 30 (Figure 36). Ask trainees to identify likely causes of each spray pattern problem depicted in Figure 36.

Cleaning and maintenance of a compressed air HVLP system is basically the same as described earlier for a conventional spray system. An overview of cleaning and maintenance procedures for typical turbine-type HVLP spray systems follows.

Step 1 When finished spraying, relieve the cup pressure through the relief valve and remove the lid from the cup. Pour any remaining coating material from the cup into a clean storage container or dispose of it as desired.

Step 2 Pour a small amount of appropriate solvent into the pressure cup. Swirl the solvent around in the cup, then empty it.

Step 3 Thoroughly clean the interior of the cup and wipe it dry.

Step 4 Remove the one-way valve on the cup lid, and clean it with solvent after each use.

> **CAUTION**
> If the valve is damaged or cannot be cleaned effectively, replace it. Also inspect the cup gasket and replace it if damaged. Do not soak the spray gun in solvent. Some solvents can damage internal seals.

Step 5 Pour a small amount of solvent into the cup and spray it through the gun to clean the fluid nozzle and needle.

Step 6 After extended use, remove the fluid nozzle, needle, and air cap and clean by hand with a soft brush.

> **CAUTION**
> Do not use a wire brush or hard tools that can damage the components.

Step 7 Clean the inside of the gun with solvent and a soft brush.

Step 8 Reassemble the gun, and test it with mineral spirits. Clean the exterior of the gun using solvent.

Step 9 Clean or replace the turbine filters. Clean the exterior of the turbine.

3.7.0 Common Spray Coating Problems

Spraying problems fall into two categories: equipment problems and application problems. Spray equipment manufacturers normally provide troubleshooting tables in their operating and service literature. If you are unfamiliar with the cause of a particular equipment problem, consult the troubleshooting table in your equipment service literature.

In addition to the problems described in the following paragraphs, there are some other problems commonly encountered with conventional and HVLP spray systems. A top or bottom heavy spray pattern (*Figure 36*) may be caused by partially clogged horn holes, a partially clogged fluid nozzle, or dirt on the air cap seat or fluid tip seat. Similarly, a heavy right or heavy left pattern may be caused by partially clogged horn holes or dirt on the sides of the fluid tip.

A heavy center pattern may be caused by a fan adjustment setting that is too low, an atomizing pressure that is too low, a coating that is too thick, too high a fluid pressure for the atomizing pressure, too high a fluid pressure for the air cap, or the wrong size tip for the coating being sprayed.

A split spray pattern is due to an improper air/fluid balance. To correct the problem, increase the fluid pressure or reduce the spray pattern width using the fan adjustment.

There are numerous causes for application problems. Following is a list of common application defects and their possible causes.

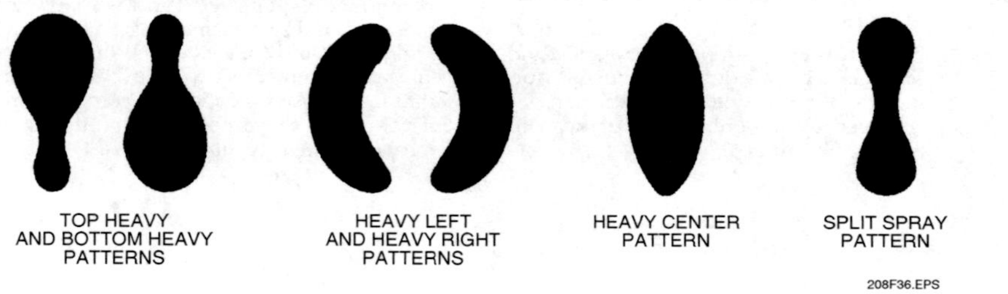

Figure 36 ◆ Conventional/HVLP spray pattern problems.

Instructor's Notes:

3.7.1 Coating Sags

If using conventional/HVLP, the coating material is applied too heavily; the trigger is not released at the end of the stroke; the gun is held too close to the surface or at an incorrect angle; the air cap or fluid cap is dirty; which may cause a distorted spray pattern; or the fluid pressure is too high or the air supply is too low.

If using airless, the coating material is applied too heavily; the trigger is not released at the end of the stroke; the gun is held too close to the surface or at an incorrect angle; the fluid tip or gun filters and/or line filters are dirty, which may cause a distorted spray pattern; or the fluid tip is worn.

3.7.2 Fingering (Streaking)

The following are reasons fingering (streaking) occurs. If using conventional/HVLP, the cap or fluid tip is dirty or worn; the gun is moved too rapidly or is too far from the surface; the air pressure is too high; or the incorrect air cap is being used.

If using airless, the gun is held at an incorrect angle; moved too rapidly; or held too far from the surface; the tip is worn or chipped; the nozzle fan slot or filters are dirty; or the coating material is too heavy.

3.7.3 Orange Peel

Orange peel can be attributed to the following causes. If using conventional/HVLP, improper or insufficient thinner has been used; the air pressure is insufficient or excessive; an improper air cap is being used; or a previously sprayed surface was overlapped before completely dry.

If using airless, the coating material atomization is poor; pressure is excessive; the tip being used is too large; the operator is moving too slowly; or the material was reduced improperly.

3.7.4 Excessive Fog

Excessive fog may result in the following situations. If using conventional/HVLP, there is too much air pressure is being used; the gun is held too far from the surface or at an improper angle; the material is too thin; or the wrong fluid tip or air cap is being used.

If using airless, the gun is held too far from the surface or at an improper angle; the fan pattern is too large; the fluid pressure is too high; or the material is too thin.

3.7.5 Lack of Material

Lack of material can be attributed to the following causes. If using conventional/HVLP, the system is out of coating material (the gun usually begins to sputter); the line is plugged; there is grit, dirt, coating, or another obstruction in the fluid tip, gun, or hose.

If using airless, the system is out of coating material; the line filters or tip are plugged; the pump capacity is insufficient; the orifice is too small.

3.7.6 Leaks from Spray Gun

Spray gun leaks may result from the following problems. If using conventional/HVLP, the needle packing is too loose or too tight, dried out, or damaged; dirt or foreign matter is in the fluid tip.

If using airless, the packing gland is too loose, too tight, or damaged; the fluid needle is damaged; there is dirt or foreign matter in the orifice.

3.8.0 Measuring the Viscosity of a Coating Material

Sometimes, the exact viscosity of a coating material must be known prior to its application. Viscosity is affected by two main factors: one is the amount of solvent (thinner) the material contains and the other is the temperature of the material. Coatings stored in the cold must be allowed to warm up before testing and use. This may be done with hot water baths or by electrical means.

A common method for checking viscosity is by timing a known amount of material while it flows through an orifice in a test container called a viscosity cup (*Figure 37*). Common viscosity cups are the No. 2 Zahn cup and the No. 4 Ford cup. The flow time through the cup orifice is measured with a stopwatch, while the temperature of the material is measured with a thermometer. Using the measured flow rate and temperature, the viscosity of the material can be determined. For example, at a temperature of 75°F, a 15-second material refers to a measured quantity of coating that took 15 seconds to flow through the viscosity cup at 75°F.

The testing and determination of viscosity should be done according to the instructions supplied with the viscosity cup being used.

3.9.0 Measuring the Thickness of a Coating Material

Coating thickness is important because either too little or too much can cause problems. A coating needs a minimum thickness to prevent corrosion, stand up to abrasion, and look good. A coating that is too thick may crack, flake, blister, or not cure properly.

Many jobs have requirements that specify the thickness of the coating that is applied to a surface or object. To ensure that the specifications are

Teaching Tip

See the Teaching Tip for Sections 3.7.1–3.7.6 at the end of this module.

Audiovisual

Show Transparency 31 (Figure 37). Ask a trainee to identify the two main factors that affect viscosity of coating materials.

Laboratory

Have trainees practice using a viscosity cup to measure and determine viscosity of various coating materials. Have trainees follow the instructions supplied with the viscosity cup.

Audiovisual

Show Transparencies 32 and 33 (Figures 38 and 39). Ask a trainee to explain the importance of coating thickness.

Laboratory

Have trainees practice measuring coating thickness using wet-film and dry-film gauges.

Homework

Assign the reading of Section 4.0.0–4.4.13 for the next session.

Materials

Ensure that you have everything required for teaching this session.

Classroom

Ask a trainee to explain the basic principles of electrostatic spray finishing. Have a trainee identify the advantages of electrostatic spray finishing.

being met, periodic checks of wet-film thickness can be made using a wet-film thickness gauge (*Figure 38*). One edge of the gauge is placed on a surface that is covered with wet-coating material. Each tooth is measured in mils or microns. The last tooth with coating on it when the gauge is removed gives the coating thickness. Multiplying the wet-film thickness by the coating's percentage of solids by volume (normally given in the manufacturer's application data) will estimate the dry-film thickness. This can be checked against the specifications for the job to see if the coating is thick enough. The manufacturer's data sheet will also list the recommended film thickness.

After a coating has dried, the thickness can be checked with a dry-film thickness gauge. There are both destructive and nondestructive dry-film thickness gauges. Nondestructive gauges measure the film thickness without damaging the surface. There are many types of nondestructive gauges, including magnetic pull-off gauges and electronic gauges (*Figure 39*). Magnetic pull-off gauges are used to test steel surfaces. Many electronic gauges can be used to measure the dry-film thickness on both ferrous and nonferrous metal substrates.

4.0.0 ◆ ELECTROSTATIC SPRAY SYSTEMS

The terms electrostatics or electrostatic spray finishing refer to a spray-finishing process in which electrical charges and electric fields are used to attract particles of atomized coating material to a target (the object to be coated). In the most common types of electrostatic spray systems, electrical charges are applied to the coating material and the

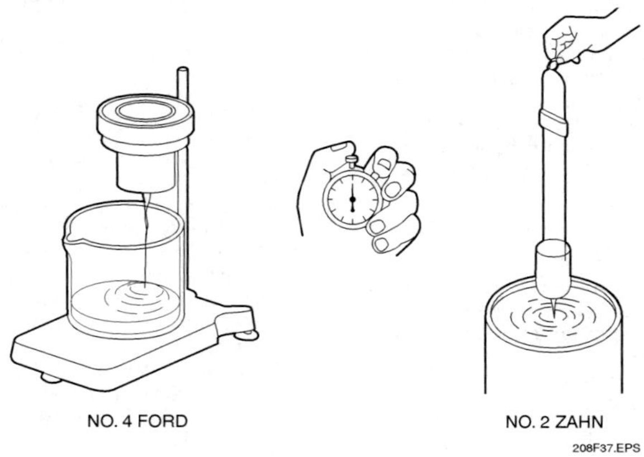

Figure 37 ◆ Measuring viscosity with a viscosity cup.

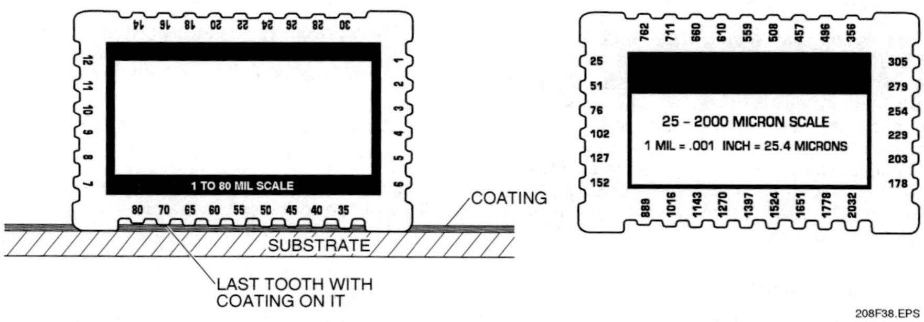

Figure 38 ◆ Wet-film thickness gauge.

Instructor's Notes:

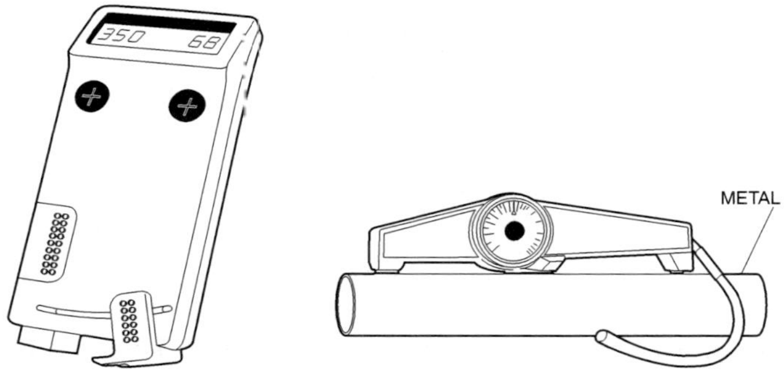

Figure 39 ◆ Nondestructive, dry-film thickness gauges.

target is grounded, creating an electric field. The charged particles of coating materials are drawn by the electric field to the surface of the grounded target because of the attraction of opposing electrical charges. Other, less common types of electrostatic systems apply an electrical charge to the target and ground the coating material, or they apply opposite electrical charges to the coating material and the target.

Electrostatic spray improves the transfer efficiency of spray-finishing equipment. The transfer efficiency improvements occur because the electrostatic forces help overcome other forces, such as momentum and airflow, that can cause the atomized materials to miss the intended target.

Electrostatic methods save coating material, time, and labor. More of the material is attracted to and adheres to the target, and less material is lost in the spray booth and the surrounding environment. Electrostatic methods also lower solvent emissions into the air since less coating material is needed to coat each part.

4.1.0 Types of Electrostatic Systems

Electrostatic spray finishing can be either automatic or manual.

The automatic electrostatic systems include:

- High-speed rotational (stationary or reciprocating) atomizers
- Reciprocating discs
- Automatic air spray, airless, air-assisted airless, or HVLP spray guns

The manual electrostatic systems include:

- Air spray electrostatic, hand-held spray guns
- Air-assisted airless, electrostatic, hand-held spray guns
- Airless electrostatic, hand-held spray guns
- HVLP electrostatic, hand-held spray guns
- Hand-held rotary atomizer systems

For the purposes of this module, only the manual electrostatic systems will be illustrated. *Figure 40* shows three of the five types of manual systems. Two of them are spray gun systems with average production capability and coating transfer efficiencies between 40% and 90%. The third system is a portable, low-speed, rotary-charged atomized system with a very high transfer capability (95% range), but with a very low application rate. It is suitable for light-duty commercial work where very little overspray is desired.

4.2.0 Principles of Electrostatic Sprayer Operation

To understand electrostatic spraying, you must first be able to identify the types of charged particles that make up matter and explain how charged particles impact one another.

Electrostatic spray charging works best when the spray particles are small and light because the electrostatic forces can more easily overcome the momentum forces on the particles. Electrostatic spray charging has the greatest positive impact in directing and controlling particles that are moving

Audiovisual

Show Transparency 34 (Figure 40). Have trainees compare and contrast the three types of manual electrostatic spray-finishing systems.

Classroom

Ask a trainee to explain the terms *stiff* and *resistive*.

Audiovisual

Show Transparency 35 (Figure 41). Ask a trainee to explain how voltage is generated in an internally powered electrostatic spray gun.

Classroom

Name the electrostatic accessories and ask trainees to explain the purpose of each. Answer any questions that trainees may have.

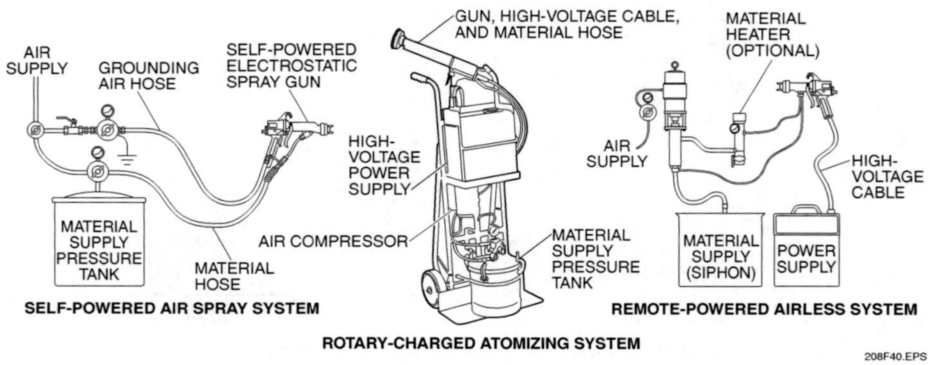

Figure 40 ◆ Manual electrostatic spray-finishing systems.

at a low velocity. Extraneous forces, such as excess air movement, should be eliminated as much as possible to aid the electrostatic process while still meeting Occupational Safety and Health Administration (OSHA) requirements. In addition, the object to be coated must be conductive or have a conductive backing or coating. The target must also be well grounded.

Safety measures are even more critical in electrostatic spray charging because of the added electrical charges. The primary hazards with electrostatic spray equipment are electrostatic shock, flash fires, unintended pressure release, and skin injection (with high-pressure systems only).

4.3.0 Key Equipment Characteristics

Terms that are often used to describe the qualities of electrostatic systems or equipment are stiff (or uninhibited) and resistive. In electrostatic applications, the terms stiff and resistive are used as antonyms (or opposites).

4.3.1 High-Voltage Power Supplies

In electrostatic spray charging, some systems may use cables to bring the power or charge to the spray gun. Other spray guns do not use cables because they have an internal power supply. In electrostatic systems, the cable types and alternatives to cables are:

- High-voltage cable
- Low-voltage cable
- Systems with internally generated power (and no cables)

In systems that use a high-voltage cable, the power supply is remote from the sprayer and uses a high-voltage cable to transmit voltage to the spray gun.

With a low-voltage cable, the power supply is remote from the sprayer, but the cable is low voltage and smaller in diameter. In these systems, voltage is increased inside the spray gun through a transformer and a cascade multiplier.

An example of an electrostatic spray gun that has internally generated power, and therefore does not require cables, is shown in *Figure 41*. Voltage is generated and multiplied through an internal power supply within the spray gun barrel. These systems use a miniature air-powered turbine to spin a miniature alternator inside the gun to generate voltage. Therefore, the device does not need cables or an external power supply.

4.3.2 Electrostatic Accessories

The accessories that are commonly used with electrostatic spray charging equipment are:

- *High-voltage probe and meter* – Measures voltage output from the spray gun
- *Coating resistivity meter* – Measures the resistivity (or conductivity) of the coating material
- *Megohmmeter* – Used to check electrical grounding, as well as electrical continuity within the system
- *Microamp meter* – Measures the current flow in the system

Instructor's Notes:

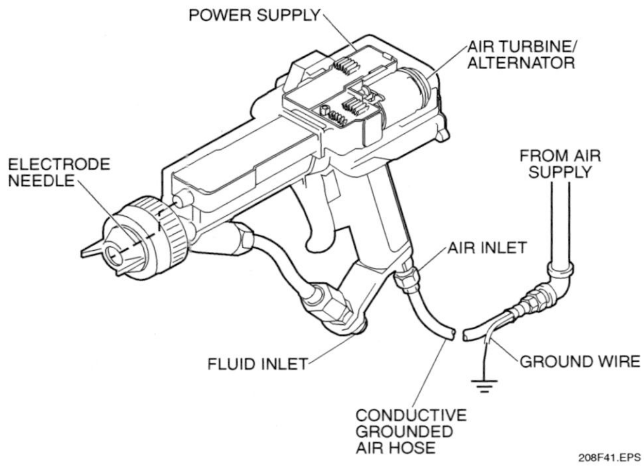

Figure 41 ◆ Internally generated power in an electrostatic spray gun.

4.4.0 Coating Materials

When the term coating materials is used in reference to electrostatic spray charging, it refers to the materials used to coat the object or surface. Like all matter, coating materials have electrical properties. The key qualities of the coatings used in electrostatic spray charging are resistivity, conductivity, and flammability.

4.4.1 Plural Component Materials and Proportioning Equipment

Material suppliers produce, package, and deliver plural component materials to the work site as two or more different component chemicals. The chemical components must be mixed together in a specified ratio at the work site for the chemicals to become a usable material. After mixing, the material is applied by spraying, dispensing, or extruding.

4.4.2 Component Chemical Reaction

A chemical reaction begins spontaneously as soon as the component chemicals are mixed. The component chemicals are transformed into a usable material in a process called curing, cross-linking, or polymerization. The reaction (called an exothermic reaction) cannot be stopped or reversed. The material increases in viscosity as the reaction continues and produces heat as a by-product (*Figure 42*).

After mixing, the material will provide good application characteristics. The period of time that the material provides good application characteristics is called working pot life. Working pot life is also known as work time when working with sealants and adhesives, or spray life when working with coatings. Working pot life, spray life, and work time are different words used to describe the same idea.

Working pot life ends when the material stops providing good application characteristics. As viscosity increases, finish quality will no longer be acceptable for coatings, sealants will not seal properly, and adhesives will not bond properly. The material is no longer usable and should be purged from the equipment before it hardens.

4.4.3 Plural Components

In a two-component material, we refer to the component chemicals in a form such as component A and component B (*Figure 43*).

One component is normally the base material. It gives the material its desired properties. For example, component A is the color in a plural component coating.

The other component (B in this example) is normally the catalyst. It starts the reactions, controls the rate of the reaction, and promotes curing, cross-linking, or polymerization.

Classroom

Ask a trainee to name the three key qualities of the coatings used in electrostatic spray charging.

Audiovisual

Show Transparency 36 (Figure 42). Ask a trainee to explain the relationship between viscosity, time, and working pot life during a chemical reaction.

Show Transparency 37 (Figure 43). Have a trainee explain the purpose of the base material and the purpose of the catalyst.

Audiovisual

Show Transparency 38 (Figure 44). Ask a trainee to explain the relationship between temperature and pot life. Have a trainee name the most critical factor that determines the physical properties of the end product.

Classroom

Provide some sample weight mix ratios. Have trainees convert them to volumetric mix ratios.

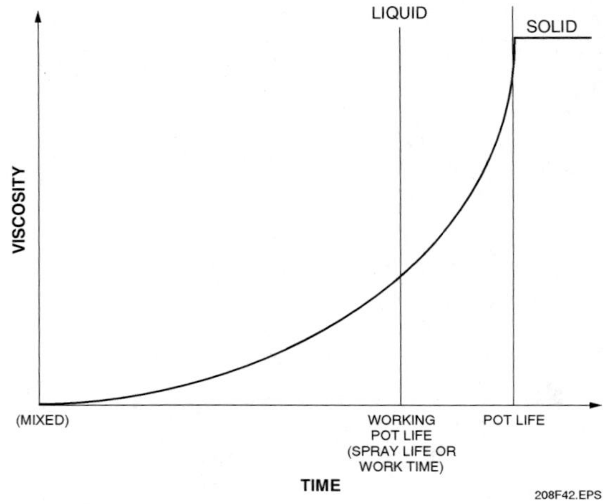

Figure 42 ◆ Plural component viscosity versus time.

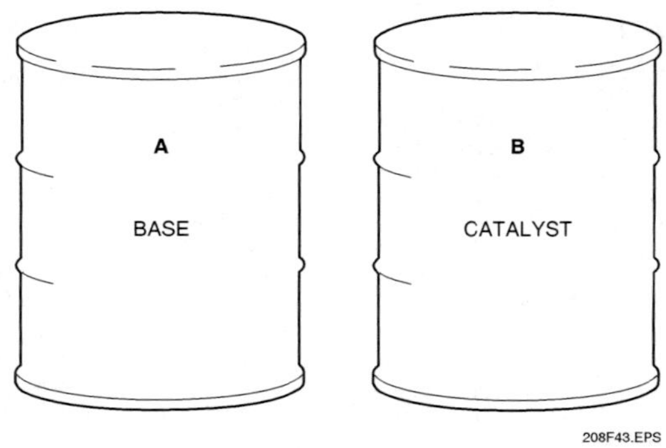

Figure 43 ◆ Plural components.

4.4.4 Mix Ratio

The mix ratio is the ratio of component A to component B that yields the best characteristics of the end product. Mix ratio is specified by the material supplier as a function of weight and/or volume. The material supplier will specify the mix ratio. The mix ratio is the most critical factor that determines the physical properties of the end product. Too much or too little catalyst may cause problems. Mix ratio also affects pot life. In addition, off-ratio materials may not cure properly. Applying heat will normally shorten pot life and speed curing (*Figure 44*).

4.4.5 Measurement of Mix Ratio

Most plural component equipment will proportion chemical components by volume. If the material supplier provides the mix ratio by weight, the following procedure be used to convert it to a volumetric mix-ratio.

> **NOTE**
> The weight/volume of A and B must be expressed in the same units.

Instructor's Notes:

For example, you are given a weight mix ratio of 16:1 (A:B). You are also given:

A = 10 lbs./gal.

B = 8.5 lbs./gal.

We can now write the equation as:

Therefore, the volumetric mix ratio (A:B) = 13.6:1.

4.4.6 Names of Component Chemicals

The names of some of the common A (base) and B (catalyst) components are shown in *Figure 45*. Some plural component materials can use three chemicals (3k): a base, catalyst, and accelerator. Some adhesives and sealants use reverse terminology. In these cases, the A component is the catalyst and the B component is the base. Learn and use the supplier's terms for each component.

4.4.7 Types of Plural Component Materials and Their Uses

Common plural component materials include epoxies, catalyzed polyurethanes, polyesters, catalyzed lacquers, polysulfides, and silicones. *Table 5* lists various uses, applications, and characteristics of epoxies. *Table 6* lists various uses, applications, and characteristics of polyurethanes.

Polyesters, catalyzed lacquers, polysulfides, and silicones are other plural component materials that have applications as decorative coatings, protective coatings, foams, sealants, and adhesives.

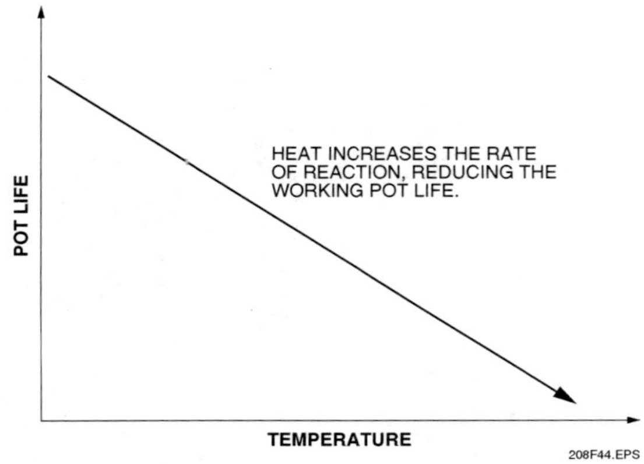

Figure 44 ◆ Effect of temperature versus pot life.

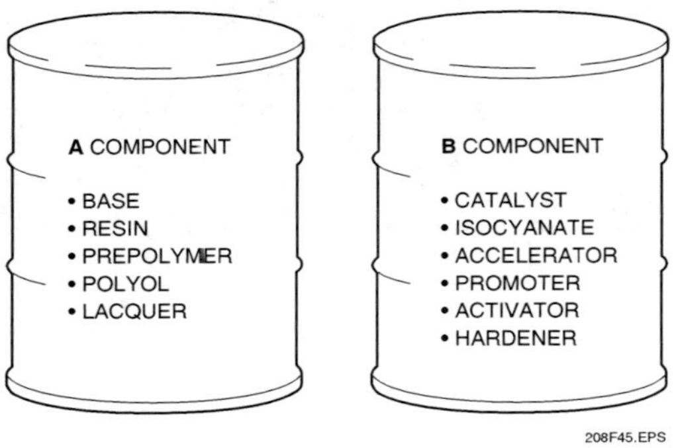

Figure 45 ◆ Various names for A and B components.

Classroom

Ask trainees questions about the information contained in Tables 5 and 6 in the Trainee Guides.

Table 5 Uses, Applications, and Characteristics of Epoxy

Uses	Typical Applications	Characteristics
Protective coatings	• Gas and oil pipe lining • Underground petroleum tanks • Chemical processing equipment	• Chemical resistance • Durability
Primer coatings	• Plastic and metal finishing	• Promotes adhesion
Sealants and adhesives	• Electrical and electronic components • Sporting equipment	• Corrosion resistance • Excellent adhesive strength • Moisture, solvent resistance • Little shrinkage • Electrical resistance

208T05.EPS

Table 6 Uses, Applications, and Characteristics of Polyurethane

Uses	Typical Applications	Characteristics
Decorative coatings	• Aircraft • Machine tools • Business machines • Railway cars • Automotive plastic • Trucks and special bodies for trucks • Clear topcoat on cars • Electric motors • Heavy machinery • Farm machinery • Wood furniture • Lorries • Plastic windows for buildings	• Flexibility • High gloss • Impact resistance • Corrosion resistance • Cures at low temperatures
Protective coatings	• Bridges • Petroleum tanks • Ships	• Chemical and abrasion resistance
Sealants and adhesives	• Composite components, aircraft, recreational vehicles • Wall supports	• Strong bond and seal • Good fatigue life
Foam	• Roof insulation • Refrigeration equipment • Truck trailers • Boats and marine equipment	• High insulation factor • High buoyancy • Sound absorption

208T05.EPS

Instructor's Notes:

4.4.8 Advantages of Using Plural Component Materials

When properly blended, plural component materials provide excellent material characteristics such as:

- Durability
- Abrasion resistance
- Chemical resistance
- Flexibility
- Adhesion promotion
- Appearance

They are also environmentally friendly:

- Low-VOC emissions
- Lower energy consumption while curing; many materials do not require heat to cure
- Mixed cured material can be disposed of as non-hazardous waste

Plural component materials are also cost effective:

- May not require ovens
- Cure quickly, reducing time on the assembly line
- May not require stainless steel equipment

4.4.9 Limitations of Using Plural Component Materials

Plural component material coating costs are high. They are typically more expensive than single component materials. In addition, incorrect mix ratios result in costly failures:

- Material characteristics do not develop properly.
- Lost production and profits are caused by stripping the material off the product and reapplying or disposing of scrapped products.
- Warranty costs are incurred due to scrap and rework.
- Production systems are required to monitor and control the mix ratio to assure quality finished products.

Exceeding the pot life of plural component materials causes costly problems:

- Equipment fails or becomes clogged when mixed materials exceed their pot life before the equipment is cleaned. You must clean, replace, or repair clogged equipment.
- Equipment must be cleaned before pot life is exceeded, generating wastes that are expensive to dispose of properly.

4.4.10 Isocyanate and Its Relationship to Fluid-Handling Equipment

Isocyanate is a plural component chemical that is often used as a catalyst. Isocyanate is sensitive to moisture, even to a slight amount of it in the air. Moisture from the air will react with the isocyanate and create a plastic film over the top of the material in the container. This plastic film can cause a variety of fluid-handling equipment problems. To prevent the plastic film from developing, a dry air blanket or an inert gas (nitrogen) is supplied over the open container of isocyanate, isolating the material from the moisture in the air. This is done using either a desiccant dryer (which uses chemical crystals to dry the air supply to the container) or a nitrogen bottle supply (*Figure 46*).

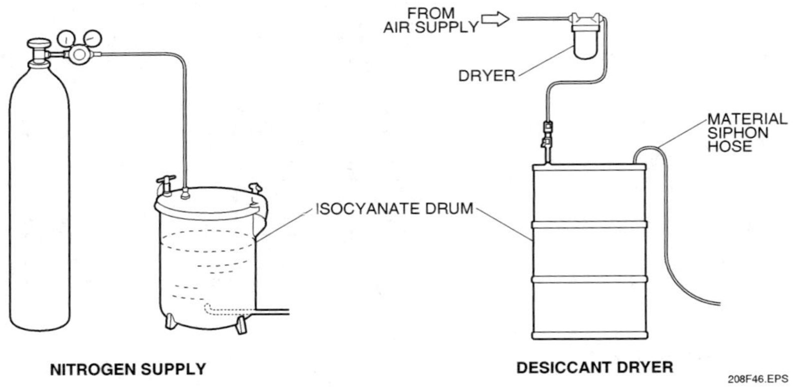

Figure 46 ◆ Moisture isolation techniques for isocyanate.

Classroom

Ask trainees to name the advantages and disadvantages of plural component materials. Have a trainee explain the consequences of exceeding the pot life of plural component materials.

Audiovisual

Show Transparency 39 (Figure 46). Ask a trainee to explain how isocyanate is used as a catalyst. Have a trainee explain why isocyanate must be isolated from the moisture in the air.

Audiovisual

Show Transparencies 40 through 42 (Figures 47 through 49). Have trainees discuss the advantages and disadvantages of plural component mixing methods.

Classroom

Ask a trainee to explain the information contained in Table 7.

Homework

Assign the reading of Sections 5.0.0–5.5.3 and the completion of the Review Questions for the next session.

Materials

Ensure that you have everything required for testing during this session.

Classroom

Ask a trainee to explain how a thermal spray system coats a surface.

Equipment failure happens when isocyanate seeps past the throat packing in a piston pump. This seepage combines with the moisture in the air and forms hard crystals. These crystals act as an abrasive to the throat seals in the pump, causing the pump to fail prematurely. Special pump oil is used in the wet cup of the pump to prevent the crystallization from occurring. The oil isolates the seepage from the moisture in the air.

Moisture in the air can pass through fluid hoses. If the incorrect hose is selected for isocyanate, the material can cure in the hose and eventually cause a hose rupture and spillage problem. Teflon® and polyolefin (Moisture Lok™) hoses will not easily absorb moisture and are recommended for isocyanate. Special safety considerations must be used when working with isocyanate materials. Be sure to refer to the material supplier's safety recommendations.

4.4.11 Advantages and Limitations of Plural Component Mixing Methods

Figures 47, *48*, and *49* briefly outline the major advantages and limitations of hand proportioning, mechanical proportioning, and electronic proportioning.

4.4.12 General Safety Guidelines for Plural Component Application

The following general safety precautions should be observed when using various application tools for plural component materials. Specific safety instructions supplied by the manufacturer of your equipment must be used. In addition, special safety precautions must be taken when working with plural component materials. Always follow the instructions given by the material supplier and refer to the MSDS for the materials.

4.4.13 Typical Plural Component Application Guidelines

A summary of the typical plural component materials, their average mix ratios, viscosities, working lives; and the tool or tools generally used for their application, is given in *Table 7*.

For spray purposes, almost any conventional air spray system, airless spray system, air-assisted airless spray system, or any of these systems equipped with electrostatic spray equipment, can be used for the application of plural component materials.

Always follow the setup, operation, shutdown, and cleanup instructions furnished by the manufacturer for the application equipment. Always make sure that both the unmixed and the mixed plural components are thoroughly cleaned from the equipment before the material reaches the end of its pot life, or damage to the application equipment will occur.

5.0.0 ◆ THERMAL SPRAY SYSTEMS

Thermal spraying (THSP) systems employ a number of different processes in which finely divided metallic or nonmetallic surfacing materials are deposited in a molten or semimolten condition on a prepared substrate to form a coating.

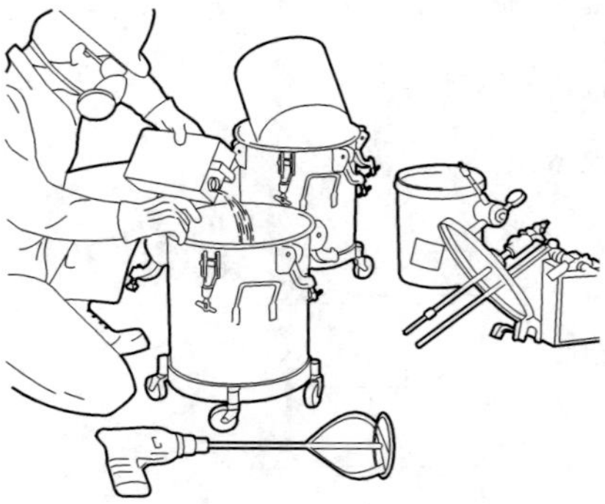

ADVANTAGES

- NO EXPENSIVE EQUIPMENT IS REQUIRED.
- SOME RATIO TOLERANCES ARE NOT CRITICAL, THEREFORE NO ACCURATE MEASURING SYSTEM IS NEEDED.
- MAY GENERATE LESS WASTE IN VERY LOW-USAGE APPLICATIONS.

LIMITATIONS

- CONTAINER DISPOSAL IS A PROBLEM.
- UNUSED MATERIAL DISPOSAL IS EXPENSIVE. LEFTOVER MATERIAL CANNOT BE SAVED.
- HIGH PERCENTAGE OF LABOR DOLLARS ARE SPENT PREPARING TO PAINT.
- QUALITY OF WORK LIFE SUFFERS BECAUSE OF HANDLING AND CLEANING TOXIC OR UNPLEASANT COMPONENT CHEMICALS.
- THERE ARE SAFETY ISSUES RELATED TO HANDLING TOXIC CHEMICALS.
- SOME MATERIALS ARE PHYSICALLY DIFFICULT TO MIX PROPERLY.
- ERRORS MAY OCCUR WHEN MIXING COMPONENTS.

208F47.EPS

Figure 47 ◆ Advantages and limitations of hand proportioning.

Instructor's Notes:

The surfacing material may be in the form of a loose powder or a powder in a plastic tube (cord), rod, or wire. The thermal spray gun uses high-temperature heat, generated by combustible fuel or an electric arc, to melt or plasticize the surfacing material. Compressed air, gas, or the products of combustion are used to accelerate and convey a confined stream of the melted or plasticized particles of surfacing material to the substrate. The particles strike the surface, flatten, and form very small, thin platelets (splatters) that conform and adhere to the irregularities of the prepared surface and to each other. As they strike the surface, they cool and build up, platelet by platelet, into a laminated structure that constitutes the sprayed coating.

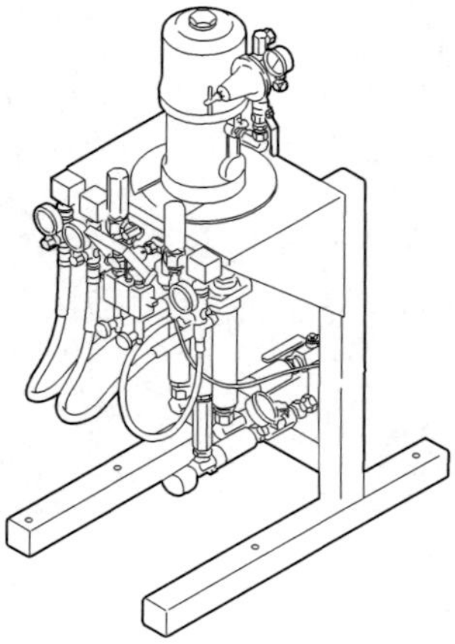

ADVANTAGES

- MECHANICAL PROPORTIONERS EFFICIENTLY HANDLE LARGE VOLUMES OF PLURAL COMPONENT MATERIALS.
- THEY ARE GENERALLY RELIABLE.
- MECHANICAL PROPORTIONERS CAN PUMP, MIX, AND APPLY MATERIALS WITH A SHORT POT LIFE VERY QUICKLY.
- THEY WORK ON DEMAND, MIXING ONLY WHAT IS NEEDED.
- THE WORK ENVIRONMENT IS CLEANER AND SAFER THAN THAT FOR MANUAL MIXING.
- THEY CAN BE INSTALLED IN HAZARDOUS AREAS WITHOUT MAJOR MODIFICATIONS.

LIMITATIONS

- THEY REQUIRE REGULAR MAINTENANCE AND REPAIRS.
- THE MIX RATIO IS AFFECTED AS MECHANICAL PARTS WEAR.
- OPERATOR ERROR CAN RESULT IN AN IMPROPER MIX RATIO.
- THEY HAVE A LIMITED RATIO RANGE.
- MECHANICAL COMPONENTS CAN CAUSE CHEMICAL BREAKDOWN (SHEARING).
- MECHANICAL PROPORTIONERS CAN CAVITATE, RESULTING IN OFF-RATIO CONDITIONS.
- CAVITATION OCCURS IF A SUPPLY PUMP MALFUNCTIONS.
- THERE IS NO INHERENT RATIO VERIFICATION OR PROCESS DATA CAPABILITY.
- CHANGING COLORS CAN CAUSE MATERIAL WASTE.
- CHANGING COLORS IS SLOW AND COMPLEX.

208F48.EPS

Figure 48 ◈ Advantages and limitations of mechanical proportioners.

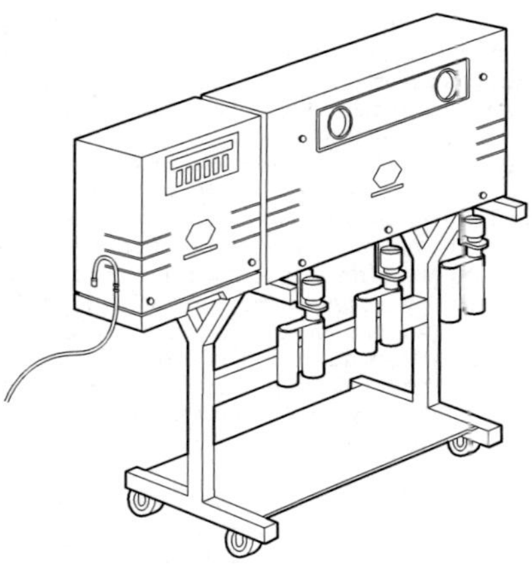

ADVANTAGES

- ACCURATELY MAINTAINS DESIRED RATIO, SELF-CORRECTING
- METERS AND MIXES ON DEMAND ONLY WHAT IS NEEDED
- DESIGNED TO MONITOR THE PROCESS EFFECTIVELY
- PROVIDES MATERIAL USAGE REPORTS
- PROVIDES CLEANER, SAFER WORK ENVIRONMENT
- HANDLES A WIDE RANGE OF RATIOS
- DISPENSES MULTIPLE RATIOS
- AUTOMATICALLY SELF-PURGES IF POT LIFE IS EXCEEDED
- SIMPLER TO MAINTAIN THAN MECHANICAL PROPORTIONERS
- GENERATES LESS WASTE WHEN CHANGING COLOR OR PURGING

LIMITATIONS

- EQUIPMENT FAIRLY EXPENSIVE
- INTIMIDATING TO LEARN AND OPERATE FOR SOME
- MUST BE OPERATED BY PROPERLY TRAINED OPERATORS ONLY
- REQUIRES REGULAR METER-CHECKS FOR ACCURACY

208F49.EPS

Figure 49 ◈ Advantages and limitations of electronic proportioners.

Audiovisual

Show Transparencies 43 through 46 (Figures 50 through 53). Ask trainees to explain the wire/rod flame spraying process.

Table 7 Typical Plural Component Application Guidelines

Materials	Mix Ratio	Viscosity Centipoise (CPS)	Working Life	Application Tool*
Catalyzed Polyurethanes				
Decorative coating	1:1 to 4:1	50 to 200	30 min. to 8 hrs.	1, 2, and 3
Protective coating	1:1 to 4:1	5,000 to 30,000	10 sec. to 1 hr	4 and 3
Adhesive and sealants	1:1	20,000 to 150,000	60 sec. to 30 min.	5
Foam	1:1	100 to 600	1 sec. to 10 sec.	4 or 6
Epoxies				
Coatings	1:1 to 4:1	1,000 to 10,000	1 min. to 20 min.	4 and 3
Adhesive and sealants	1:1 to 2:1	1,000 to 10,000	1 min. to 10 min.	6
Polyesters				
Decorative coating	1:1	50 to 200	10 min. to 30 min.	1, 2, and 3
Laminate systems	51:1 to 85:1	Not reduced	12 min. to 37 min.	8
Vinyl Esters				
Laminate systems	43:1 to 71:1	Not reduced	12 min. to 37 min.	8
Polysulfides				
Protective coating	1:1 to 3:1	100 to 500	1 hr. to 8 hrs.	7 and 3
Adhesive and sealants	7:1 to 12:1	5,000 to 300,000	1 hr. to 3 hrs.	5
Silicones				
Protective coating	10:1	100 to 2,000	1 min. to 5 min.	7 and 3
Adhesive and sealants	1:1 to 10:1	5,000 to 30,000	10 min. to 30 min.	5
Foam	1:1	100 to 2,000	5 min. to 15 min.	7 or 6

*NOTES: 1 = Conventional air spray equipment
2 = Air-assisted airless spray equipment
3 = Electrostatic equipment to be combined with primary application equipment
4 = Heated airless spray equipment
5 = Extrusion equipment
6 = Dispensing equipment
7 = Airless spray
8 = Airless spray with catalyst injection

208T07.EPS

5.1.0 Types of Processes

Thermal spray systems use one of two general processes to apply a coating. The first uses combustible fuel and oxygen or air as the heat source. The second uses a low-voltage, high-amperage electric arc as the heat source.

5.2.0 Flame and Wire Arc Spraying System Equipment

Of the various thermal spraying processes described in the previous paragraphs, wire-fed or rod-fed flame spraying systems or wire arc spraying systems are the most commonly used for field applications. The following paragraphs describe the equipment typically used for these systems and the operation of some of the associated spray guns.

5.2.1 Wire/Rod Flame Spraying System Equipment

Wire/rod flame spraying guns, equipped with a wire/rod tip up to 3/16" in diameter, can spray anywhere from 6 to 60 lbs. of material an hour depending on the material. *Figure 50* shows a typical diagram of a wire flame spraying system that uses compressed gases as the fuel source. In this system, wire from a reel is supplied to the flame gun along with a fuel gas and oxygen to establish and maintain a melting flame for the wire. Dry, clean, compressed air is also supplied to the gun to atomize the melted wire material.

Figure 51 illustrates a wire flame spraying gun, along with optional nozzle extensions used with the gun, for a gas-fired, wire flame spraying system. The wire is drawn into the rear of the gun by drive rolls mounted on the gun that are powered by an integral electric motor, air motor, or air turbine. As shown in *Figure 52*, the wire is fed into a coaxial flame created by the combustion of an oxygen/fuel gas mixture. The end of the moving wire is continuously melted by the flame, and the melted material is then atomized and carried to the substrate by a shaped stream of compressed air or inert gas.

Figure 53 illustrates a type of high-velocity-oxygen fuel (HVOF) flame spray gun that uses wire instead of powder material. This particular gun has no integral wire-feeding device. Wire is supplied to this gun through a hollow cable from a remote wire feeder sometimes called a *pusher feed*.

Instructor's Notes:

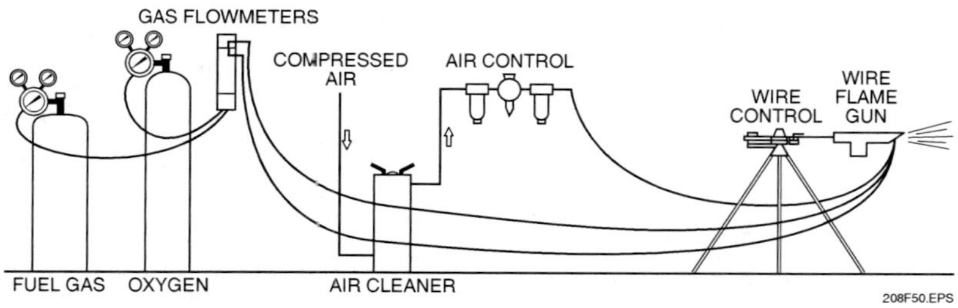

Figure 50 ◆ Typical diagram of a wire flame spraying system.

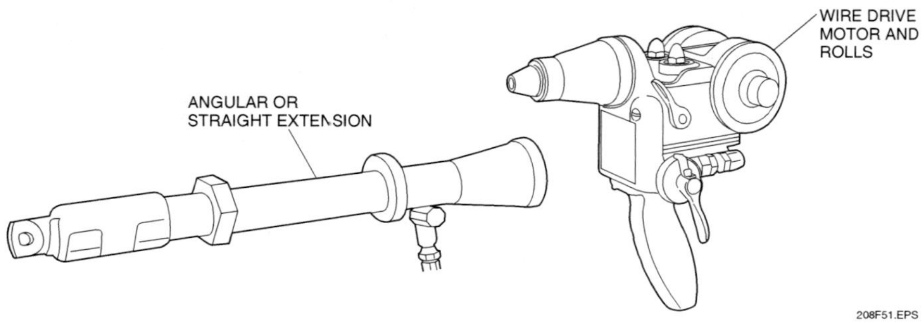

Figure 51 ◆ A wire frame spray gun with optional nozzle extensions.

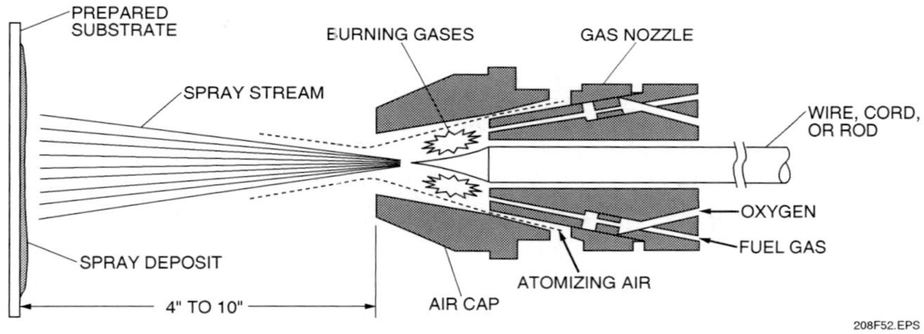

Figure 52 ◆ Cross section of a typical wire flame spray gun head.

Audiovisual

Show Transparencies 47 through 49 (Figures 54 through 56). Have trainees explain the typical wire arc spraying system.

5.2.2 Wire Arc Spraying Equipment

These systems, equipped with wire tips up to 3/16" in diameter, can spray anywhere from 7 to 85 lbs. of material an hour depending on the material. Of the various thermal spray processes, the highest production is usually obtained from wire arc spraying systems. *Figure 54* shows a diagram of a typical standard system. In this system, two wires (each up to approximately 1/16" diameter) are supplied from wire reels through insulated, flexible, hollow tubes to the arc spray gun. Low-voltage direct current (DC) (18 to 40 volts DC) at high amperage (50 to 400 amps) is supplied from a remote DC power supply/control panel to the arc spray gun to establish and maintain the electric arc that is used to melt the wires. Dry, clean, compressed air is also supplied to the gun to atomize the melted wire material.

Figure 55 illustrates a typical standard gun along with an optional nozzle extension that can be used with the gun. The two consumable wires are drawn into the rear of the gun from the insulated hollow tubes by insulated drive rolls mounted on the gun that are powered by an integral electric motor, air motor, or air turbine. In some systems, where a long distance exists between the wire reels and the gun, an insulated wire feeder (pusher feed) at the wire reels is synchronized with the drive rolls in the gun in order to provide a smoothly moving wire supply to the gun. As shown in *Figure 55*, the wires are fed through insulated wire guides into insulated copper contact tubes that supply the low-voltage, high-current electric power to the wires.

Figure 56 shows a cross section of a typical wire arc spray gun head. The two wires are accurately and continuously fed through the angled contact tubes and wire tips so that they converge at a precise point where the electric arc is struck and maintained between the wires. The intense heat generated by the electric arc melts and, in some cases, partially vaporizes the wire material, especially aluminum or zinc. As the ends of the moving wires are continuously melted/vaporized by the arc, the melted material is atomized and, along with

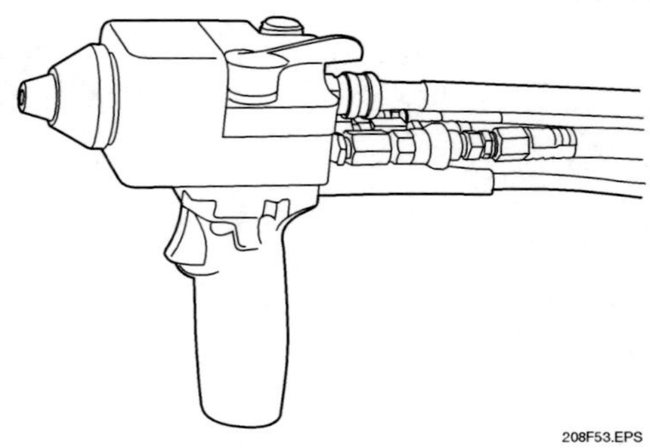

Figure 53 ◈ Wire-fed type of HVOF flame spray gun.

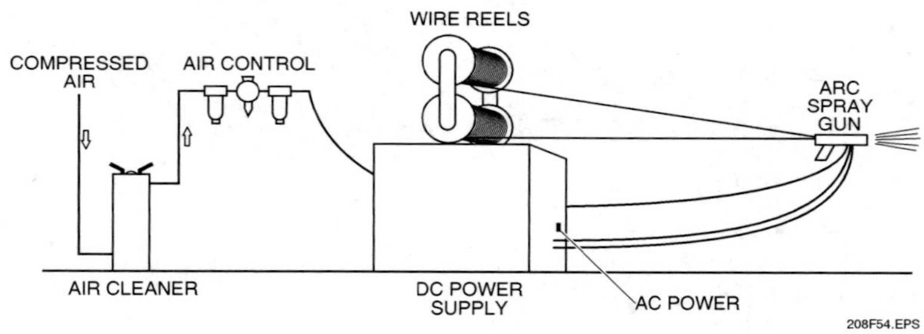

Figure 54 ◈ Typical wire arc spraying system.

Instructor's Notes:

any vaporized material, is carried to the substrate by a shaped stream of compressed air or inert gas. The high particle temperatures, along with the condensation of any vaporized material on the surface of the substrate, causes minute metallurgical weld spots that greatly increase the bonding of the sprayed material. An arc shield at the front of the gun or gun extension significantly reduces the intense visible light, ultraviolet light, and infrared heat radiation exposure to the operator.

Figure 57 illustrates a high-output system (600 amps or more) which uses a compact, pusher-fed, wire arc spray gun that does not have an internal wire-feed drive. The two wires, each up to 3⁄16" in diameter, are supplied to this gun through insulated, flexible, hollow tubes from a remote wire feeder. In some pusher-fed guns, the low-voltage, high-amperage DC power is coupled to the wires at the wire feeder instead of at the gun.

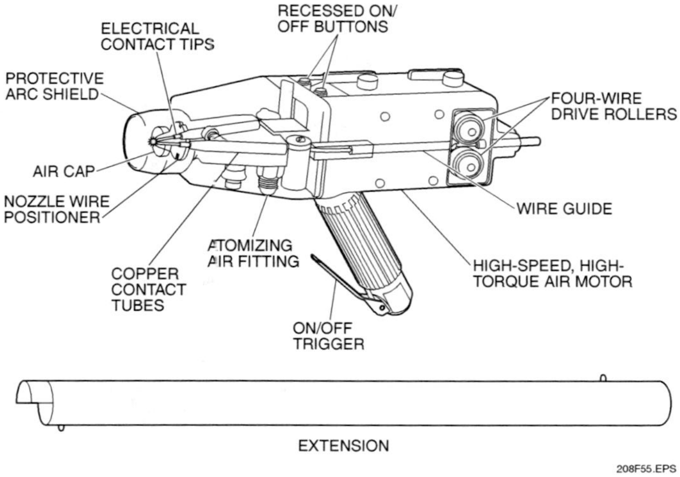

Figure 55 ◆ Wire arc spray gun with optional extensions.

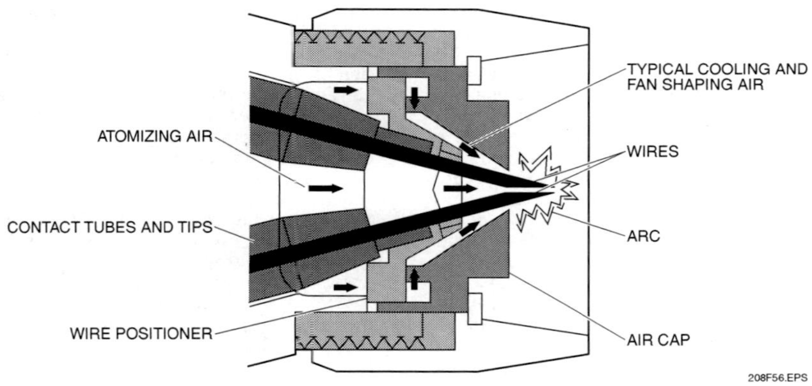

Figure 56 ◆ Cross section of a typical wire arc spray gun head.

Audiovisual

Show Transparency 50 (Figure 57). Have a trainee explain the differences and similarities between a regular wire arc system and a high-output wire arc system.

Laboratory

Have trainees list the various types of thermal spray coatings and characteristics of each. Discuss each type of thermal spray coating.

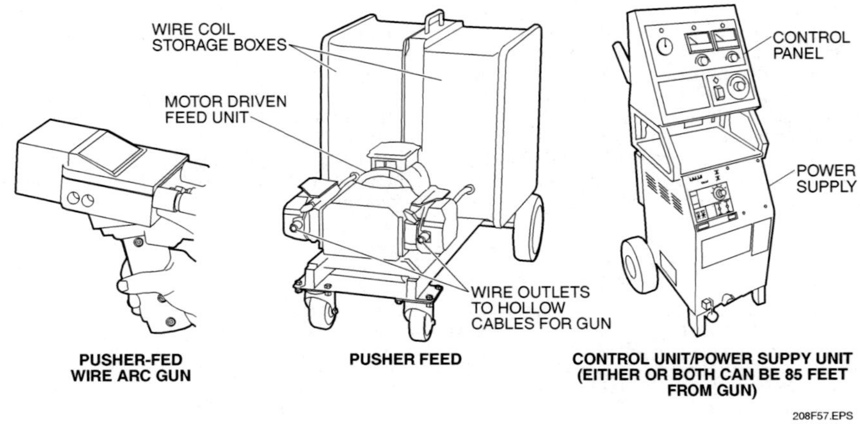

Figure 57 ◆ A pusher-feed, wire arc spray system.

High-output systems are capable of supplying thermally sprayed corrosion protection coatings at production rates that are close to those achieved with conventional airless coating sprayers. Typically, a high-output system using 1/8" diameter wires can apply coatings 3- to 4-mils thick per pass at fan widths of up to 8" (20 cm). Using 3/16" diameter wires yields coating thicknesses of 4- to 6-mils per pass at fan widths of 12" (30.5 cm) or more. These systems typically require a minimum of 45 amps of 460 volts (V), three-phase power and 70 CFM of clean air at 100 psi (1,982 L/m at 6.9 kiloPascals [kPa]) for operation.

5.3.0 Thermal Spray Coating Materials

A great many materials, both metallic and nonmetallic, may be applied to the surface of a substrate by thermal spray systems for a wide variety of purposes. However, for this module, only corrosion protection coatings will be described.

Thermal spray coatings are used for protection of iron and steel in a wide range of corrosive environments. When properly sealed with a compatible sealer, long-term effectiveness (over 20 years) in both industrial and marine applications has been documented. Austenitic stainless steels, aluminum bronze, and cobalt and nickel base alloys, when sealed, offer varying degrees of atmospheric protection in certain specific applications. However, zinc, aluminum, and alloys of both provide the broadest atmospheric protection. They are more corrosion resistant than steel and iron, and they are also anodic to steel and iron. This means that they will serve as a sacrificial anode in an electrolytic solution and will be consumed over time, thus preventing corrosion of the ferrous substrate (reserve cathodic protection). This protection will occur even if the coating coverage is incomplete or suffers mechanical damage. In addition, they form an oxide film which resists corrosion.

Aluminum corrodes less rapidly than zinc in acidic conditions (below a pH of 6). Zinc performs better than aluminum in alkaline conditions. For the protection of steel in gas and chemical plants or for other applications where the temperature of the coating will exceed 390°F (200°C), aluminum is used. However, spraying of aluminum coatings should be avoided whenever there is a fire or explosion hazard (mines or oil rigs) because of the possibility of particle impact sparking. In these cases, zinc coatings are used because they are nonsparking.

Zinc coatings are also preferred for use in fresh, cold, hard water; but in soft water, salt water, or any water above 150°F (65°C), aluminum is used. When zinc is alloyed with aluminum (85% Zn, 15% Al), the zinc-rich coating has the attributes of both. The zinc provides cathodic protection, and the aluminum strengthens the coating providing abrasion and erosion resistance.

Coating thicknesses are determined based on the degree of protection required. Coating life is proportional to the coating weight per unit area. For very long life or in highly corrosive conditions, the coating thickness should be increased to enhance protection. Zinc coatings can be applied

Instructor's Notes:

in thicknesses ranging from 2 mils to over 20 mils. Aluminum coatings can be applied in thicknesses ranging from 4 mils to 10 mils.

Thermally sprayed zinc and aluminum coatings have extremely good adhesion to fresh, clean, grit-blasted steel. Once applied, maximum service life is achieved by covering the surface with a 5- to 8-mil layer of suitable sealer. Typically, this may consist of a phosphate primer sprayed directly on a freshly sprayed zinc or aluminum surface followed by either a vinyl co-polymer (colored or aluminum pigmented) or a coal tar epoxy (the latter for severe environments).

Other low-viscosity sealers based on phenolic, modified epoxy-phenolic, polyester, silicone, sodium silicate, ethyl silicate, anaerobic methacrylate, polyurethane, or epoxy-based resins may also be used. Sometimes, the addition of aluminum flake can improve the protection of a sealer because the aluminum penetrates the pores of the zinc or aluminum coating and retards further aqueous penetration. Chlorinated rubber has proven to be an unsatisfactory sealer for sprayed zinc or aluminum coatings.

5.4.0 Thermal Spraying Guidelines for Zinc and Aluminum Coatings

To provide maximum bonding when zinc coatings up to and including 9 mils thick or aluminum coatings up to and including 7 mils thick are to be applied, the surface of the substrate should be freshly cleaned to a white metal blast finish (*SSPC-SP5/NACE 1*). The surface must also be grit profiled to produce a 3-mil anchor pattern using one of the following:

- Aluminum oxide or silicon carbide, mesh size SAE G-25 to G-40
- Hardened steel grit, mesh size SAE G-25 to G-40
- Angular silica sand, garnet, flint, or crushed slag, mesh size SAE G-25 to G-50

When zinc coatings greater than 9 mils thick, or aluminum coatings greater than 7 mils thick, are to be applied, the surface must be white metal blast cleaned and grit profiled to provide a 5-mil anchor pattern using one of the following:

- Aluminum oxide or silicon carbide, mesh size SAE G-18 to G-25
- Hardened steel grit, mesh size SAE G-18 to G-25
- Angular silica sand, garnet, flint, or crushed slag, mesh size SAE G-18 to G-25

> **NOTE**
> Steel shot should not be used when profiling a surface for zinc or aluminum coating application.

Blast cleaned and profiled surfaces can be held within containment areas or confined areas by keeping the humidity level below 50%.

After the substrate has been properly cleaned and profiled, areas that are not to be coated must be masked. After proper masking is accomplished, the desired thermal spray system may be set up and placed into operation in accordance with the manufacturer's instructions. However, before beginning spraying operations, make sure that the proper ventilation and dust collectors for metallic dust have been emplaced and turned on.

Thermal spraying techniques are generally the same as those used for conventional or airless spraying. The standoff distance of the gun to the surface of the substrate varies from 4" to 10", depending on recommendations given on the manufacturer's MSDSs for the material being sprayed. It is critical for proper layering of the coating, and to reduce dusting, that the gun be held as perpendicular to the substrate surface as possible when spraying. However, any areas of discontinuity or insufficient coating thickness can be simply rectified by additional spraying at any time before sealing.

Arc spraying with a high-output system (600 amps or more) is generally the fastest method and inherently produces high bond strengths due to the high temperatures used. In any case, the best bond strengths are achieved by making the first pass as hot as possible to produce large particles (low air pressure and a 2" [50 mm] standoff distance). This establishes conditions for super-heated particles to adhere tightly. Zinc and especially aluminum have good bonding properties under these conditions. The adhesion plus the rough profile obtained with this type of first pass makes the resulting surface ideally suited for subsequent coating buildup. After thermal spraying operations are complete, the surface should be sealed as appropriate for maximum service life.

5.5.0 Dry-Powder Coating Application Processes

A number of processes are used to apply dry-powder coatings to objects that can tolerate being heated to several hundred degrees Fahrenheit.

Classroom

Ask a trainee to explain how to prepare the substrate to provide maximum bonding when zinc coatings up to 9 and aluminum coatings up to 7 mils thick are applied.

Classroom

Have trainees discuss the advantages of using dry-powder coatings. Ask trainees if they have any questions.

Audiovisual

Show Transparency 51 (Figure 58). Have a trainee identify the components of the powder-bed fluidizing tank.

Show Transparencies 52 and 53 (Figures 59 and 60). Ask a trainee to explain how a powder-bed fluidizing tank works.

The most commonly used processes can be categorized as follows:

- Fluidized bed process
- Electrostatic fluidized bed process
- Electrostatic powder spray process

Each process has particular advantages for specific coatings applied to workpieces that have been prepared for final coating by power tools, blast cleaning, or pickling as described in previous modules. These processes are only suitable for use in fixed or mobile shops. The most common advantages for the use of dry-powder coatings are as follows:

- Minimal material loss (up to 98% of most dry-powder coating particle residues can be recovered and reused)
- Usually a one-coat coverage process (no primer coats required)
- No need to mix, use, recover, and dispose of solvents which, in some cases, can be toxic
- No need for solvent flash-off before curing booths/ovens
- Curing booths/ovens require no outdoor makeup air because no solvents are used (the air is filtered to remove any dust and then returned to the shop)

5.5.1 Fluidized Bed Process

This process is the oldest and simplest method of applying a dry-powder coating. It is used mainly to apply thick coatings of thermoplastic resins for functional rather than decorative purposes. However, it can be used with thermosetting materials when very high film build is required or when the workpiece cannot be electrostatically sprayed for one reason or another.

The equipment usually consists of a preheating booth or box, a powder-bed fluidizing tank and recovery system, and (if necessary) a post-heating booth/oven or box to complete the melting and flowing of the coating. In this application, the preheating booth is used to heat the workpiece to the point (250°F to 450°F) (121°C to 232°C) at which the desired amount of powder coating will adhere to the piece and melt during a specific time period and before the temperature of the piece drops below the fusion point of the powder-coating material. Heating can be accomplished by exposure of the workpiece to infrared heat sources, high-power electrical induction heat, or convection heat using circulating hot air heated by fuel gas or electric heaters.

Once heated, the workpiece is automatically or manually transferred and immersed for a specific amount of time in the fluidized powder within the powder-bed fluidizing tank. The powder-bed fluidizing tank (*Figure 58*) consists of an upper and lower chamber separated by an air-penetrable membrane or plate. Pressurized, clean, dry air is applied to the lower chamber (air plenum). The air passes up through the membrane or plate, creating a lifting and agitating effect that causes the powder in the powder bed of the upper chamber to expand and act like a fluid (*Figure 59*). As air pressure is increased, the fluidized powder level rises in the tank. Vibration of the powder bed is

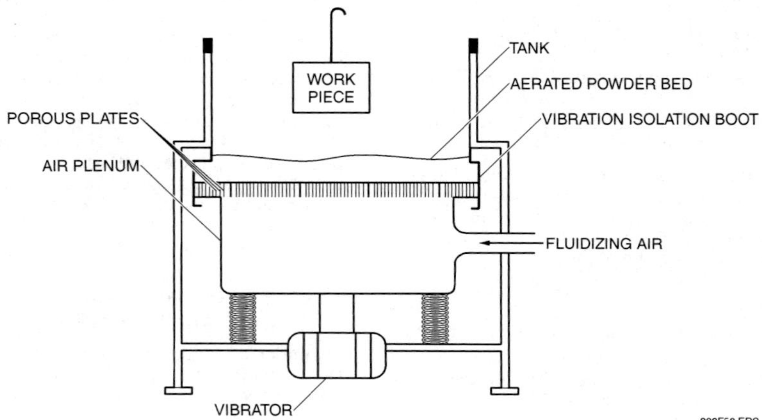

208F58.EPS

Figure 58 ◆ Typical powder-bed fluidizing tank.

Instructor's Notes:

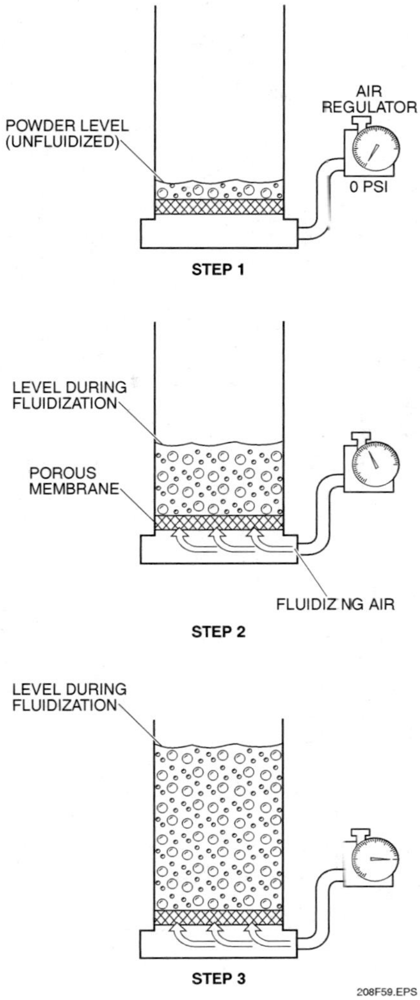

Figure 59 ◆ Powder-bed fluidizing under air pressure.

sometimes used to increase the lifting and agitating effect, and to help prevent stratification caused by the settling of heavier powder particles.

A typical fluidized bed system is shown in *Figure 60*. In this system, stray powder is recovered and reused along with new powder. In addition, powder near the bottom of the tank is drawn off, mixed with new and recovered powder, and recirculated to help prevent stratification.

Once the powder is fluidized to the proper point, the heated workpiece is immersed for a specified period of time, as mentioned previously. It is then withdrawn and, if necessary, post-heated (usually in the same manner as it was preheated) to cause additional flow-out of thick thermoplastic coatings.

5.5.2 Electrostatic Fluidized Bed Process

In this process (which is not as widely used as other processes), the combination of a fluidized bed and electrostatic electrodes creates a cloud of charged powder particles above the powder bed. The fluidized bed is of less importance in deposition of powder in this process than in the regular fluidized bed process. Basically, the electrostatic field produced in this process is used to apply the powder material to the workpiece.

The equipment consists of an electrostatic power supply and control unit, an electrostatic powder-bed fluidizing tank, and a post-heating booth/oven or box. In a few cases, a reheating booth/oven or box may also be used. The tank (*Figure 61*) is similar to the tank described previously for the fluidized bed system, except that the pressurized air passing through the porous membrane or plate is ionized by charging electrodes. The ionized air, passing up through the powder bed, charges the powder particles. The charged particles repel each other and move upward. Aided by the fluidizing air, a cloud of charged powder forms above the upper surface of the powder bed. Grounded workpieces can be conveyed or immersed into this cloud.

This process is only suited for workpieces that have a small vertical dimension, such as flat sheets or wire mesh The process offers the advantage of being able to coat objects such as flat sheets on one side only, instead of having to dip the whole part. Note that the workpiece does not require preheating. The electrostatic field created between the electrodes and the workpiece will

Audiovisual

Show Transparency 54 (Figure 61). Ask a trainee to explain the electrostatic fluidized bed process. Have a trainee identify the advantage of the electrostatic fluidized bed process.

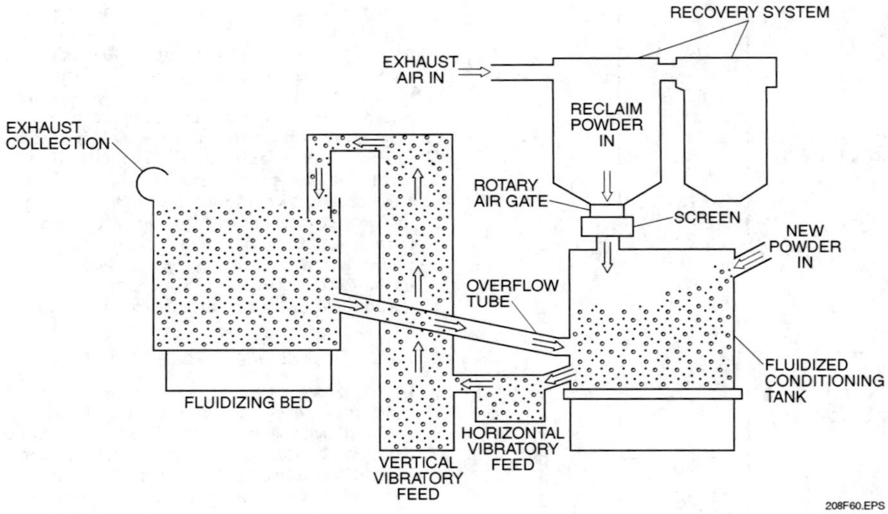

Figure 60 ◆ Typical fluidized bed system.

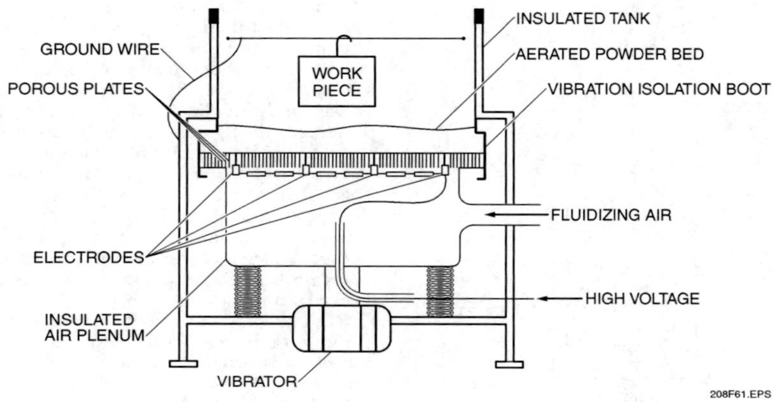

Figure 61 ◆ Typical electrostatic powder-bed fluidizing tank.

Instructor's Notes:

cause charged, powder-coating particle transfer to the workpiece regardless of whether the piece is hot or cold. If the workpiece is hot, the deposited powder will melt and flow on contact. If the piece is cold, the charged powder particles will adhere until the piece is removed and post-heated to cause flow-out of the coating.

The electrostatic powder-bed fluidizing tank uses a power supply and control unit similar to those used with manual electrostatic spray systems described earlier in this module. Because the powder cloud density is very high, an output voltage of approximately 100 kilovolts (kV) is required from the power supply. The short-circuit energy must not be greater than the ignition energy of the powder material. In addition, if workpieces are preheated, it is important with this process that they be kept below temperatures that can cause explosive gases/vapors to be released from the powder material. Explosions can result if sparking occurs across air/gas mixtures within the powder cloud.

5.5.3 Electrostatic Powder Spray Process

Electrostatic powder spraying, using automatic or manual guns, or rotary disks, is the most widely used method of applying powder coatings. There are many varied types of electrostatic powder spray coating systems available. They generally consist of five basic pieces of equipment

- Powder-feed unit
- Electrostatic powder spray gun or rotary disk
- Power and control unit
- Powder recovery units
- Spray booth

In this process, powder is typically supplied to the spray gun from a powder-feed unit (*Figure 62*). The unit illustrated uses a siphon-lift venturi (*Figure 63*) to supply a fluidized powder/air mixture to the gun. Some units are also equipped with a cyclone reclaimer and/or sieve to allow recovery of expended sprayed powder and/or powder vented from the hopper of the feeder unit. In some cases, a simple gravity-fed hopper mounted over a venturi unit is used to fluidize and supply powder to a gun.

The general construction and principles of operation for an electrostatic powder spray gun or rotary disk are very similar to the equipment described in previous sections of this module and are also used to apply conductive liquid coatings. Some of the guns (called *electro-gas-dynamic [EGD] guns*) use internal instead of external corona charging. With this type of gun, the internal electrodes (positive and negative) charge the particles as they move past the electrodes in the airstream, and little or no field is established between the electrodes and the workpiece. The deposition field around the workpiece results from the field between the charged particle cloud and the workpiece. However, the transfer efficiencies are basically the same as for an externally charged gun. Safety and operating requirements for an electrostatic powder spray gun are basically the same as those for guns used for conductive liquid applications.

Several typical spray booths and recovery systems for powder coating are shown in *Figures 64* and *65*. In one system (*Figure 64*), the oversprayed powder is reclaimed by allowing gravity to carry the falling powder directly into a hopper. Any remaining airborne powder is removed by a cyclone separator/reclaimer and final filter. In the other system (*Figure 65*), the oversprayed powder is drawn down to a moving, fabric-filter belt by a low-vacuum, high-air volume exhaust system. The powder, captured by the belt, is removed by a low-volume, high-vacuum pick-up system and reclaimed by a cyclone separator for return through a sieve to the powder-feed unit. New powder is also picked up by a cyclone separator and supplied to the powder-feed unit through the sieve.

Classroom

Ask a trainee to explain the precaution that must be taken with the electrostatic fluidized bed process when workpieces are preheated.

Audiovisual

Show Transparencies 55 and 56 (Figures 62 and 63). Ask a trainee to explain the electrostatic powder spray process.

Show Transparencies 57 and 58 (Figures 64 and 65). Ask a trainee to explain the purpose of spray booths and recovery systems for powder coating. Have a trainee compare and contrast the two systems illustrated in Figures 64 and 65.

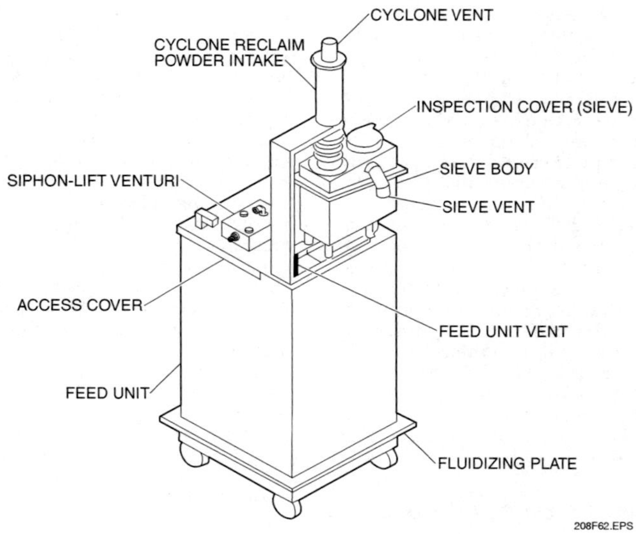

Figure 62 ◆ Powder-feed unit.

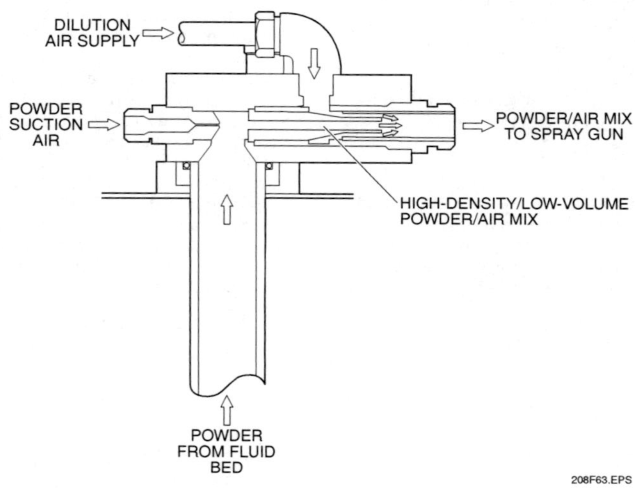

Figure 63 ◆ Siphon-lift venturi.

Instructor's Notes:

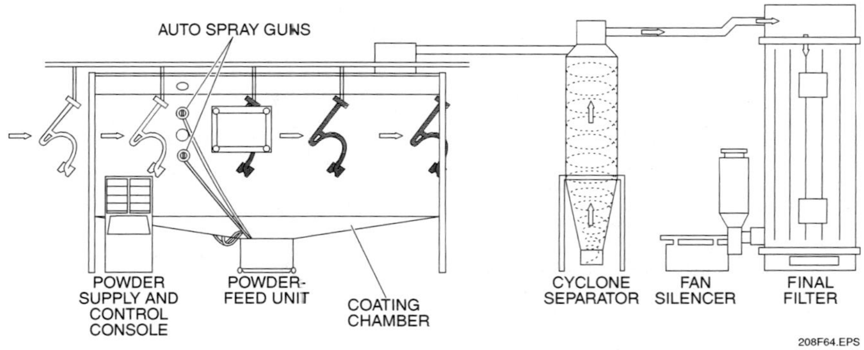

Figure 64 ◈ Typical spray system with gravity-feed recovery.

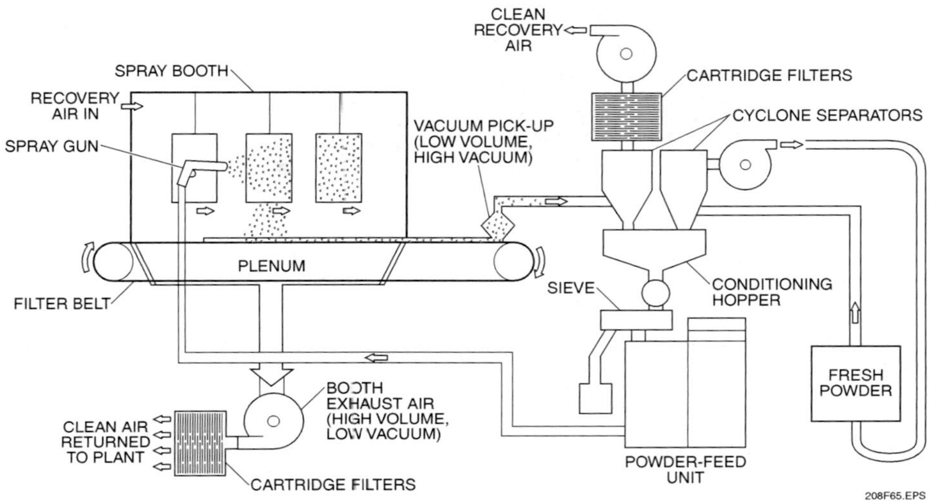

Figure 65 ◈ Typical spray system with vacuum filter-belt recovery.

Performance Testing

Have each trainee perform Performance Tasks 1, 3, and 4 to your satisfaction. Fill out a Performance Profile Sheet for each trainee.

Classroom

Go over the Review Questions for this module, and answer any questions trainees may have.

Summary

For environmental reasons, new coatings have been developed to provide efficient protection for metal while getting rid of harmful compounds such as VOCs. This is particularly true for the new water-based, high-performance industrial coatings. All coatings contain some or all of four basic ingredients: solvents, resins, pigments, or additives.

Coatings are formulated and selected for particular applications and requirements. Some of these factors are the type of substrate to be coated, operating conditions, ambient conditions during application, government and environmental regulations, budget, time, and surface preparation. The selection of the right coating, and proper mixing and application are critical if the coating is to perform the function for which it was designed. Proper surface preparation is perhaps the most critical step because the surface must provide the anchor to hold the coating.

Systems used to apply coatings include air spray systems, airless spray systems, HVLP spray systems, electrostatic systems, and thermal spray systems. Each system must be used in accordance with manufacturer operation specifications. Each system must also be used according to company and NIOSH safety procedures. Personal protective equipment (PPE) is required for most spraying operations. These may include respirators, goggles, gloves, and other equipment. Ventilation must be adequate for spraying operations.

Common spray-coating problems include both those associated with equipment and those associated with application. Application problems include coating sags, fingering, orange peel, and excessive fog and are caused by not using the equipment or coating material properly. Machine problems should be addressed with reference to the manufacturer's service and troubleshooting directions. All completed coatings must be inspected and tested to ensure that they meet the specification and quality requirements.

Review Questions

1. Coatings have been re-engineered to eliminate or reduce the use of _____.
   a. thinners
   b. solvents
   c. VOCs
   d. oil

2. Pigments give a coating its _____.
   a. smoothness
   b. fluidity
   c. durability
   d. color

3. Solvents are _____ liquids used to dissolve the solid pigment.
   a. volatile
   b. thermoplastic
   c. inorganic
   d. viscous

4. Films formed by oxidation are called _____.
   a. deformable
   b. thermoset
   c. thermoplastic
   d. resilient

5. The process by which molecules join together is called _____.
   a. emulsion
   b. dispersion
   c. oxidation
   d. polymerization

6. Inorganic zinc-rich binders are always _____.
   a. silicates
   b. vinyls
   c. epoxies
   d. urethanes

7. The degree of sheen is determined by the proportion of _____ in the coating.
   a. pigment
   b. resin
   c. binder
   d. solvent

8. _____ refer to a change in the operation that could damage or destroy a coating.
   a. Specification changes
   b. Upset conditions
   c. Exposure limits
   d. Ambient conditions

Instructor's Notes:

9. Thinning a coating is sometimes necessary to obtain the required wet film _____.
   a. volume
   b. coatibility
   c. adhesion
   d. viscosity

10. Minimum chemical, corrosion, and fire resistance are determined when the coating is _____.
   a. inspected
   b. selected
   c. tested
   d. dried

11. Pressure-feed flow is dependent upon the _____.
   a. atomization
   b. spray pattern
   c. pump pressure
   d. viscosity

12. Airless spraying is best suited for jobs where _____ areas must be covered quickly.
   a. medium to large
   b. small to medium
   c. confined space
   d. complex and intricate

13. There are two main types of HVLP systems: _____ and _____.
   a. airless; turbine
   b. compressed air; turbine
   c. compressed air; siphon
   d. siphon; pressure-feed

14. The air cap, fluid tip, and needle must be selected as a matched set, commonly called the _____.
   a. triplet set
   b. nozzle combination
   c. nozzle suite
   d. spray ensemble

15. The most common mistake in spraying is moving your arm in a(n) _____ motion.
   a. arcing
   b. straight line
   c. circular
   d. up and down

16. To adjust the speed of HVLP application, the _____ can be changed.
   a. air pressure
   b. nozzle size
   c. air cap size
   d. fluid pressure

17. In the most common types of electrostatic spray systems, electrical charges are applied to the _____.
   a. sprayed surface
   b. sprayer
   c. target
   d. coating material

18. Terms that are often used to describe the qualities of electrostatic systems are _____ and _____.
   a. fixed; portable
   b. inhibited; resilient
   c. stiff; resistive
   d. flexible; reactive

19. Isocyanate is a plural component chemical that is often used as a _____.
   a. sealant
   b. binder
   c. catalyst
   d. base

20. _____ is the oldest and simplest method of applying a dry-powder coating.
   a. Electrostatic powder spraying
   b. Thermal spraying
   c. Electrostatic bed spraying
   d. Fluidized bed process

Examination

Administer the Module Examination. Be sure to record the results of the Exam on Craft Training Report Form 200 and submit the results to the Training Program Sponsor.

Performance Testing

Ensure that all Performance Tasks have been completed and Performance Profile Sheets for each trainee are filled out. Be sure to record the results of the testing on Craft Training Report Form 200 and submit the results to the Training Program Sponsor.

Instructor's Notes:

Glossary

This glossary will help you learn some terms specific to applying coatings using spray applications. Trade Terms introduced in this module appear in italic text.

Bond/adhesion: The attachment of a coating to a substrate. Adhesion is the key to coating, effectively allowing the coating to become part of the substrate. The three types of adhesive bonds are chemical, polar, and mechanical.

Bounceback: The atomized particles that rebound or deflect from the surface.

Cigarette wrap: Application technique for cut lengths of sheet coating. The width of the sheet must exceed the circumference of the pipe to provide an overwrap. The sheet is placed longitudinally and wrapped no more than twice around the pipe. Ending seams should be alternated along the pipe as to not be continuous. Each sheet should overlap the previous. This method is not commonly used, with the exception of heat shrink sleeves.

Coal tar: A tar distilled from bituminous coal used to coat piping. Usually a paper wrap is used over coal tar.

Confined space: A space that: 1) Is large enough and so configured that an employee can bodily enter and perform assigned work; and 2) Has limited or restricted means for entry and exit (e.g. tanks, vessels, vaults, etc.); and 3) Is not designed for continuous employee occupancy.

**Cracking (of coating):* Breaks in a coating that extend through to the substrate.

Dry Film Thickness: The thickness of a coating after it has completely dried or cured, usually measured in mils (1 mil = 0.001 in.).

Fusion bond: Application of a powder coating to a heated pipe, performed by electrostatic methods. The powder gels, flows with the heat, then cures and hardens during cooling. The process creates a tight physical bond between the coating and metal.

Hand cleaning: The process of removing burrs, slag, rust, mill scale and other contaminants from steel pipe. Burrs or slag may be removed by grinding.

***Holiday:** A discontinuity in a protective coating that exposes unprotected surface to the environment.

Mastic: An asphalt-like mixture used to coat piping.

Opacity: A coating's ability to cover or hide the underlying surface.

Orange peel: Hills and valleys in a coating that resemble the skin of an orange.

Orifice: The small, specially shaped hole or opening in the spray gun tip that controls the flow of fluid.

Overlap: An amount of coating covering the previous row of coating in a row-to-row sequence.

Over spray: A dry, flat, pebbly surface resulting from paint particles falling outside the spray pattern.

Permit-required confined space: A confined space that has one or more of the following characteristics: 1) Contains or has a potential to contain a hazardous atmosphere; 2) Contains a material that has the potential for engulfing an entrant; 3) Has an internal configuration such that an entrant would be trapped or asphyxiated by inwardly converging walls or by a floor which slopes downward and tapers to a smaller cross-section; or 4) Contains any other recognized serious safety or health hazard.

Pre-heat: Pre-heating of the pipeline metal prior to application of certain coatings. Heat shrink sleeves and several other coatings require the pipe to be pre-heated prior to applying the coating. Caution: Excessive heat can damage the coating.

**Primer:* A coating material intended to be applied as the first coat on an uncoated surface. The coating is specifically formulated to adhere to and protect the surface as well as to produce a suitable surface for subsequent coats. [Also referred to as Prime Coat.]

Refractive index: Ability to reflect light.

Sags: Also called runs or curtains, these coating defects are caused by excess flow of coating material.

Shrink sleeve: A conformable, wrap-around sleeve designed to protect below-ground and submerged metal surfaces from corrosion and electrolysis. The sleeve consists of two overlapping pieces or a single tube surrounding the pipe. Heat is applied to the sleeves until an adhesive material in the sleeves bonds them to the pipe and joins the overlapping sections to form a tight bond.

Thermoplatic: The property of becoming soft when heated and hard when cooled.

Thermoset: The property of becoming hardened when cured and not softening or deforming upon being heated.

Transfer efficiency: The amount of coating sprayed that actually goes on the surface as compared to the amount used.

Volatile: The property of being evaporated or being converted to vapor.

Wax: The wax present in petroleum crudes primarily consists of paraffin hydrocarbons known as paraffin wax and naphthenic hydrocarbons.

Wet film thickness: The thickness of a fresh coating before it dries, usually measured in mils (1 mil = 0.001 in.).

Wrinkling: A coating defect causing a rough, crinkled surface.

Instructor's Notes:

MODULE 61208-02 — TEACHING TIPS

The following are suggested activities or instructional methods to help you teach the material in this AIG.

General

When you call on someone to answer a question, the rest of the class relaxes or even tunes out because they expect that the question and answer will take place only between you and the trainee you called on. Instead, use this technique to involve more trainees in answering questions and to keep them on their toes.

1. Ask trainees to define a term or explain a concept.
2. After one trainee has answered, ask a trainee seated nearby if the answer is right. Then ask whether a trainee in the back of the room agrees.
3. Ask trainees to explain why they think an answer is right or wrong.
4. Use the session to clear up incorrect ideas and encourage trainees to learn from their mistakes.

Section 2.7.0 *Powder Coatings*

This Quick Quiz will familiarize trainees with the various types of powder coatings. You will need photocopies of the quiz provided on the following page. Trainees will need pencils. If you allow trainees to use the Trainee Guide, decrease the amount of time you give them to complete the quiz.

1. Make as many photocopies as you will need of the quiz.
2. Give trainees between 5 and 10 minutes to complete the quiz.
3. Go over the answers to the quiz.
4. Ask trainees whether they have questions about powder coatings.

Answers to Quick Quiz

1. c. Nylon
2. e. Polyester TGIC
3. a. Acrylic urethane
4. d. Polyvinyl chloride
5. b. Polyethylene

Quick Quiz *Powder Coatings*

For each description listed, identify the powder coatings that the text best describes. Write the corresponding letter in the blank provided.

_____ **1.** This is a tough thermoplastic coating with excellent abrasion, wear, and impact resistance.

_____ **2.** This thermoset coating provides a good combination of sharp-edge coverage with a good exterior durability.

_____ **3.** This thin-film, exterior-durable thermoset coating is used extensively in the appliance industry.

_____ **4.** This thermoplastic coating bonds well to most metals and can withstand bending and embossing.

_____ **5.** This thermoplastic coating has excellent chemical resistance and has many uses in the coating of laboratory equipment.

a. Acrylic urethane
b. Polyethylene
c. Nylon
d. Polyvinyl chloride
e. Polyester TGIC

Sections 3.4.0–3.4.4

Spraying Techniques

Trainees will practice using conventional spray equipment. Trainees will need appropriate personal protective equipment and their Trainee Guides. You will need to arrange access to a spray system and provide necessary materials and equipment. Alternatively, you may arrange for trainees to observe an expert use spray equipment or have trainees simulate the use of the equipment. Allow 1 to 2 hours for this exercise.

1. Review the safety guidelines for using a spray system.
2. Have trainees put on their personal protective equipment.
3. Have trainees use the manufacturer's data to apply the proper settings for the spray system.
4. Supervise trainees as they practice using the spray equipment. During the exercise, encourage trainees to experiment by holding the gun closer and further away from the surface. Have trainees purposely move their arm in an arcing motion while spraying to see the undesirable results. Then have them practice the correct motion.
5. Have trainees inspect each other's work and provide suggestions. If possible, save samples of trainees' work for a future exercise on common spray coating problems. Ask trainees if they have any questions.

Section 3.4.5

Cleaning and Maintaining Conventional Spray Equipment

This exercise will allow trainees an opportunity to practice cleaning and maintaining conventional spray equipment. Trainees will need appropriate personal protective equipment and their Trainee Guides. You will need to arrange access to a conventional spray system that needs to be cleaned and provide necessary materials and equipment. Alternatively, you may arrange for trainees to observe an expert clean and maintain spray equipment or have trainees simulate the process. Allow 1 to 2 hours for this exercise.

1. Have trainees review the manufacturer's operation and service manuals to become familiar with the cleaning and maintenance guidelines.
2. Remind trainees to dispose of all used solvent waste in accordance with all laws.
3. Supervise trainees as they follow the procedure for cleaning and maintaining a gun and pressure container or pressure cup. Ensure that they follow local and state laws regarding VOCs and avoid soaking the spray gun in solvent.
4. Ask trainees if they have any questions.

Sections 3.6.0–3.6.4

Using, Cleaning, and Maintaining HVLP Spray Equipment

This exercise will allow trainees an opportunity to practice using, cleaning, and maintaining HVLP spray equipment. Trainees will need appropriate personal protective equipment and their Trainee Guides. You will need to arrange access to an HVLP spray system and provide necessary materials and equipment. Alternatively, you may arrange for trainees to observe an expert operate HVLP spray equipment or have trainees simulate the process. Allow 2 to 3 hours for this exercise.

1. Have trainees review manufacturer's instructions for preparing, operating, cleaning, and maintaining the HVLP spray equipment.
2. Ensure that trainees wear appropriate personal protective equipment.
3. Supervise trainees as they practice spraying with the HVLP system. Remind trainees to slow down and not increase production or material output.
4. Have trainees clean and maintain the equipment, following all laws concerning release of VOCs and proper disposal of waste solvents.
5. Inspect trainees' work and answer any questions they may have. If possible, save samples of trainees' work for a future exercise on common spray coating problems.

Sections 3.7.1–3.7.6

Common Spray Coating Problems

In this exercise, trainees will learn to recognize spray pattern problems and their causes. Trainees will need pencils, paper, and their Trainee Guides. You will need to provide samples of each type of spray coating problem. You may use trainee practice samples from prior exercises, obtain samples from the field, or obtain photos of spray coating problems. Also include samples of coatings that do not have problems. Alternatively, you may simply describe each spray coating problem. Allow 2 to 3 hours for this exercise.

1. Set up several stations at which you have placed numbered examples of sprayed coatings. Divide the class into small groups and assign each group to a station.
2. For each example, have trainees determine whether the coating has a problem and, if so, identify possible causes. Trainees can refer to their Trainee Guides for assistance. Ask trainees to write down their findings on paper.
3. Review trainees' reports. Discuss their findings. Ask trainees if they have any questions.

MODULE 61208-02 — ANSWERS TO REVIEW QUESTIONS

Answer	Section
1. c	1.0.0
2. d	2.1.0
3. a	2.3.0
4. b	2.5.1
5. d	2.5.3
6. a	2.5.4
7. b	2.6.3
8. b	2.8.0
9. d	2.10.0
10. b	2.11.0
11. c	3.1.1
12. a	3.2.0
13. b	3.3.1
14. b	3.4.2
15. a	3.4.4
16. b	3.6.3
17. d	4.0.0
18. c	4.3.0
19. c	4.4.10
20. d	5.5.1

CONTREN™ LEARNING SERIES — USER UPDATES

The NCCER makes every effort to keep these textbooks up-to-date and free of technical errors. We appreciate your help in this process. If you have an idea for improving this textbook, or if you find an error, a typographical mistake, or an inaccuracy in NCCER's Contren™ textbooks, please write us, using this form or a photocopy. Be sure to include the exact module number, page number, a detailed description, and the correction, if applicable. Your input will be brought to the attention of the Technical Review Committee. Thank you for your assistance.

Instructors – If you found that additional materials were necessary in order to teach this module effectively, please let us know so that we may include them in the Equipment/Materials list in the Instructor's Guide.

Write: Curriculum Revision and Development Department
National Center for Construction Education and Research
P.O. Box 141104, Gainesville, FL 32614-1104

Fax: 352-334-0932

E-mail: curriculum@nccer.org

Craft ______________ Module Name ______________

Copyright Date ______________ Module Number ______________ Page Number(s) ______________

Description ______________

____

____

____

(Optional) Correction ______________

____

____

(Optional) Your Name and Address ______________

____

____

Index

Index